The Grid

The Grid

Biography of an American Technology

Julie A. Cohn

The MIT Press
Cambridge, Massachusetts
London, England

This book was set in ITC Stone Serif by Scribe Inc., Philadelphia, PA.

Library of Congress Cataloging-in-Publication Data

Names: Cohn, Julie A., author.
Title: The grid: biography of an American technology / Julie A. Cohn.
Description: Cambridge, MA: The MIT Press, [2017] | Includes bibliographical references and index.
Identifiers: LCCN 2017011363 | ISBN 9780262037174 (hardcover: alk. paper)
ISBN: 9780262537407 (paperback)
Subjects: LCSH: Electric power distribution--United States--History.
Classification: LCC TK3001 .C58 2017 | DDC 333.793/20973--dc23 LC record available at https://lccn.loc.gov/2017011363

ISBN: 978-0-262-03717-4

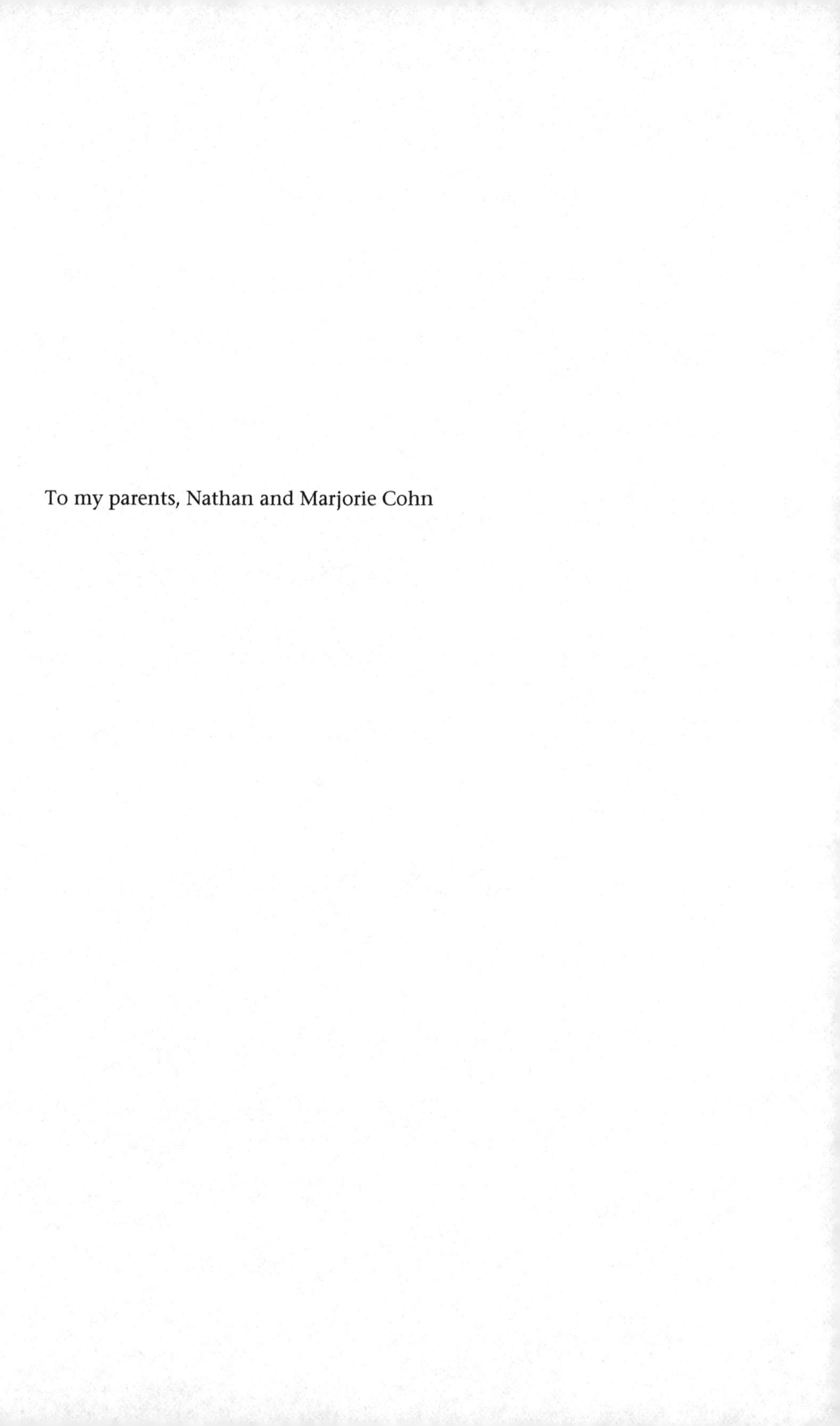

To my parents, Nathan and Marjorie Cohn

Contents

Preface

As an elementary-school-age girl on vacation in Florida, I found the insider's view of a power plant control room to be a very dull side trip. The noisy buzzing of a transformer station did not stand up to the allure of beach sand, waves, and sunshine. The droning voices of men admiring meters and charts held little interest. And yet when my father, Nathan Cohn, explained the control of generation and power flow on interconnected systems—a string of words that would normally induce sleep in any child—his passion for his work was palpable. After all, he was discussing a collection of technologies known as the world's largest machine: the interconnected power system that kept the lights on and industry moving all across America. It was—and still is—a network that connects nearly every living soul on the continent. The side trips, the buzzing, the meters, the droning voices, and the long explanations all were part of the story of North America's power grid.

Looking back from the perspective of the twenty-first century, "the grid" seems a logical outcome of economic and technological trends under way at the start of the twentieth century. Many studies of electrification end in the 1930s, as if the grid was a fait accompli by that decade. But if the completion of a national-scale grid was indeed assured in the 1930s, why were the engineers so passionate about their work and their accomplishments in the decades that followed? Looking at the record of the work of these hands-on power experts, one learns that the grid was not at all the logical outcome of a natural process of building networks. Instead, the grid grew out of the aggregation of hundreds of different projects initiated by an equal number of different entities, over decades, and with numerous setbacks and side trips. And even today, "the grid" is a misnomer because there are actually four grids in North America, although there are links between them. While there were analogs for an interconnected power system—roadways, railroads, telegraph networks, telephone networks, and early urban gas networks—the physical properties of electricity created unanticipated challenges for those trying to control interconnected systems. At the same time, the economic and political relationships that framed electrification added complexity to the process of control. In the absence of a central authority directing the size, shape, direction, and standards of electrification, it was the control room operators, manufacturers, utility managers, and academic engineers who imagined, experimented, and cooperated to build today's grid. How and why they pulled this off merits a good retelling.

This story begins in the late nineteenth century, so my research began with perusal of the major industry journals that documented electrification from its start: *Electrical World* and the *Electric Journal*. *Electrical World* in particular offered a weekly review of everything electrical in the United States and abroad. Within its pages, it was possible to trace the first experiments in interconnection, the reasons companies chose to interconnect, and the alternative technologies that offered similar benefits. For the period covering the late 1880s to 1920, these journals and other contemporaneous publications provided the basis for piecing together the origins of the grid. Secondary literature from many fields—including corporate histories, business history, political history, environmental history, urban history, rural history, and regulatory history—provided additional insights into the context in which advances in electrification took place. By 1920, the concept of interconnection had taken hold across the industry, and for this story, it was necessary to find a means of tracing further development beyond that date that was both compelling and universal to electrification.

The passion of power systems engineers offered me a clue. I chose to use the challenges of power control—both conceptual and technical—as the point of entry for the more detailed portion of this project. While there are myriad entities involved in electrification, varied technologies for power generation, and significant regional differences in how interconnection developed, all the stakeholders had to wrestle with the same physical characteristics of electricity and the related problems of control on their systems. A narrative that documented development of a single power pool would diminish the scope of the entire project of building a grid. At the same time, a step-by-step description of each new transmission line and interconnection from the East Coast to the West, beginning in the 1890s and ending in the 1990s, would bore both the researcher and the reader. By focusing on control techniques, I was able to follow the story of interconnections across the continent and through the century, engaging a number of economic, organizational, and political trends that shaped the process.

My father entered the power industry in 1927, the same decade in which utilities began to define key physical control problems. He worked for Leeds & Northrup Company, an instrument manufacturer involved in addressing these control problems, until 1972, after which he continued to consult for the industry until his death in 1989. This span of time roughly corresponds with the decades during which control issues dogged system operators across North America. Nathan Cohn's technical papers, professional publications, and personal memorabilia thus provide a good starting place for delineating the technical challenges and control experiments that ensued.

The Nathan Cohn collection at the Massachusetts Institute of Technology archive, early records of the Leeds & Northrup Company at the Hagley Library, additional personal collections of power system engineers on loan to the North American Electric Reliability Corporation (now housed at Hagley Library), and photocopies of records from the Consolidated Edison of New York archives (on loan from historian Joseph Pratt) comprised primary archival material for my research. In addition, I consulted myriad technical journals, textbooks and handbooks, and trade publications and met with several engineers who worked actively in the power industry on system control challenges during the twentieth century. This research allowed me to follow the fraught path of power control innovation from the 1920s to the 1980s. The secondary literature on topics ranging from energy and cultural history to cybernetics offered contextual clues that allowed me to frame the technical history within broader trends.

Acknowledgments

Many thanks are due to numerous individuals and institutions for providing insight, critique, resources, and support over the past several years. First, I must acknowledge my mentors at the University of Houston, Kathleen Brosnan (now at the University of Oklahoma), Martin Melosi, and Joseph Pratt (recently retired). Together these three individuals steered me through graduate training, dissertation writing, professional development, and the project of putting together a book. It would be hard to imagine a more supportive, insightful, critical, and encouraging team. I am indebted to numerous other colleagues who read, advised, discussed, and encouraged my work, including but not limited to Kristen Contos-Krueger, Matt Evenden, Debbie Harwell, Bernice Heilbrunn, Richard Hirsh, Paul Hirt, D. C. Jackson, Chris Jones, Kairn Kleiman, Marc Landry, Bob Lifset, Cyrus Mody, David Nye, Isabelle Parmentier, Adam Rome, Jimmy Schafer, Rebecca Slayton, Stephanie Staatz, Joe Stromberg, and Jason Theriot. Several engineers who worked in the power industry during the latter half of the twentieth century, including some who continue to keep our lights on today, took time to explain operations, demonstrate technologies, and correct my misunderstandings. I thank especially John Adams, B. H. Behroon, Joel Mickey, N. D. R. Sarma, and the rest of the team at the Electric Reliability Council of Texas (ERCOT) for their time, their tours of the control room, and their clarifying e-mail exchanges. I greatly appreciated the opportunity to meet with Ralph Masiello, Jim Resek, and Walt Stadlin, all former Leeds & Northrup employees, who discussed grid operations during recent decades and further amplified explanations in follow-up e-mail exchanges. In addition, Walt Stadlin read and advised on technical chapters of my dissertation, the precursor to this book. Bob Cummings at the North American Electric Reliability Corporation (NERC) provided details, explanations, and great images. I owe a deep debt of gratitude to David Nevius, formerly with NERC, for introducing me to these outstanding individuals in the power industry, for providing me with archival material, and for always agreeing to answer questions and offer further explanations. To Dave and his colleague Jim Robinson, both of whom read early complete drafts of the book and helped enormously with technical insights, thank you for your enthusiasm for this project—it made all the difference! Numerous other individuals at ERCOT, the IEEE History Center, NERC, and several federal agencies and power companies provided valuable assistance in tracking down data, resources, and images.

It was a privilege to share portions of this work at several professional conferences, including annual meetings of the American Society for Environmental History; Electric Worlds/Mondes électriques, hosted by the Committee for the History of Electricity and Energy of the EDF Group; the "Green Capitalism?" conference, cosponsored by the Hagley Library Center for the History of Business and the German Historical Institute; "Le pouvoir des riverains" colloquium, cosponsored by l'Université catholique de Louvain and l'Université de Namur; annual meetings of the Society for History of Technology and the Special Interest Group for Computers, Information and Society; and the Second World Congress of Environmental History. I am also grateful to organizations that provided material and institutional support for my research and conference travel, including the Center for Public History at the University of Houston; the IEEE Life Members' Fellowship in Electrical History; the Special Interest Group for Computers, Information and Society; and the University of Houston Department of History. Without archivists, archives are nothing but a collection of old stuff. Many thanks especially to Max Moeller and his colleagues at the Hagley Museum and Library (Lucas Clawson deserves special acknowledgement for locating and preparing digital files of key images included in this book); Alex Simons and the Interlibrary Loan staff at the M. D. Anderson Library at the University of Houston; Myles Crowley, Nora Murphy, and their colleagues at the MIT Institute Archives and Special Collections; staff at the National Archives; and Julie Kersson at the Seattle Municipal Archives for assisting with the location of particular documents and collections. To the anonymous reviewers of my book manuscript, your critical reading helped me extend the breadth and depth of this project. I offer special thanks to Katie Helke and Justin Kehoe at the MIT Press for their patience, assistance, and most excellent advice throughout the project.

Friends and family members who ask questions, offer to read chapters, insist on diversions, and provide entertainment balance the solitary work of writing. With love and deep gratitude, I thank the Algers, Susan Cohn, the Ebys, Alysse Einbender, Janet Hamel, Becky Grant, Jeanne and

Mike Maher, Karen and Dave Pernell, Sally McNagny, my yoga circle, the Heights Literary Guild, the dog-walking neighbors, and numerous other dear friends for many years of fun and moral support. An extra thank you to Lori Hedrick for proof reading the entire manuscript. Gayle Hoffer and Joy Stapp are due special appreciation for reading portions of the book and providing wise counsel. My extraordinary siblings, Ted (in memoriam), David, Anna, and Amy, and their spouses, children, and grandchildren inspired me. My mother, Mimi, kept me honest at the Scrabble board. My mother-in-law, Mary Ann Connor, never failed to ask, "How's the book coming along?" My gratitude extends to my extended Cohn and Connor clans. To my two closest and dearest family members, John Connor and Miriam Connor, thank you for coming along with me on this journey and making sure there was laughter every step of the way.

While I appreciate the aforementioned assistance and support, I take full responsibility for any errors or shortcomings within these pages.

1 Introduction

Each moment of the day, millions of Americans turn on lights, computers, coffeemakers, vacuum cleaners, cars, conveyer belts, and other devices using electricity. Without having to know what electricity is, where it is generated, or how it gets from the power plant to the switch, Americans rely utterly on the everyday miracle of electrification. Most Americans may not realize that the generators, transformers, transmission lines, distribution lines, motors, chargers, plugs, and bulbs that comprise our interconnected world all must operate at the exact same speed—60 hertz (Hz)—for the system to work.[1] Those who have been intimately familiar with this collection of technologies we call "the grid" often describe it as the largest interconnected machine in the world. Unless every part of the power grid cycles in near-perfect unison, the system will experience disturbances. Of course, minor disturbances occur all the time, because each flick of a power switch causes a slight change in the demand for electricity, and power networks must adjust. But with control technologies, power companies seek to guarantee that the disturbances will be so small as to escape notice, and Americans continue with their activities unconcerned about the stability and reliability of the grid, except during a blackout.

Within a growing body of literature about electrification in North America, the grid gets short shrift. Scholars focus on inventors, manufacturers, utilities, cities, regions, customers, dams, nuclear plants, environmental battles, and politics—all critical elements of the history of power systems.[2] But with only a handful of exceptions, scholars tend to gloss over the process of building interconnections. They take a leap of faith that once conceived, the networks were as good as built. Some of the historical evidence supports this view. In 1912, a reporter commenting on linked systems in California said, "If three or four stations must be operated together in defiance of all precedents, in go the switches, and the plants operate as if they had worked together from the very beginning."[3] Fifteen years later, a senior engineer advised a new MIT graduate to pass over a job with a power control instrument manufacturer, saying, "The companies now existing are very old and have practically all their methods, etc, and well standarized [*sic*] to a cut and dry proposition."[4] Despite these optimistic outlooks, the challenges of power control on interconnected systems dogged engineers for decades across the mid-twentieth century. Although first articulated in the 1910s, the dream of a national grid—or, more accurately, a single power network serving the entire continental United States—was not even close to a reality until 1967 and then was abandoned in 1975. Both the nature of electricity and the structure of the industry complicated what appears to be a simple matter—linking two or more power companies together.[5] This book argues that the question of control, both social and technical, was central to the project of building the grid and reflected the peculiarly American technology that today powers our economy.

What is "the grid"? The grid is a misnomer. In fact, there are four major networks carrying electricity to the vast majority of power users in North America, (these are shown in figure 1.1). But in casual use, the grid refers to all four networks together.[6] An English engineer, Charles Merz, coined the term "grid" in the 1910s to describe a transmission network that carries electricity from generators to consumers.[7] The US Energy Information Administration describes the grid as "the network of nearly 160,000 miles of high voltage transmission lines" that moves electricity from power plants to substations and eventually to consumers.[8] The North American Electric Reliability Corporation (NERC) uses similar phrasing: "The network of interconnected electricity lines that transport electricity from power plants and other generating facilities to local distribution areas."[9] A key concept in the NERC definition is the word "interconnected," which typically means that the entire network operates in synchrony. Other definitions for the grid extend the network concept to include power stations and transformers.[10]

While these definitions capture in nontechnical terms what the grid is and does, the essence of the grid is both more and less than these explanations offer. The grid is more in that it is a crucial lifeline in the modern world's energy-dependent economy. Nearly four trillion kilowatt-hours of electricity travel across transmission lines each year. Further, in the United States, electric power accounts for 40 percent of the energy used annually. The interconnected power system allows North Americans to rank among the top energy consumers in the world.[11] The grid is less in that it is, to a degree, a phantom and a hodgepodge. NERC, a single private electric reliability organization, certified by both the US federal and the Canadian provincial governments, oversees reliability of the system in the United States. Eight different regional

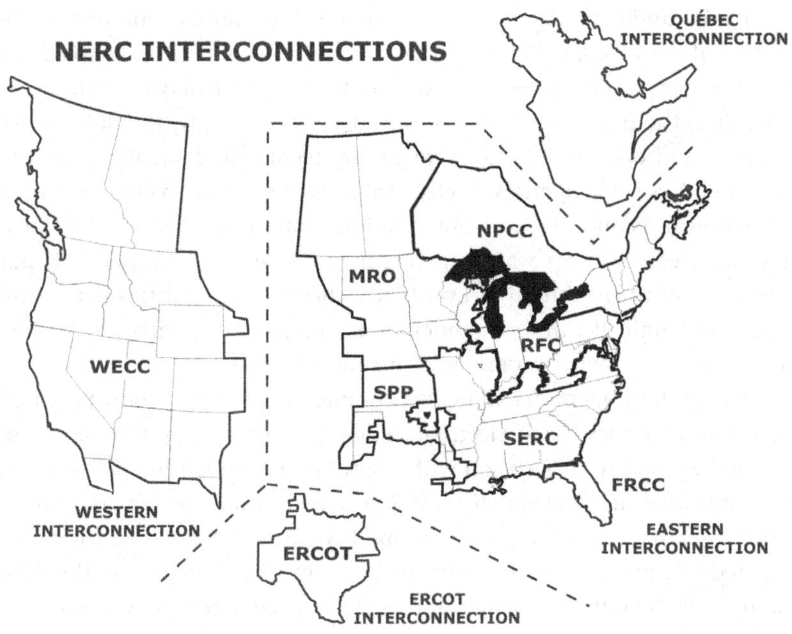

Figure 1.1
Map of North American interconnected systems.

reliability entities, however, monitor and enforce reliability compliance; dozens of state and provincial agencies regulate segments of the grid in the United States and Canada; and thousands of entities own the bits and pieces of the networks that comprise the grid.[12] In fact, when a speaker refers to the grid, he or she may be talking about the entire system that reaches from coast to coast and across international boundaries or any one of the subsystems that serves a region. Finally, except when we look at a line of poles holding up power lines, we barely notice that the grid exists at all.

With a backward glance from the twenty-first century, it seems logical and natural that electricity travels through a networked system to reach customers. From telegraph wires, to railroads, to gas pipelines, to highway systems, to the Internet, networked technologies have heralded modernity for the past two centuries. Electrification, from the very beginning, was nothing if not modern. Yet the process of building a grid was more complex, more contested, more haphazard, and more uniquely American than is generally understood. For technical, political, economic, and even environmental reasons, a collection of public and private entities cobbled the grid together piecemeal over decades. With each new network and every expansion, the power system operators faced increasingly difficult control problems. On one level, these were strictly technical challenges resulting from the physical properties of electricity. But on other levels, these were also political, economic, and organizational problems, because in North America, electrification was neither strictly a capitalist proposition nor solely a government initiative. Eventually, thousands of entities—large and small, economically interdependent and autonomous, investor owned and government owned—built and operated the grid.

Power system operators addressed complex control problems posed by the physical properties of electricity and compounded by the variety of stakeholders involved. To begin with, there is no good method for generating a large quantity of electricity and then storing it for later use. To provide a useful service (or, arguably, a commodity), a power company must be prepared to deliver electricity at the exact moment of demand, in the right quantity, at the right frequency, and of the expected quality. Second, electrical current on a power network flows according to the laws of physics, not the desires of human operators. Thus, on a network of power lines, electricity generated at one spot and used just a few blocks away might travel miles to get there, depending on other conditions affecting the system. Third, in North America, we use alternating current, a dynamic form of energy. The electric current on our power lines is actually moving back and forth at a very high speed. Constant changes in generation and demand compound the dynamic behavior of alternating current. Over the past century, as individual power companies built links to share power, they encountered challenging nuances in the physical behavior of electricity. On an organizational level, two or more interconnected companies acted as fully autonomous economic entities and navigated social interactions that were sometimes as complex as the electricity itself.

This book examines how the power system experts assembled the world's largest interconnected machine, despite the myriad complications that militated against success. This is a long story that covers a broad geography. To grasp the scale and significance of America's grid, it is important to consider the process across the entire continent. No single company or type of company, government agency, engineer, or regional network is likely to serve as an effective stand-in for the system as a whole. But it is essential to find a comprehensible point of entry. The technologies of control serve that function. The need for physical control of electric power appeared at the very start of electrification—for example, through the use of speed governors on the generators at Thomas Edison's Pearl Street Station. It continues to challenge system operators even today, as they worry over how to handle the variability and intermittency of wind and solar power on the grid. In addition, physical control of moving electric power offers a nice metaphor for social control of electrification, and this has likewise interested power producers and users from the very start. Finally, the topic of control reflects tensions that cross multiple aspects of electrification: is it a commodity or a service, is it in the public domain or the private, is it a unified system or a collection of autonomous operators, is it robust or fragile, and does it foster conservation or consumption or both?

Over the course of nearly a century, engineers, system operators, utility managers, and academics collaborated on the problems of control. Despite the competing interests of employers, each of these experts had compelling reasons for sharing ideas, testing equipment, and maintaining collegial relations as they built the grid piece by piece. The failure of one link spelled problems for all the others; thus interconnections bound the public and private power companies together. Successful operations, on the other hand, promised real cash savings, more effective use of coal and falling water, and a more reliable service for customers. For most of the period covered by the book, the relationships among the experts were informal, the experiments in control took place in real time on the power systems themselves, and the adherence to standard practice was entirely voluntary on the part of each company.

The development of particular practices and devices form the narrative spine of *The Grid*. Frequency control, load control, tie-line bias, economy dispatch, system security, and digital solution of power networks are by no means the only crucial technologies of interconnection, but they are essential to modern power networks. These technologies of control, addressed in this book, are well hidden from most power users, instrumental in facilitating the development of the grid, and emblematic of the nature of America's power system. Often the technical innovations followed the social organization of the industry. Further, the apparatus used to control power on interconnected systems was common to most of the grid, while other materials and technologies—from primary energy resources, to types of generators, to applications of electric power—varied greatly across the continent. Leeds & Northrup Company (L&N) was one of many control instrument manufacturers, including the better-known Westinghouse Electric Corporation and General Electric Corporation. During the 1920s, L&N became deeply involved in solving control problems newly defined by the power companies, and by 1949, 90 percent of the industry used L&N load and frequency control instruments on interconnected systems. Thus the narrative will feature the work of L&N as a means of linking the diverse entities that produced, delivered, deliberated, and regulated power, much as the grid linked them physically. At the same time, the book will engage the work of a wide array of engineers, utilities, manufacturing companies, regulators, and others who made equally important contributions to the development of today's power network.

By definition, the North American interconnected electric power system is a network. It is a physical network made up of wires and connection nodes over which electrical energy moves from a point of generation to multiple points of use. It is also a virtual network of individuals and organizations that coordinate to make and move and use electricity. In *Networks of Power*, a seminal work for the history of electrification, historian Thomas Hughes used the term "network" in multiple ways when referring to power systems.[13] As Olivier Coutard explained, Hughes used "network" as a synonym for the physical infrastructure, as a metaphor for the political relationships among people and organizations involved in electrification, and as a reference to the external events and trends that linked power systems across the globe.[14] In the field of infrastructure studies, scholars underscore the importance of investigating more than the physical components of a network: "Beyond bricks, mortar, pipes or wires, infrastructure also encompasses more abstract entities, such as protocols (human and computer), standards, and memory."[15] This comports with an approach to the history of technology that emphasizes the role of society in defining the meaning and purpose of physical technologies.[16] The power grid can be usefully compared to other energy, transportation, and communication network infrastructures in the United States, but it differs in significant ways.

While electrification began in the United States as a competitive private sector enterprise, legislators, managers, and civic leaders eventually determined the power system was a natural monopoly. When weighing regulatory changes in the 1990s, economists Paul Joskow and Roger Noll drew parallels among power networks, telephone networks, and other systems that fell into this category. Natural monopolies comprise technology components that "exhibit economies of scale over a range of output or capacity that is comparable to the magnitude of demand," and the industries "require some standardization and coordination . . . to operate efficiently."[17] Crucially, the transmission element of power systems was both a transportation network that moved power and also a coordination system that integrated multiple facilities over a wide region.[18] Additionally, the power grid is a type of communication network in that the flip of the on/off switch at the user end signals the status of demand to the generator end. In the United States, private interests built, owned, and operated numerous natural monopoly networks, while different governments regulated price, entry, and later other aspects of the industry such as worker health and safety.[19] Joskow and Noll theorized that elements such as transmission lines continued to be most effective when under monopoly control and close government regulation, while other elements, such as generating plants, might benefit from market competition. Like telephone, telecommunications, natural gas, and rail infrastructure, the natural monopoly network of the grid is mostly under private control, but as this history indicates, public sector investment in electrification, in addition to regulation, played an important role in its development.

The intentionality of certain networks differentiates them from the power grid. Historian Richard John argues that in the case of the telephone, it was the purposeful creation of a national interconnected long-distance network that brought the system as a whole into the category of a natural monopoly.[20] As John describes, William Forbes, president of the Bell Telephone Company between 1879 and 1887, decided to invest in an interconnected long-distance telephone network as a hedge against the eventual expiration of Alexander Graham Bell's patents. Early on, the network was a drain on revenues, yet it had "epochal significance," dominating American telephony for one hundred years.[21] There is no comparable individual entrepreneur in the history of electrification who conceived and pursued construction of a single national long-distance transmission system for power in the United States. Instead, beginning in the 1910s, many individuals in different sectors discussed the benefits of a national grid, but none undertook to create it single-handedly. The telephone systems differed from the power network in two other significant ways. As John explains, the early long-distance telephone network did not enjoy network effects—that is, it did not increase in value as more people used it—and in fact, intensive use caused congestion on long-distance lines, which was problematic. Further, in the earliest years, Bell subsidiaries viewed business communications as the primary market and did not equate expanded service areas with increased profit. As *The Grid* will illustrate, power companies quickly found advantages in widespread markets and pursued interconnections in part to benefit from network effects.

The builders of the Internet also intentionally created a national network. As historian Janet Abbate explains, the Advanced Research Projects Agency Network (ARPANET), a group funded through the Department of Defense, designed a network to facilitate rapid digital communication between certain locations across the continent.[22] The original links of the ARPANET system formed the spine to which other smaller networks, both government-sponsored and commercial, later attached themselves. Abbate asks, "Is there something unique about the Internet's seemingly chaotic development?"[23] While every infrastructure network is arguably unique in terms of its technology, the material it transports or transmits, and the social organizations that surround it, the history of the grid suggests that it shares an element of chaos with the Internet. Rather than a backbone to which other networks connected themselves, the grid is composed of many, many smaller networks, which grew regionally and eventually interconnected with each other. But in the case of both the Internet and the grid, the trajectory of development unfolded without blueprints for the systems as they now exist.

Networks in general are negotiated in two senses. First, the human participants who develop and use the infrastructure find ways to resolve competing priorities—that is, to make compromises and trade-offs. Second, as many historians of technology might argue, the developing infrastructure itself influences the organizations and broader social environment in which it functions and is used.[24] Historians Paul Edwards and Geoffrey Bowker argue that tensions between different stakeholders are endemic to the process and must be resolved to move toward an "operational, maintainable, robust infrastructure."[25] These traits are evident in the history of the interconnected power system. Public and private entities, academic engineers and control room operators, politicians and industry leaders, fishermen and dam builders all negotiated the particulars of electrification. They also strove to ensure that the networks operated together smoothly. Borrowing from computer science, the notion of interoperability captures the sense of both linking and operating smoothly. Technically, interoperability means the ability of two or more systems to exchange information and to use the information that has been exchanged.[26] As legal scholars John Palfrey and Urs Gasser explain, interoperable systems harness the benefits of interconnection while protecting core societal values. They argue, "Even while the best systems and people are enabled to work together, the powerful force of diversity can be preserved."[27] In this sense, interoperability was crucial to successful interconnection of power systems. System experts developed technical and organizational solutions to interconnection challenges that also protected the integrity of economically autonomous power systems.

In the twenty-first-century United States, an aging transmission infrastructure supports complex power transactions among every possible type of generating facility, from a hydroelectric plant more than one hundred years old, to a nuclear plant licensed in the 1960s, to the newest solar panels atop a residence. System operators rely on techniques validated in the 1950s to support markets created after the 1990s using computers manufactured in the 2010s. The hierarchy of organizations that control the grid dates back decades yet reflects a postderegulation and postblackout legislative world. By investigating the social and technical underpinnings of American interconnected power systems, it may be possible to better understand how future energy decisions will be framed by the opportunities and the limitations of the grid.

The narrative of *The Grid* is structured to follow the development of the grid much as one might follow the biography of a living person—from birth to a natural stopping point preceding major change. The story begins with the earliest links between power stations in the 1890s and culminates with the closure of ties between the eastern and western interconnected systems in 1967, the first and brief moment in which the North American grid truly existed. The book follows the further development of power networks after 1967, when conceptual and technical advances led to fairly mature and stable system operations. Between 1967 and 1990, however, economic and political trends laid the groundwork for major restructuring of the industry and its regulators. The closing portion of the book illustrates the centrality of the grid to the changing power system, the longevity of the grid's systems of control, and the ongoing paradoxes inherent in the grid's organization and operation. The chapters also underscore key themes that characterize the grid and the evolving context in which it developed.

Chapter 2, "The Birth of the Grid, 1899–1918," describes the origins of the grid, from the earliest links between power companies to the interconnections that fueled war industries during World War I. Diversity within the industry, competing technologies, a fraternity of experts, and early conservation initiatives all played a role in furthering interconnections. Chapter 3, "Contests for Control, 1918–1934," describes the emerging control challenges as interconnection became more popular. Despite public approbation for interconnection, the stakeholders, including utility owners, politicians, and consumers, could not reach consensus about who should own, regulate, and benefit from expanding power networks. At the same time, the operators in control rooms met with difficulty as they attempted to match generation and demand on linked systems. Chapter 4, "Balancing Reliability and Economy, 1930–1940," illustrates how one solution precipitated the next problem and how the fraternity of experts navigated the challenges through both collaboration and competition. This unfolded in the context of a national economic depression, government infrastructure investment and rule making that favored interconnections, and the development of many different arrangements for power sharing. Importantly, while retrospectively the expansion of interconnections appears inevitable, when looking ahead from the system control rooms of the 1930s, the process of operating in concert with neighboring systems probably appeared challenging.

Chapter 5, "Power Transformations on the Home Front, 1935–1950," examines the effects of World War II on electrification in the United States. Wartime accelerated the growth of interconnections, magnified some control problems and precipitated solutions to others, and validated the benefits of interconnection—all in the context of unprecedented cooperation between the public and private sectors. Chapter 6, "Nuances of Control in an Increasingly Interconnected World, 1945–1965," illustrates how the growth and increasing complexity of power pools deepened the challenges of interconnection at the very moment that industry experts began to plan explicitly for a coast-to-coast grid. Chapter 7, "Drifting 'Lazily' into Synchrony: From Blackout to Grid, 1965–1967," describes the 1965 Northeast blackout and the 1967 East-West Closure, two seminal events in the history of electrification. The blackout marked the first time both the expertise and the organization of the industry faced major critical public scrutiny. The closure, much less widely observed, represented the completion of a "national" grid, albeit for only a few years. Chapter 8, "Reaching Maturity: Integration, Security, and Advanced Technologies, 1965–1990," describes the maturing of the grid under renewed efforts to bring about stability and reliability following both the blackout and the closure. Engineers conceptualized power networks as integrated human/machine systems and introduced new digital technologies for planning and controlling interconnections. At the same time, engineers acknowledged the regional and local diversity, and hence autonomy, of subsystems in both planning and operating activities. Chapter 9, "Deregulation and Disaggregation: A Brief Overview, 1980–2015," describes the regulatory changes, corporate restructuring, and technical innovations that resulted in a reshaped power industry. The book concludes in chapter 10 with a visit to a twenty-first-century control room, offering comparisons between historical and contemporary challenges of the grid.

The origins of the grid are hidden in thousands of pages of trade journals published at the turn of the nineteenth to the twentieth century. In 1897, a journalist reported on a test of two power lines joined together. An 1899 story told of a surprisingly long transmission line. Another year or so later, there was a mention of various power plants running together in parallel.[28] Power system operators were testing, extending, and incorporating new technologies on the fly and laying the groundwork for a decades-long project of interconnection. The young fellows charged with keeping the lights on in, say, Virginia in 1899 probably couldn't imagine that their experiment linking a steam plant with a hydroelectric plant "to take up generation when there is a dry spell" was a precursor to a key engineering marvel of the coming century.[29] For the next hundred years, the elements of the grid appeared more and more clearly in maps, photographs, anecdotes, technical arguments, legislative proposals, meeting transactions, and eventually full-blown strategic plans—and, of course, across the landscape. This is a story about the loose network of people, ideas, and systems of ownership and governance that led to the construction of America's tightly coordinated machines of electrification.

2 The Birth of the Grid, 1899–1918

In 1899, the San Gabriel Electric Company in Southern California built a link to the Los Angeles Railroad Company, and the two companies exchanged power. The San Gabriel Electric Company relied primarily on a hydroelectric plant on the San Gabriel River in Azusa Canyon to provide power to customers in Los Angeles, twenty-three miles away.[1] The Los Angeles Railroad Company, owned by Henry E. Huntington, relied on steam-powered generators. When the river ran high in the winter, the railway was the "dumping ground" for surplus power.[2] When the river ran low, the power company purchased electricity from the railroad. San Gabriel "gives or takes a 500-volt direct current from the Los Angeles Railroad Company . . . the very quintessence of novelty, originality and boldness of design."[3] The trailblazing connection between the San Gabriel–Los Angeles transmission line and the Los Angeles Railroad circuit allowed two separate companies to equalize their loads, provide each other with backup reserve power, and manage the use of water and hydrocarbon resources. Within a few years, the owners of the electric company and Huntington consolidated their interests in the Pacific Light and Power Company. The Los Angeles Express described Huntington's ventures in electric rail expansion as "figuring upon a complete gridironing of the country tributary to L.A. with electric lines," a phrasing that anticipated the future notion of the power grid.[4]

The San Gabriel Electric–Los Angeles link embodied the rationale for and the effect of building interconnections at the start of the twentieth century. For these two companies, it was both an economic and a resource management decision. To realize a satisfactory return on its investment in a hydroelectric dam, the San Gabriel Electric Company had to find customers for its maximum power-generating potential throughout the year. But the seasonal variation in river flow left the company with too much power in the winter and not enough in the summer to meet the needs of its customers. The railroad, on the other hand, relied on fossil fuels to generate electricity for its railcars. Hydroelectric power, when available, was cheaper. The railroad's generating plant, however, when running with the greatest efficiency, produced excess electricity. Through the link, both companies made the most cost-effective use of their power-producing investments. At the very same time, they maximized use of the river and conserved fossil fuels, fundamental concepts embraced by conservationists at the turn of the century.

Links between companies conferred additional benefits. In the late nineteenth century, Henry Sinclair, a local entrepreneur, invested in numerous power operations in the Los Angeles region, each functioning as an autonomous company.[5] Sinclair formed Redlands Electric Light and Power Company in 1892 to provide power to a local ice-making facility. In 1896, Sinclair and another prominent business investor, Henry Fisher, established Southern California Power Company to build a hydroelectric dam on the Santa Ana River. Southern California Power transmitted power from the headwaters of the Santa Ana River to Los Angeles on an eighty-three-mile line, deemed the world's longest in 1899.[6] A short segment of this line is illustrated in figure 2.1. Southern California Power also employed a temporary interconnection with the Redland Electric Light & Power Company to provide backup power during construction of a new powerhouse. With the link, Southern California Power accessed power during planned and emergency construction outages. By 1905, Southern California Power operated four hydroelectric plants and three steam-powered plants as "a coherent group connected to a common network."[7] Southern California Power's network formed one of the "three great systems" in the country, the others operating in the San Francisco Bay area and Salt Lake City, Utah.[8] The system allowed the separate plants to "help and receive help," share in the general average of supply, balance out different stream flows, render multiple flows more valuable, provide each other with "virtual reservoirs," and "duplicate" each other.[9] As a result, the power company enjoyed economy of operations, enhanced safety from interruptions, and a wider and more profitable market.

These two examples reflect the concerns that led early power developers to interconnect. Investor-owned power companies, especially, looked for economic benefits, including a diverse customer base, a lower cost per unit of electricity generated, and the least expensive option for backup power. Alternative technologies, such as storage batteries, offered similar advantages. Only interconnections provided the added benefit of allowing power companies to manage primary energy resources. Key to interconnection, however, was the notion that each system in the network operated autonomously,

Figure 2.1
The Santa Ana River 33 kV Line.

Source: Southern California Edison Photographs and Negatives, the Huntington Digital Library, used with permission of Edison International.

meeting its own economic objectives while sharing responsibility for keeping the links stable. For the San Gabriel–Los Angeles link, this was not difficult. Because the companies shared direct current (DC), the operators simply sent electricity across the power line in one direction or the other according to plans and needs. For networks over which companies transmitted alternating current (AC), things quickly became more complex.[10] To understand why companies first chose alternating current and then chose to interconnect, it is important to look back to the origins of electrification in America.

The Building Blocks of Interconnection

In 1882, Thomas Edison flipped a switch at his brand-new Pearl Street central generating station and illuminated four hundred lamps in downtown Manhattan, changing the world's understanding of electrification and its practical use. His direct current system replicated the benefits of existing urban gas lighting, in which a central station provided energy through lines connected to the points of illumination. But Edison's approach also added the advantages of fire safety, cleanliness, improved quality of light, and the "wow" factor of electrification.[11] Within a very short time, Edison systems, in competition with both gas systems and other types of electric lighting systems, appeared in cities all over the world.[12] By 1890, there were one thousand central station plants in the United States alone.[13]

Before 1882, experiments in electrification had captivated the public from Europe to California. Theaters in Paris, London, and New York employed lighting for special stage effects; arc lights illuminated central cities; generators powered lighthouses; and several inventors had introduced incandescent bulbs.[14] Charles Brush installed the first true central station service for arc lighting in San Francisco in 1879. During the mid-nineteenth century, applications of electricity to motor force drew the interest of scientists and inventors as well. The first invention of a dynamo electric machine appeared in 1860, and Nikola Tesla defined the challenge of developing a simple induction motor in 1877. In 1879, English physicist and engineer William Ayrton forecasted the advantages of central station service for a combined lighting, motor power, and heating system built on the model of gas service. The beauty of Edison's lighting system lay in the integration of generators, distribution lines, and long-lasting incandescent bulbs in a single, and potentially profitable, DC network.[15]

Within a very short time, however, operators recognized the limitations of the Edison, Brush, and other lighting systems. Three issues dominated discussions about future directions in electrification: (1) longer-distance transmission of electricity, (2) electrification of trains and other forms of transportation, and (3) application of electricity in manufacturing shops and plants. In Edison's DC system, the generators and the load (lamps or motors) operated at the same voltage, carried over a three-wire copper line. As the electricity moved across the line, some of the energy dissipated. Thus, to reach a lamp or motor at a farther distance, the generator had to supply extra electricity to compensate for the line losses. And to carry more power over a longer distance, the copper line had to be ever larger. The high cost of copper made it prohibitively expensive to locate generators more than one mile away from the load. The limited area served by Edison's Pearl Street Station in the 1880s is indicated in figure 2.2. The distance limits could be overcome if the transmission line carried a higher voltage of power, resulting in smaller line losses, but in the 1890s, no practical technology existed for changing the voltage up and down on a DC system. Line losses are inversely proportional to the voltage and are smaller when the voltage is higher on a given power line.[16] Further, the Edison system did not have the flexibility to support motors at higher or lower voltages. George Westinghouse and others who operated alternating current lighting systems, in competition with Edison, faced problems of their own, chiefly the lack of a practical motor. Edison overcame multiple shortcomings in his central station service during the 1880s and even experimented with AC systems. But he still promoted DC as the safest and most technically feasible approach to providing electrical service. Edison and Westinghouse engaged in a period of bitterly fought publicity contests and patent litigation on this matter. During this time, inventors outside the Edison system resolved the challenge to provide power economically beyond a one-mile radius.[17]

The key to building a more flexible system hinged on using alternating current rather than direct current, converting it to high voltage as it left the generator for economical long-distance transmission, and then reducing the voltage when it reached its point of use. While several inventors experimented with the use of alternating current for motors, Nikola Tesla achieved the most successful conceptualization of the problem and the solution. In May 1888, Tesla patented the first alternating current induction motor and polyphase system, which proved crucial for twentieth-century electrification. A polyphase system has three or more conductors, or wires, carrying alternating current. In Tesla's system, an alternating current generator created a rotating magnetic field. Rotors placed in this magnetic field then whirled to produce mechanical power. Transformers "stepped up" the voltage for long-distance transmission and then "stepped down" the voltage for a wide variety of uses at the customer end. By July 1888, George Westinghouse purchased rights to Tesla's patents, hired Tesla to collaborate with in-house engineers, and began experiments to implement the Tesla system for long-distance transmission of power.[18]

The first major practical demonstration of long-distance power transmission in the United States occurred in 1893 at Chicago's Columbian Exposition.[19] The Westinghouse Corporation introduced an integrated AC

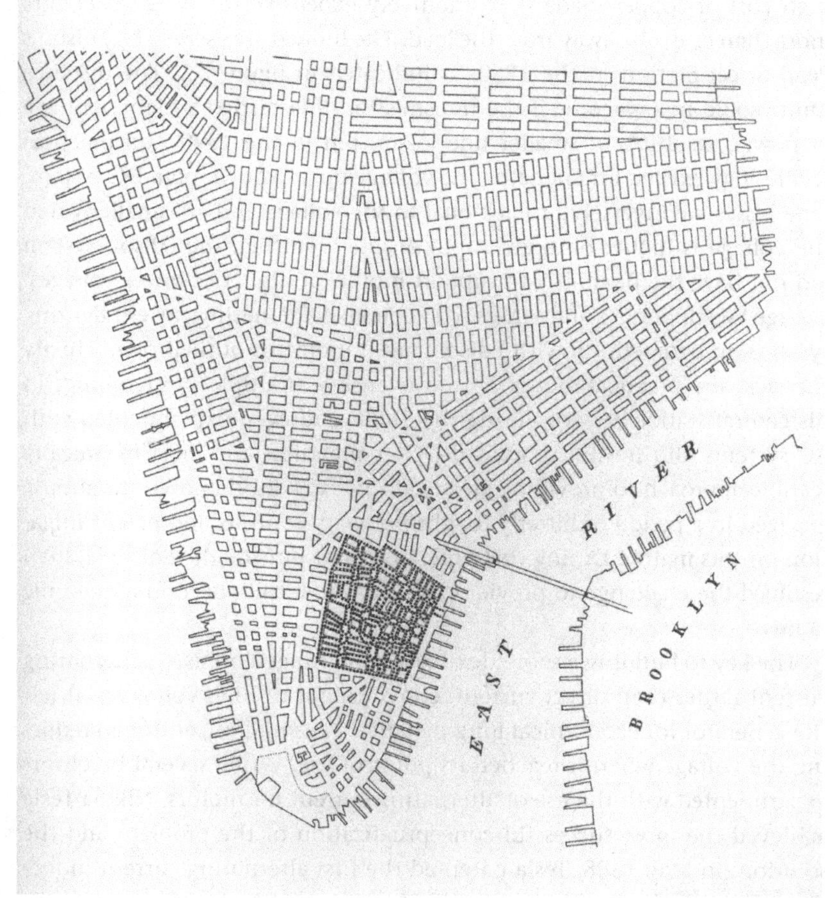

Figure 2.2
Pearl Street Station distribution area, New York, ca. 1883.
Courtesy of the Smithsonian Institution Electricity Collections.

electrical system capable of powering lights as well as multiple types of motors at a distance. George Westinghouse had previously pioneered a series of devices that shared several key characteristics: (1) transmission over a distance, (2) crucial linking mechanisms, and (3) mechanisms providing feedback to allow regulation of the system.[20] In the mid-1880s, Westinghouse installed a natural gas well and a patented pipeline that provided the most compelling model for his later AC power system. Natural gas entered the pipeline under high pressure at the well. The pressure pushed the gas through narrow pipes over a long distance. Finally, the gas moved through widening pipes to a lower-pressure level safe for use at the consumer end. Over the course of several years, Westinghouse aggregated inventions, inventors, and patents to create an analogous system for electric power. Inventors introduced transformers, rotary converters, and motors that formed the basis of the Westinghouse system. The Tesla induction motor and polyphase system lay at the heart of the Westinghouse approach.

Electricity experts immediately recognized the advantages of a universal system. Within the year, the managers of the very high-profile project to build a power plant at Niagara Falls contracted with Westinghouse to design an AC powerhouse and transmission system. The Westinghouse installation at Niagara Falls is shown in figure 2.3. Shortly thereafter, Chicago Edison installed one of the first Westinghouse AC rotary converters in a commercial electric system. Utility developers now imagined reaching larger territories with central station service, waterpower investors envisioned moving energy from the mountains to the cities, and manufacturers

Figure 2.3
Westinghouse umbrella type 5000 HP waterwheel type generators, Niagara Falls, New York, 1895.

Courtesy of the Smithsonian Institution Electricity Collections.

located factories closer to labor and market sources while transmitting motor power from farther away.[21]

The successful demonstration of the Westinghouse universal AC power system encouraged a number of electric companies, in addition to power developers at Niagara Falls, to build long-distance transmission lines. Early adopters included the Telluride Power Transmission Company in Utah, the Bay Counties Power Company in Northern California, and the Southern California Power Company in Southern California.[22] In these mountainous regions, long-distance transmission lines carried electricity from waterpower sites to consumer centers. In Boston, Chicago, Philadelphia, Montreal, and Hartford, installation of larger generators and long-distance transmission lines facilitated urban expansion.[23]

Power company investors and managers took advantage of the combined attributes of central station service and long-distance transmission, along with continuous improvements made to generators, power lines, bulbs, and motors, to spread electrification in the United States. As the century turned, the young power industry grew rapidly, as shown in the bar chart in figure 2.4. The nascent art of interconnection emerged during these first decades of power system expansion.

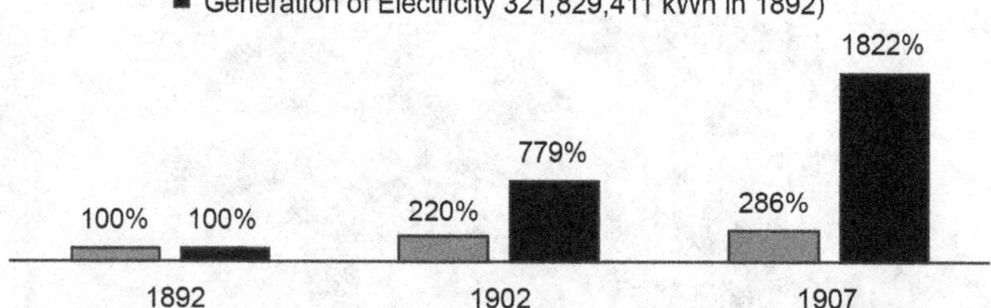

Figure 2.4
Comparison of percentage growth in number of power stations and quantity of electricity generated, 1892–1907.

Sources: Bureau of the Census, *Central Electric Light and Power Stations, 1902* (Washington, DC: US Government Printing Office, 1905), 7; Bureau of the Census, *Historical Statistics of the United States, 1789–1945* (Washington, DC: US Government Printing Office, 1949), Series G 200–204, 156–157.

What It Means to Interconnect

An interconnection exists when two or more discrete entities build a link, called a "tie line," "tie," or "intertie," across which they share electric power. The entire connected system then operates together. On an alternating current system, the generators turn at the same frequency as the electrons traveling across the power lines and at the same frequency as the devices using the power. Organizationally, each participant in an interconnected system is autonomous. In the example of the Southern California Power Company link with the Redlands Electric Light and Power Company, one investor, Henry Sinclair, had interests in both. But each company had its own financial obligations to investors and its own customer base. Power sharing thus involved both technical and financial arrangements.

In the late 1890s and early 1900s, companies pursued multiple approaches to electrification and expansion. In competition with central station service, for example, electrical equipment manufacturers sold small generating equipment to industrial customers, department stores, and apartment buildings. Industrial manufacturers in particular favored installation of on-site generating facilities over connection to central station service until the mid-1910s. Figure 2.5 illustrates the dominant use of isolated plants in manufacturing facilities until 1915. Power companies also used different strategies to handle the variations in customer demand and the threat of power failure. Storage batteries proved to be increasingly popular in these early years. Companies installed storage batteries adjacent to generating stations, charged them up during periods of low customer demand, and used them during periods of peak demand or for emergency backup power. In 1907, the storage battery was termed "the watchdog of electric service," and experts prophesied increased usage.[24] Investor-owned companies also built networks without interconnections. For example, the Commonwealth Edison Company in Chicago and Boston Edison both expanded by purchasing small neighboring utilities and repurposing their generating plants into substations.[25] The Commonwealth Edison networked facilities all served customers of a single company. None of these technical innovations called for links among autonomous companies.

The number of interconnections started to grow after 1899. In 1901, the Bay Counties Company built a connection with the Standard Electric Company in Northern California in order to "achieve greater than 200 miles of long distance transmission from Colgate Power House to Burlingame."[26] During its first six months of operation, "the experience . . . of the extreme long distance transmission service . . . is that no further question need be

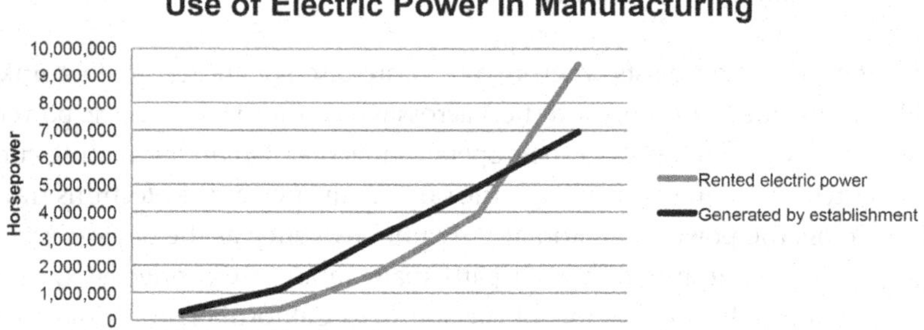

Figure 2.5
Trends for use of electric power in manufacturing, 1899–1919.

Source: Fourteenth Census of the United States Taken in the Year 1920, vol. 8, Manufactures (Washington, DC: US Government Printing Office, 1919), table 27, 122.

raised concerning the reliability of the service."[27] This complex network, deemed to be a commercial success at the time, is depicted in figure 2.6. Within a few years, "several others are feeding in at various points along the line, generating, not only at different voltages, but at a different number of phases, which are made so they can be paralleled by transformation."[28] Power systems run "in parallel" when they operate at exactly the same frequency. In Utah, where river flow caused variability in the amount of electricity available from any one plant, companies linked across watersheds. According to one report, "The throwing together in parallel a number of them greatly increases the reliability of current supply. The plan works well."[29] The editor of *Electrical World* touted interlinked systems as the key for realizing the benefits of water storage reservoirs in both economical use of generating plants and the elimination of waste.[30] By 1904, engineers calculated that interlinking would allow plants to take on a "surprising amount of load," and plants that would be uneconomical on their own could be profitable when part of a network.[31] *Electrical World* lauded interconnection as the wave of the future: "Probably the most important present tendency in power transmission is the union of several plants in feeding a great network."[32] The topic of interconnection reached an international stage at the end of 1904.

Power planners from around the world met in St. Louis at the Louisiana Purchase Exposition in 1904 (see figure 2.7). A full session was devoted to the now rapidly advancing technologies of long-distance transmission and regional systems, marking a turning point in the industry. The majority of

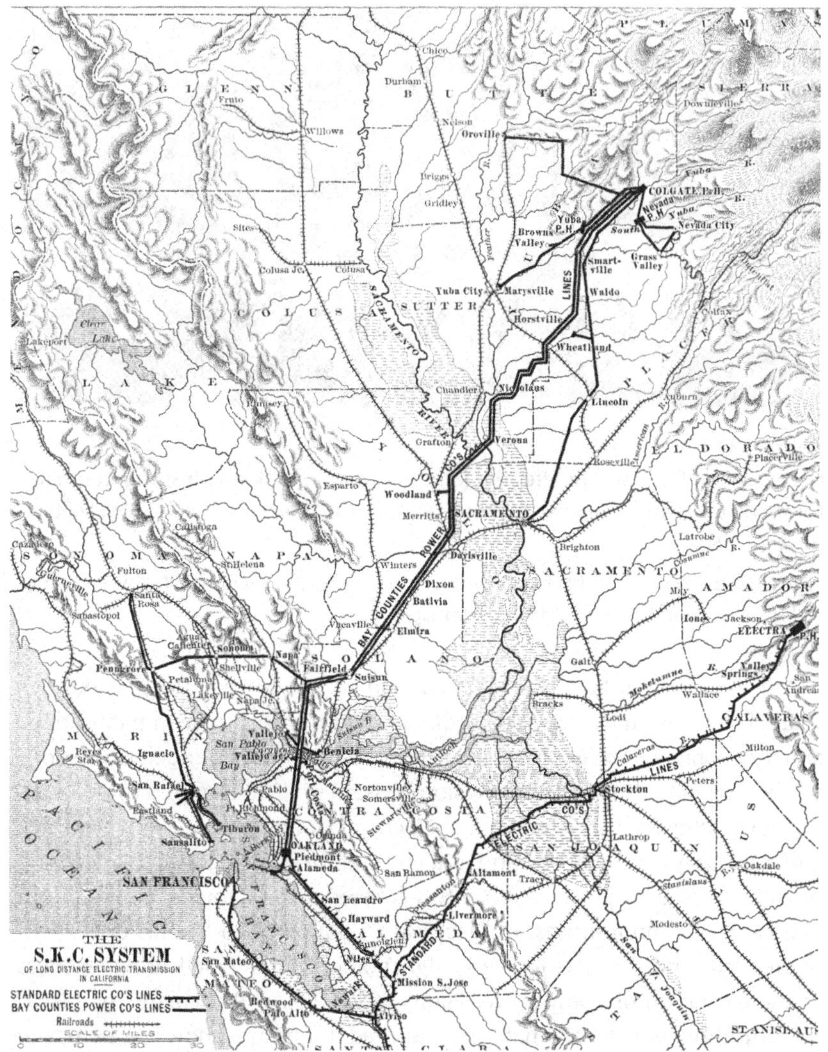

Figure 2.6
Map of the Bay Counties Power Company system in California, 1902.
Source: "Electrical Transmission on the Pacific Coast," *Electrical World* 39, no. 1 (1902).

Figure 2.7

Group before Festival Hall after the joint meeting of Section F, Institution of Electrical Engineers and American Institute of Electrical Engineers, Wednesday, September 14, 1904.

Source: Transactions of the International Electrical Congress, St. Louis, 1904, vol. 2 (St. Louis, MO: J. B. Lyon Company, 1904), frontispiece.

advances in long-distance transmission had occurred in just the previous five years. The focus on electrification coincided with a brief spike in publications about the state of the electrical arts, including the first extended discussions in print of long-distance transmission of power.[33] The Electric Power Transmission Section of the International Electrical Congress conferred over three and a half days at the exposition and covered the pressing issues concerning movement of bulk power.[34] Presenters covered sixteen topics, ranging from "The Use of Aluminum as an Electrical Conductor," to "Notes on Experiments with Transformers for Very High Voltages," to "Practical Experiences in the Operation of Many Power Plants in Parallel."[35] Participants pondered the stunning possibility that five hundred thousand kilowatts of power could travel more than five hundred fifty miles in the foreseeable future. In 1904, transmission of twenty-five thousand kilowatts over two hundred miles was considered "barely feasible."[36] Notably, the attendees at the St. Louis meeting displayed a great deal of curiosity about the experiences of presenters who had operated stations in parallel—a requisite of interconnection.

Bringing two or more AC systems into parallel and then keeping them in synchrony represented a key challenge of interconnections. On a system using alternating current, not only does the electric current reverse direction at a given frequency; it also has a particular wave pattern. If one system's wave pattern is out of synchrony with the connected system's wave pattern, movement of electricity will be disrupted and become unstable. As noted in the case of the Bay Counties Company interconnection, reliable operations were crucial to a successful project. Frederic A. C. Perrine, founder of Stanford University's Department of Electrical Engineering, explained to his international colleagues in St. Louis, "In this country, the plants that are commercially the most successful, are those having more than one generating station, and in which the generating stations are inter-connected, running in parallel and aiding each other. In a few parts of the country, this has even been done by rival plants."[37] He claimed that the very complexity of the power systems ensured their success. Regarding leading or lagging currents, he said, "The system is fed at many points and supplying current at many points as in this case, such difficulties disappear."[38] Parallel operations, and hence system stability, made interconnection profitable.

Until the 1904 meeting, most of the reporting on interlinked systems described the power plants and the related challenges of building them but gave short shrift to operations. According to engineer R. F. Hayward, in the hands of a skillful writer, "it would be a tale of fights with the forces of nature in the great valleys and mountains of the West; fights against ice, snowslides, floods and rockfalls, brushfires, windstorms and lightning, where time was always on the side of the enemy."[39] Hayward had worked in the power industry for twenty years by this date and, as electrical engineer for Utah Light & Railway Company, directed the development of interconnections in Utah.[40] At the 1904 St. Louis meeting, Hayward reported on the operating success of the combined system of the Telluride Power Company and Utah Light & Railway Company. He identified seven key points that reflected the practices and technical innovations of the system. First, he underscored the importance of hiring and organizing an ambitious and well-prepared staff of engineers and "artisans" and noted that "character is as important as skill."[41] Second, he promoted installing a private telephone service between all powerhouses and substations, laid out with as much care as the transmission lines themselves. Third, system operators had to be attentive to, and address, the load factor of the system as a whole:[42] "By a proper combination of water-power plants (some with storage, others without) and a steam plant, all proportioned properly in relation to one another and to the load factor, it is possible to utilize every cubic foot of the water in the minimum season and to obtain an economy that is greater than either by steam alone or water alone."[43] Fourth, Hayward reminded the assembled engineers, "Good speed regulation is absolutely essential to the success of a large power system."[44] In other words, maintaining steady frequency was essential on these networks. Fifth, proper voltage regulation assured that the system could provide good service to a combination of customers for multiple uses. Sixth, Hayward called for careful location and construction of the physical elements of a network to minimize breakdowns on the transmission lines. Finally, he addressed technologies that enabled networks to withstand lightning strikes and other types of disturbances. Hayward confidently projected, "It would be possible to-day to operate a string of steam and waterpower plants in parallel from the Atlantic to the Pacific Coast and to supply power to trunk railroads with so few interruptions that train service could be as punctual as it is today on steam roads."[45]

Fairly early in the process of electrification, company managers realized how much reliability mattered to their clients. As one noted in 1897, "With the growth of systems, the service which was in its early days rather fitful and liable to shutdowns at any time has become more reliable, and the feeling has grown up that the shutting down of a line is something almost criminal."[46] Early in the 1900s, another offered that "continuity of service has been one of the prime requisites of commercial success," and with growth, it "has become imperative."[47] In 1905, the editor of *Electrical World* mused, "One of the kinds of information about power transmission most needed is exact knowledge of the degree of continuity of services which is now attained."[48] The growing focus on reliability reflected a tension among power companies: did they offer a commodity or a service?

In the earliest years of electrification, Edison and others sold a commodity to customers capable of paying a high price. By the early 1900s, some utility managers realized they offered a service critical to a modernizing country. With interconnections and other advances in the technologies of electrification, power producers brought the per-unit price of electricity down quickly, as illustrated in figure 2.8. In just the first few years of the century, the cost to consumers fell by almost 50 percent. In response, more customers electrified their homes and businesses, and rates of use rose. Power companies began to reframe their product as a service: "Every step upward in the overall efficiency means a chance for a more economical supply and a larger market. . . . It should be possible to make electrical supply a necessity and not, as it now is in many instances, rather a luxury."[49] Customers increasingly expected reliable service, and this added both value and risk to the process of interconnection. On the one hand, companies built links to have access to emergency backup power without having to invest in larger generating plants or storage batteries. On the other, the link

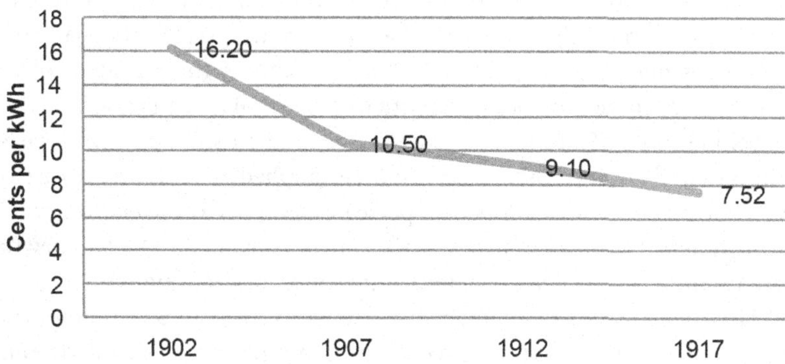

Figure 2.8
Declining cost of electricity from 1902 to 1917.

Source: Historical Statistics of the United States, Colonial Times to 1970 (Washington, DC: US Government Printing Office, 1975), Series S 108–109, 827.

itself might pose a greater possibility of a power failure. Through informal exchanges of information and experience, power system operators learned how to improve system reliability.

The Role of Experts

The group assembled at the St. Louis Exposition represented a growing fraternity of experts in electrification.[50] Hailing from Canada, South America, Europe, India, and Australia, these men worked at the leading edge of innovation in power system development. Not seen in the photograph from the meeting, in figure 2.7, are the additional thousands of engineers, system operators, manufacturers, and academics who joined professional associations, read trade journals, attended conferences, and visited each other's facilities. Nearly all men, these individuals together established the body of knowledge that facilitated the reliable generation and delivery of electricity across the globe. In lieu of designated governing authorities, the information exchange networks created by the experts provided opportunities for collaboration, critique, and the development of voluntary standards, especially for power control and reliability, followed by most of the industry.

The notion that a group of power system experts represents a fraternity of sorts provides a useful way to understand the role of these individuals in advancing the art of interconnection. Across a wide variety of fields, the term "fraternity of experts" describes practitioners who have reached a level of both capability and shared experience that sets them apart from the more general population. In 1897, for example, the president of the Institute of Actuaries encouraged participation in his organization, saying, "To belong to this Fraternity of Experts as a student should now prove an inspiration."[51] In 1902, the same moniker applied to fashion: "Insist on seven buttons, for some tailors who persist in ignoring Poole and the whole British fraternity of experts are offering them only six."[52] And somewhat more recently, it was used to refer to specialists researching social welfare and labor legislation: "The fraternity of experts, in contrast to any other segment of society, was distinctive for its combination of knowledge and allegiance to the broad community interest."[53] While the word "fraternity" can be defined as a group sharing common interests or professions, it has a particularly male slant. The root word *frater* specifically means "brother." Although a number of past references to fraternity of experts address the involvement of women in activities ranging from golf, to material preservation, to the corporate boardroom, the male connotation is particularly apt for the specialists in electrification.[54] This was, and still is, a notoriously male-dominated industry. Only a tiny minority of women appeared in the power system control rooms through most of the early years of electrification.[55] The bylines of journal articles, authors of papers presented at professional conferences, and textbook editors were by and large male. According to Amy Sue Bix, men in the engineering professions "deliberately cultivated a macho image for their field."[56]

These predominantly male experts working on power systems engaged in collegial exchanges, another attribute of fraternities. Not only did they share a special language that described how generators, power plants, transmission systems, and electrical appliances worked, but they also shared knowledge deliberately and widely. Of the journals targeting experts in the fields of electrification, lighting, power station operations, and general advances in electrical technologies, *Electrical World* was one of the preeminent publications at the turn of the last century.[57] *Electrical World* began publication as *Operator and Electrical World* in 1874 and, over the years, merged with numerous other periodicals that carried similar stories. Articles addressed telegraphy, telephony, electricity, lighting, power, automobiles, and transit systems. In addition, *Electrical World* published a weekly digest of news from other related periodicals—including those from foreign countries—thus offering the reader broad coverage of technical, regulatory, and operating advances in the electric power field from around the world. Visits by groups of engineers from one locale to see power systems in another locale were especially highlighted.[58] Every issue listed recent patents and business formations and contained photographs, diagrams, maps, mathematical equations, and drawings to aid in the dissemination of knowledge. Despite the heated competition among electrical equipment manufacturers, the system operators made a point of describing in detail their techniques and innovations for efficient power production and distribution. Beginning in mid-1908, each issue of *Electrical World* included a section called "Central Station Management, Policies and Commercial Methods," with tips and queries from operators on everything from maximizing boiler efficiency to establishing rate structures to attracting new business and displacing isolated plants.[59] *Electrical World* articles and editorials seldom included a byline, thus most contributors, presumably knowledgeable about electrification and the utility business, remained anonymous.

In addition to the pages of *Electrical World*, the fraternity shared its expertise in many other fora. The American Institute of Electrical Engineers (AIEE), founded in 1884, began publication of meeting transactions that year. Professional associations of the other subdisciplines of engineering—from mechanical to military—also addressed electrification. By 1884, numerous other journals featured papers on electrification, including, for example, *Cassier's Magazine, Electrical Record, Electricity, Engineering and Mining Journal, Electrical Review, Transactions of the Institution of Civil Engineers, Journal of the Franklin Institute, Telephone Journal and Electrical Review*, and *Engineer*. Regional publications followed.[60] In 1885, arc lighting companies organized a trade association, the National Electric Light Association (NELA). Dominated by executives in the lighting business, the organization prominently lobbied for and against regulatory measures and disseminated technical, policy, and economic information across the industry.[61] In 1933, NELA reorganized as the Edison Electric Institute, which functions as an association of investor-owned electric companies and continues to publish widely distributed technical journals, books, maps, and industry data. Thomas Edison and his associates organized the Association of Edison Illuminating Companies to "facilitate the 'interchange of opinion'" between the Edison manufacturing company and central station franchisees.[62] The "Westinghouse men" began publishing the *Electric Club Journal*, later the *Electric Journal*, in 1904 to "inspire young men to greater endeavor, to give all electrical engineers a fuller knowledge of their craft, and to raise thereby the standard of the electrical profession."[63] Across the engineering professions and in academic settings, the experts met regularly to discuss advances in generation, lighting, transit, distribution, and motor technologies.

The multiple types of organizations and media through which the power experts communicated suggest the informality of their interactions. The associations, such as the American Society of Civil Engineers, the American Institute of Mining Engineers, the American Society of Mechanical Engineers, and the AIEE, provided the platform for debating and establishing professional standards, ethics, and policy positions for their respective membership.[64] Historian David Noble argues that engineers ensured the stability of corporate capitalism specifically through the efforts of these organizations to set technical standards, reform patent law, promote industrial and university research, and transform engineering education.[65] Between 1880 and 1930, engineers also worked through a variety of committees within the associations to set technical standards. Historian Andrew Russell illustrates that the American political culture of this era "stressed the progressive potential of engineering and the value of associations."[66] In addition to these more formal entities, power system engineers and system operators exchanged information through multiple avenues to support the practical challenges of everyday work, regardless of the larger socioeconomic and political issues at stake. As in other industries, when a problem arose at more than one location across the country, the operators, manufacturers, academics, and consultants collaborated to find a solution.[67] Further, these power system experts were often both the designers and the users of technical innovations elaborated through their formal and informal networks of communication.[68]

The crucial importance of widespread sharing of information emerged as power companies chose to interconnect. Whether they operated under a single holding company, as friendly but autonomous neighbors, or as rivals, the connected entities were bound to find common techniques for sharing power successfully. Especially with regard to frequency, companies had to agree on voluntary standards or interconnections were impossible. As linked systems expanded to include more companies and wider geographies, the value of voluntary standards grew as well. Some in the industry debated the merits of formal standards. In 1900, the editor of *Electrical World* pled for standardization.[69] In 1902, the experts examined a "Standardization Report" at the AIEE annual meeting.[70] The following year, the AIEE prepared rules for standardized electrical machinery, as did the parallel organization in Germany.[71] The editor of *Electrical World* implored engineers to establish and adopt international rules that addressed terminology and definition while acknowledging that "local custom and commercial precedent would probably render international uniformity impossible" on matters of "minute industrial detail."[72] Yet only a year later, experts mused that standards tended to inhibit innovation.[73] Though they questioned the merits of formal standards for certain aspects of electrification, the experts used the informal exchanges of technical meetings and publications to converge on practices that ensured their ability to operate power networks.

The ability of engineers and manufacturers to set standards outside the authority of the state may be considered a unique characteristic of American industries. Earlier in the nineteenth century, for example, American railroad managers developed standards in order to operate a unified network with consistent track gauge, easy interchange of trains and cars, coordinated timetables, and common accounting practices.[74] They coordinated through regional professional associations, in their technical journals, and as a result of business deals. In some instances—for example, in establishing and enforcing standards for safe braking—federal agencies did play a significant role.[75] In many European countries, by contrast, the state took direct ownership of transportation and communication networks. Russell explains that associations and special committees, peopled by "men of rank," took on industrial standardization in the nineteenth century.[76] The standards committee of an association such as the AIEE gave engineers from different companies leeway to address technical issues and "avoid the spectre of monopoly."[77] In addition, this type of collaboration strengthened the sense of unity among geographically dispersed power system experts. In the twenty-first century, international standard-setting organizations continue to rely on voluntary consensus, for which there is a long precedent.[78] In the nineteenth century, for example, engineers began working through the International Electrical Congress to establish standard electrical terminology.[79] Relative independence from government authority also characterized the development of American interconnected power systems, especially as they crossed state and international boundaries. The grid works only because autonomous operators voluntarily adopted standard practices and equipment for certain key control activities.

Expansion for Conservation

Companies increasingly built high-voltage transmission lines and interconnections across the United States in the first two decades of the twentieth century (see figures 2.9 and 2.10). In 1908, distinct systems appeared in Washington State, northern and southern California, western Michigan, the Niagara Falls region, and the Southeast. By 1918, those systems had

Figure 2.9
Map showing high-voltage transmission lines and related networks in 1908.

Source: Report on the Status of Interconnections and Pooling of Electric Utility Systems, Edison Electric Institute, 1962.

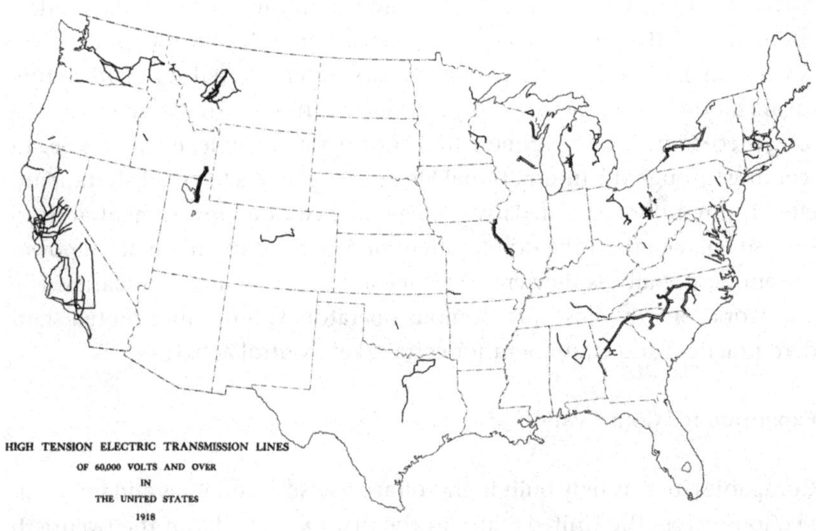

Figure 2.10
Map showing high-voltage transmission lines and related networks in 1918.

Source: Report on the Status of Interconnections and Pooling of Electric Utility Systems, Edison Electric Institute, 1962.

grown and others had developed in the Rocky Mountain regions, near Chicago and along the Mississippi River, in New England, in Texas, and in the eastern Midwest. Engineers at the Southern Power Company in North Carolina, for example, had anticipated the benefits of interconnection in the late 1890s. Contemplating the potential of the Catawba River to power textile mills and other development in the region, Southern Power built four interlinked hydroelectric plants between 1900 and 1911.[80] In 1912, Southern Power interconnected with Carolina Power and Light Company, and soon thereafter with the Georgia Railway and Power Company. By 1914, *Electrical World* dubbed this the "Great Southern Transmission Network."[81] Seven large systems across four states shared electricity, though not full time. It was theoretically possible to operate a motor in Nashville, Tennessee, with power from Rockingham, North Carolina, that traveled more than one thousand circuit miles across the lines of six companies. Within the network, "the various systems are controlled by separate and distinct syndicates," underscoring that interconnection was not synonymous with consolidation.[82]

In other regions, power companies followed similar patterns for various reasons. In the Willamette Valley of Oregon, for example, "the final step of building interconnecting lines is soon to be taken and thenceforward a great network will be built up to do for Western Oregon what the two immense plants to the southward have done for California."[83] The high cost of coal was the "greatest incentive" for expansion of the Bay Cities networks in Northern California.[84] Companies in central Colorado combined steam and hydroelectric systems in order to transmit electricity across three hundred miles.[85] Interconnection and consolidation in Ohio facilitated increased electricity output, expanded market territory, spurred modernization of equipment, and facilitated combination of hydroelectricity and steam in a single network.[86] While each network differed in configuration, they all joined a trend to "do in efficiency and reliability what no one plant can reasonably hope to accomplish."[87]

Between the turn of the century and the advent of World War I, electric power expansion took place against the backdrop of increasingly influential conservation movements and perceived coal shortages.[88] By the late nineteenth century, coal and falling water provided the foundation for most systems of electrification, and these resources were at the heart of conservation debates. The movements grew out of several concerns, including the loss of monumental places of natural beauty; the wasteful use of nature's bounty; the need for large-scale irrigation to develop arid regions of the country; inequities resulting from private monopoly control of resources; and the lack of orderly, scientific, and efficient procedures in manufacturing.[89] Conservation leaders advocated full development of river basins for irrigation, flood control, and power production. They argued that hydroelectric power generation could displace, if not fully replace, coal-fired power plants. Some deemed electricity the ideal alternative to other forms of energy used in manufacturing, for both its increased efficiency and its flexibility. These utilitarian conservationists focused on the wise use of natural resources so that the resources would be available to future generations. Others were not so sanguine about the power companies. Preservationists, nature lovers, and local populations argued that hydroelectric dams marred the scenic beauty of remote mountain areas, destroyed fisheries, and flooded upstream landscapes. In some areas, residents protested that power lines threatened human health, safety, and aesthetic preferences. Urban activists complained that coal-fired generating stations caused noise and smoke pollution, and antimonopolists objected to increasing consolidation in the private-sector portion of the industry. During this era, however, conservationists focused little, if at all, on reducing consumption, and utilities actively promoted increased use of electricity.[90]

As companies justified their plans to interconnect, they often outlined advantages that aligned with emerging ideas about resource conservation. By building links, the experts claimed, they saved coal, maximized the use of falling water, and reduced pollution in urban areas. Early on, they focused on economy and waste:

> To-day with the resources of electrical power transmission at hand, enabling half a dozen plants to be linked together and utilized as a unit, and allowing power to be economically transmitted a hundred miles or more, every power has a potential far greater than ever before. A little skillful storage and the interlinking of stations so as to distribute the load in the most advantageous way, works wonders in the economy of transmission as a whole. The art is young yet, spanning scarcely a decade, and there are many things to learn, not the least of which is the economical employment of water.[91]

The linking together of waterpower systems represented the "last great step in [the] advance toward complete utilization of the hydraulic resources of the country."[92] Americans had recklessly wasted energy resources, especially coal, in the prior century.[93] At the same time, allowing water to flow unused into the sea was also a "sinful waste."[94] With hydroelectric plants and coal-fired plants on common networks, power companies addressed both types of waste. In 1909, for example, three companies initiated plans for a transmission line, termed a "trunk line," to link systems in Ohio, Indiana, Illinois, Michigan, New York, and Ontario.[95] Both hydroelectric and steam-powered plants would feed into the network as the transmission lines passed through different regions. They claimed, "The conservation of resources and energy is the key to the work being done along the line from the Hudson to Chicago."[96] Long-distance transmission and interconnection also promised reduced urban air pollution. By locating a large generating plant outside the city, for example, a power company gained greater popularity by burning less coal within the city.[97]

Over time, discussion of waterpower development and resource management converged on the idea that interconnection could achieve conservation goals. As early as 1908, engineers observed that in regions like the Pacific Northwest, with roughly one-fourth of the United States' potential waterpower, "great combinations" would provide "general and conclusive value . . . in electrical generation and transmission" and that an interconnected network could "do in efficiency and reliability what no one plant can reasonably hope to accomplish."[98] Even if the aggregation of hydroelectric plants gave the appearance of an emerging power trust, interconnection tended "distinctly toward conservation of . . . natural resources."[99] In praising the interconnection of the "Great Southern Transmission Network," a journalist noted that it was the means by which each plant "could be utilized to its greatest advantage and the waste of water . . . could be averted."[100]

America's power companies had a vested interest in the resource-management side of conservation. Their product had peculiar characteristics rendering it useful only at the instant of demand. They faced long-term amortization of their capital investments, and they had to attract a widely diverse market over many years to both achieve operating economies and repay investors. While power companies had direct control over their own generators, their primary energy supplies depended on the vagaries of nature and fluctuating resource markets.[101]

Engineers, system operators, and manufacturers found themselves at the center of conservation debates in other ways. Leaders of the utilitarian conservation movement called for experts to assess and plan for thoughtful resource development. At the nation's first Conservation Congress in 1908, Gifford Pinchot, then President Theodore Roosevelt's chief of the US Forest Service, posited, "There is no body of men as intelligently posted on the subject of the natural resources of the country as the engineers."[102] One year later, AIEE president Lewis B. Stillwell explained to colleagues at a joint meeting of civil, mechanical, electrical, and mining engineers that conservation of natural resources was "essentially an engineering question."[103] Later, Stillwell pushed electrical engineers to engage more deeply in conservation efforts: "It is also the duty of the technical electrical engineer to furnish the leadership and sinews of war in conquest of the fields where waste and useless dissipation of energy still prevail."[104] As conservation leaders thrust the power industry into the spotlight, engineers and system operators adopted the language of conservation to describe their quotidian work. Before 1900, journal articles routinely reported on reduced operating costs, lowered capital investment (termed "avoided first cost" at that time), load factor, losses on the transmission line, and profitability. Those concerns did not disappear, but increasingly after 1900, success was also weighed in terms of coal saved, pollution reduced, waste averted, and water used. For example, one engineer reported to the AIEE that, among other benefits, a large system "solves the smoke problem by wiping out small coal plants."[105] Another told his colleagues, "The transmission of electrical energy generated by water-power which would otherwise be wasted is the most important work engineers of the present day have accomplished."[106] Engineers estimated that one horsepower of hydroelectricity displaced the burning of twelve tons of coal, and in 1916, anywhere from thirty million to two hundred million horsepower of water sat unused in the United States.[107]

Some engineers began to envision national interconnections. A speaker at the 1911 AIEE annual meeting in New York proposed, "It is to be expected that in time the country will be covered with high-tension lines."[108] One year later, the head of one of the country's largest holding companies explained, "It is possible to-day to cover the entire area of the United States with a network of high-tension lines connecting together, with efficient distribution, all of the waterpowers capable of development."[109] At his annual lecture in Chicago, famed General Electric engineer Charles P. Steinmetz predicted a coming era of cooperation among power companies and "a network of energy transmission wires covering the country."[110] Early conceptualizations of a national grid coincided with conservationist themes in the public discourse.

Interconnections neatly addressed all the areas of intersection between conservation and electrification. With interlinking power systems, utilities more closely managed coal and water resources and also operated more efficiently. Utilities that participated in full river development projects carried electricity to distant customers with interconnected long-distance transmission lines. Experienced power system engineers sat at the central controls of interlinking networks. Interconnections allowed a utility to move smelly, polluting, coal-fired plants away from urban centers and closer to either coal mines or industrial centers. Finally, the Progressive Era conservation movements focused on producers, not consumers.

The industry, despite aligning with conservation initiatives in many respects, never sought to reduce the use of energy resources in absolute terms and instead promoted increased consumption. Whether owned by investors or a local government, power companies were in business to sell electricity. Herein lies a paradox of the power industry's relationship with conservation. With interconnections, utilities achieved conservation at the producer end of the power lines while encouraging use at the customer end. In the early 1900s, the power industry found itself at the start of a lengthy and complex courtship with conservation in the United States.

Interconnections and War

Interest in electrification and the advantages of interconnection intensified during World War I. When the war broke out in Europe in the summer of 1914, the power industry focused initially on foreign markets for electrical manufacturers, not on the size and shape of electric systems at home. With a healthy and growing lighting market in the United States, central stations anticipated little change resulting from conflict abroad. Equipment manufacturers, on the other hand, felt the pinch of limits on the export of electrical equipment to Europe. Electrical manufacturers had enjoyed a healthy sales business in European countries, but access to these markets was terminated once the war began. Industry experts viewed this as a technical and commercial setback. They turned their attention to South America, which was nearby and relatively undeveloped as an export market for electrical equipment. By September 1914, *Electrical World* introduced a weekly section on "Prospects for Domestic and Foreign Business" documenting industry adjustments to the wartime economic environment. The war did eventually affect the central stations and domestic utilities—as demand for war materiel rose, so did demand for power.[111]

As a consequence of the uptick in war manufacturing, central station managers no longer competed with isolated plants for the industrial market. In the early 1900s, manufacturing establishments tended to prefer replacing waterwheels and steam engines with in-house electric generators. Factory owners believed in-house electricity was more affordable and more reliable than central station service. However, efforts to compare the economics of in-house power generation with central station service for manufacturers led some to promote the latter approach. This was described as the "Industrial Power Problem." Some speculated that the solution would involve placing a power plant at a coal mine and transmitting electricity to industrial centers. The debate continued for several years.[112] Central station managers eyed the market share of manufacturers with envy. Over the twenty years from 1899 to 1919, the amount of electricity used in manufacturing grew fortyfold. This presented a huge opportunity because utilities that succeeded in selling "rented power" to manufacturers added an excellent customer to the consumer mix. Manufacturers often demanded power during periods when residential, traction, and commercial customers did not. For example, most factories operated during the day, when lights in homes were out and traffic on electric rail lines was light. Some factories even introduced round-the-clock operations once they used electric lighting, thus benefitting the utilities even more. Central station managers actively exchanged ideas for luring industrial customers in order to expand service areas. But utilities competed directly with their own potential customers, and the latter tended to choose to install isolated plants. As the *Abstract of the Census of Manufactures* reported, "Electric motors . . . are in most cases owned by the establishments using the power."[113]

When utilities finally added industrial customers in large numbers toward the end of the 1910s, the benefits of interconnection grew more attractive. The isolated plants began to lose ground to central station service after 1910, as shown in figure 2.5. In 1912, industry journals began to report on mine-mouth plants that offered economical electricity to manufacturers. By 1913, engineers considered the comparative economy of an isolated plant to be a "moot point."[114] According to the trade journals, central stations used a third of the fuel burned by isolated plants, conserving 1,750,000 tons of coal per year. As utilities succeeded in bringing down the cost of electricity, manufacturers began to shift to central station service. During World War I, many industries faced coal shortages and found they could not generate electricity on-site. By the end of the war, the majority of manufacturers shifted to "rented electric power," and the trend continued thereafter. Postwar proposals for large-scale interconnected systems specifically focused on delivering electricity to the growing industrial market for power.[115]

The coal shortages experienced by manufacturers during World War I were greatly magnified in certain regions, and as a result, government and power industry leaders collaborated to expand interconnections. The problems began in 1917 at Niagara when Canadian authorities curtailed hydropower exports to the United States in order to meet demand at home. Niagara was considered a "war load center."[116] Manufacturers in the region had been overloaded with orders for wartime production from the English, French, and Russian governments, heavily taxing the existing generating capacity. Especially icy conditions in the winter of 1917–1918 further slowed power production. This hindered the electrochemical industry along the river and ultimately affected production of war materiel. In retrospect, Lt. Col. C. Keller of the Army Corps of Engineers reported to the secretary of war, the president, Congress, and the public, "We were taken by surprise by the shortages that finally became evident and we were without really effective means for curing them."[117] The two countries successfully negotiated a plan to curtail hydropower delivery for nonessential purposes; increase reliance on steam power on the interconnected system, even when uneconomical; accelerate the construction of new steam plants; and increase coal deliveries to Canada.[118]

Other regions, however, also suffered energy shortfalls. Before the war ended, New England, New Jersey and eastern Pennsylvania, North Carolina, South Carolina, Georgia, Alabama, eastern Tennessee, Pittsburgh and eastern Ohio, western New York, and the entire Pacific Coast experienced shortages. Residents in urban centers faced a long, cold winter with insufficient food and fuel while the Wilson administration ordered industry shutdowns and established rules for limiting the use of artificial lighting.[119] It might have been desirable to build new power-generating facilities close to the centers of defense manufacturing, but it was not feasible. Utilities measured the time from design of new plants to full operations in years rather than months, and the demand for power was immediate. Together, industry and government leaders developed strategies for providing power where and when it was needed. The federal government found three approaches especially important for war production: (1) direct defense production to parts of the country with excess electrical capacity and relatively low demand; (2) ration and manage coal shipments, giving priority to areas with war industries; and (3) call on utilities to interlink stations and run them at maximum load and efficiency.[120] The energy shortages highlighted the "desirability of interconnection" as some sections of the country formed plans to distribute power over large areas with a "maximum of economy and reliability."[121] Interlinked transmission lines were key to conserving fuel and ensuring sufficient electric power for wartime loads. Notably, ten utilities serving New England, including Boston Edison, reported improved reliability as a result of interconnected and enforced operating efficiency.[122]

In the months immediately after the war ended, industry leaders pled the case for building a power network across the country. Guy E. Tripp, chairman of the board of Westinghouse and assistant chief of ordnance during the war, favored "one reservoir" of power. He noted that the industry was hampered by fuel waste, poor loading of stations, and the inefficiency of small power utilities. Commonwealth Edison engineer Rudolph F.

Schuchardt predicted, "Ultimately the country will have a network of transmission lines."[123] Noting the lessons of the war, he advocated for universal interconnection and a common frequency of sixty cycles, and he urged Congress to stop wasting waterpower.[124] Lt. Keller, in his report on the power situation during the war, offered a plan for preventing future wartime energy shortages that highlighted the importance of linking adjoining systems with long-distance power lines. On the eve of the Roaring Twenties, the idea of large-scale interconnected power systems crossing North America fully captured the interest of engineers, executives, and government officials alike.[125]

3 Contests for Control, 1918–1934

During the early 1910s, the West Penn Power Company of Pittsburgh and the Central Power Company of Canton decided to invest together in a new generating plant. Engineers had foreseen an "almost desperate power situation developing at Canton," with steel manufacturing on the rise, increasing electrification of area factories, a growing urban population, and demand for electricity outstripping the capacity of existing plants.[1] The two companies had weighed building separate facilities but found they were able to cut costs in half by building a single larger facility. Located adjacent to abundant coalfields in Wheeling, West Virginia, their new Windsor power plant showed "every promise of becoming the most economical electric generating station ever built."[2]

Construction began in 1915 and continued for two years. With long-distance transmission lines operating at up to 130,000 volts, the companies planned to link the plant with systems in Ohio, Pennsylvania, and West Virginia.[3] Access to coal supplies had been problematic for central stations during the winter of 1917. By locating the plant at the mouth of the coal mines, the project "removed at one stroke the danger of coal shortage."[4] In addition, once the plant was fully operational, it released four thousand rail cars back to the railroads, freeing the attendant locomotives, terminals, and tracks.[5] Every major electrical manufacturer and two high-profile consulting engineering companies played roles in the design and construction of the plant.[6] Engineers praised the project in extravagant prose:

> As a fitting start in its field of service to man, this mighty energy that has lain dormant in the West Virginia hills for ages is shot at lightning speed to the top of a 165-ft. tower, across that picturesque yet industrious, complacent and docile yet at times turbulent and unmanageable Ohio River, climbing steep rocky hills, down over fertile valleys, through farm lands, woodlands, towns and villages to a city teeming with industry, there to serve its master, man, at the touch of a button to perform wonders undreamt of by our fathers who fought the Indians and cleared the forests of this great section of our country.[7]

The power experts involved in the Windsor project anticipated that it would benefit the region's participation in the war effort. The first unit began test operations in August 1917, just four months after the United States declared war on Germany.[8] By autumn, however, the city of Pittsburgh complained to the secretary of war that power shortages were crippling the region.[9] In the heart of the country's steel industry and the contiguous coal-mining territory, wartime production demand "had completely exhausted the power resources of the district."[10] The War Department, working through a regional power section, called together the local companies to address the critical challenge and devised a program that established priorities for power service, a schedule for repairs of aging equipment, access to coal, and a "comprehensive plan for interconnection and new construction."[11] Despite these efforts, the Pittsburgh area experienced a power shortage in 1918, though not as great as it might have been without Windsor station and the coordination efforts.[12] After the war, the companies continued to pursue coordinated operations with marked success. In 1921, due to interconnections, Windsor station produced "thirty to fifty per cent more power per generator than similar undertakings at present serving the larger cities."[13] The Windsor station links formed the kernel of what eventually became the largest power pool on the continent, the Interconnected Systems Group (ISG).

During the interwar years, power companies such as West Penn and Central Power expanded transmission networks from a handful of locations across the country to entire regions. Figures 3.1 and 3.2 illustrate that the fragments of networks visible in 1918 grew dense by 1934. As in the case of the Windsor station interconnection, which can be seen as a dark line crossing the Pennsylvania–West Virginia–Ohio borders in figure 3.1, much of this development took place on a piecemeal basis. Further, and perhaps just as importantly, the system operators were overwhelmed by the difficulty of controlling the flow of electricity on these expanded networks. The war department worked closely with companies in particular regions in 1917 and 1918 to develop plans that would increase power production and transmission specifically to support defense manufacturing. The war ended, but the plans remained as blueprints for expansion. Immediately after World War I, multiple proposals for large-scale interconnection emerged, all featuring coordination by a central authority. These schemes languished. The fraternity of experts, meanwhile, pursued techniques for physical control of electricity through competition and collaboration. The concentration of high-voltage power lines in California, New England, and the central Midwest, seen in figure 3.2, belies the lack of both a coordinating plan and coordinating techniques for building a grid.

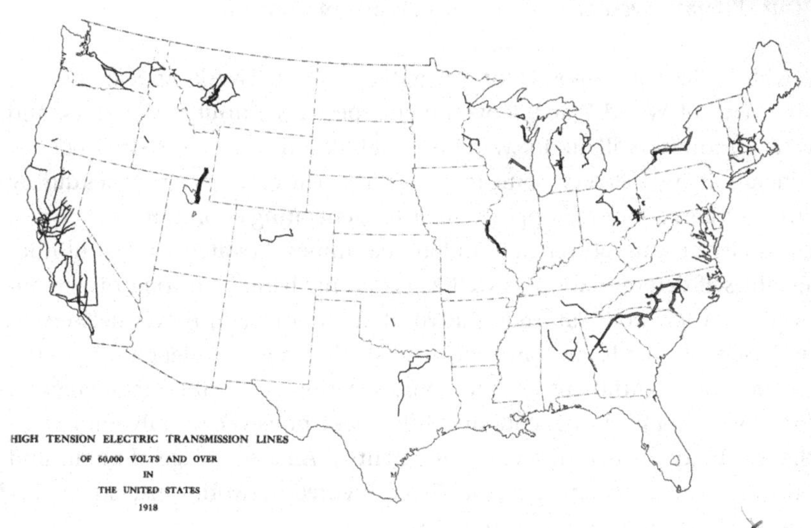

Figure 3.1
Map showing high-voltage transmission lines and related networks in 1918.

Source: Report on the Status of Interconnections and Pooling of Electric Utility Systems, Edison Electric Institute, 1962.

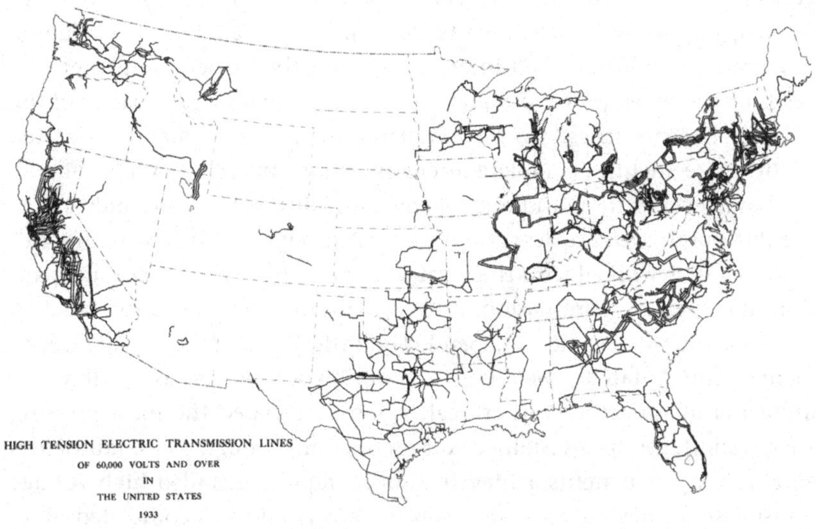

Figure 3.2
Map showing high-voltage transmission lines and related networks in 1933.

Source: Report on the Status of Interconnections and Pooling of Electric Utility Systems, Edison Electric Institute, 1962.

"Our Ultimate System": Competing Visions of Control

A wide variety of proposals for integrated power networks appeared in the aftermath of World War I. Government specialists, utility executives, and college professors all put forward ideas for building unified systems of power generation and delivery. Some focused on averting energy shortages during crises. Others addressed opportunities for generating enormous amounts of electricity at waterpower sites and at coal mines. Privately owned utilities sought self-determination as well as economic benefits through interconnection. A few political leaders advocated for equity in power delivery to underserved populations and regions. All the strategies depended on the creation of a central entity that would determine the size and shape of a future grid. Political and economic differences, however, heavily influenced the feasibility of creating a controlling entity. An assemblage of public and private utilities, operating regionally configured networks, instead characterized electrification in the United States.

The central coordination of energy resources during World War I was a highly attractive model to both utility executives and politicians for future electricity growth, despite variable success. In Niagara and New England, both private power companies and industrial manufacturers generally cooperated with the War Department in planning activities.[14] At Niagara, through "curtailments, economies, and additions," including new transmission lines and interconnections, the power companies successfully increased power production to match the unused capacity of the area defense industries.[15] In New England, the private companies boasted of their agreement to complete interconnections in eastern Massachusetts and save seventy thousand tons of coal annually.[16] The power industry in the Pittsburgh area was less cooperative. Although the War Department's representative worked with the local companies to devise plans for expansion and greater coordination, "the conclusion was reluctantly reached that local initiative could be depended on little, if at all."[17] The War Department's representatives met with investor-owned power companies and industrial manufacturers in critical regions, alleviated the most pressing energy shortages, and sustained war production through 1918. Journalists emphasized the benefits achieved through rapidly installed high-voltage transmission lines. Construction was less costly and was completed more quickly than new generating capacity, and power moved efficiently from areas of high production to areas of greatest need. Interconnection represented "our ultimate system" for postwar expansion and preparation for future engagements.[18]

In July 1918, before the war's end, Ross McClelland, chief engineer of the Electric Bond and Share Company (EBASCO), outlined one of the first large-scale interconnection proposals.[19] As the subhead to McClelland's article stated, "Shortage of power a major war problem—critical railroad situation calls for generation of energy near mines and transmission electrically—interconnection of central-station systems in war industries districts will greatly increase capacity." McClelland noted that projects already under way in Britain and France called for comprehensive increase, interconnection, and centralization of electrical supply. To keep pace, he proposed that the United States centralize the power supply in large and efficient generating plants built in the mining districts, fully exploit waterpower, electrify steam railroads, electrify coal mining, and interconnect. He called for "a comprehensive and rational policy for future power supply for maintaining our industrial standing after the war."[20]

More ideas appeared in the postwar years. Officials from the Army Corps of Engineers reported that the status of the power industry "shows clearly the need of adopting a comprehensive policy with definite plans for the construction of unified power systems over large areas, many of which are interstate in context."[21] The Smithsonian Institution offered a vision in which a network of transmission lines would serve as a common carrier to facilitate the development of mine-mouth and waterpower site plants and thus alleviate the transportation burdens experienced by the country.[22] Utilities in the Pacific Northwest echoed the growing national interest in regional planning and interconnection. Electrical engineer and University of Washington professor Carl Edward Magnusson proposed developing a 220,000-volt interconnection to link power generators and consumers from British Columbia to Southern California.[23]

In 1919, William S. Murray, a pioneer in high-tension rail electrification, and E. G. Buckland, president of the New York, New Haven, & Hartford Railroad Company, urged Franklin Lane, secretary of the Department of the Interior, to survey energy sources from Maine to Washington, DC. Murray, as the lead promoter of this project, envisioned a plan much like the one proposed previously by McClelland. The "Superpower" system encompassed development of trunk lines connecting power plants along the Eastern Seaboard, electrification of the rail system, and construction of new hydroelectric plants.[24]

At the outset, Superpower received wide support from the business community and notably from then Secretary of Commerce Herbert Hoover. Numerous engineering and utility associations, including the National Electric Lighting Association (NELA), the American Institute of Electrical

Engineers (AIEE), state and regional engineering societies, the US Chamber of Commerce, and many local chambers of commerce endorsed the plan. With Superpower, the United States could provide adequate power to meet growing industrial demands for electricity. Murray explained, "The enormous development of war industries had created an almost insatiable demand for power, a demand that was overreaching the available supply with such rapidity that had hostilities continued it is certain that we should now be facing an extreme power shortage."[25] The Superpower plan also addressed conservation of natural and labor resources: "Such a comprehensive system of power supply, making use as it would of unutilized or undeveloped waterpower and of fuel now wasted at the mines, will result in large savings in coal."[26] Decreased reliance on coal equated to a reduced need for labor at the coal mines. For the power companies, this also tempered concern about labor strikes that might interrupt access to primary energy resources.[27] The project would save an estimated fifty million tons of coal annually while maximizing the preexisting installed electrical generating capacity in the northeastern states.[28]

Opponents, however, felt that the Murray proposal fell far short of addressing the growing menace of holding companies, unequal access to electricity, and the proper role of government in providing utility services. Governor Gifford Pinchot of Pennsylvania, formerly chief of the US Forest Service under President Theodore Roosevelt, and a key leader of the Progressive Era conservation movement, promoted an alternative proposal, "Giant Power," focused on his own state.[29] While similar in many respects to the Superpower system, the Giant Power plan defined a power district entirely within Pennsylvania and specifically addressed several matters, including rural electrification, byproduct recovery at mine-mouth plants, stringent state regulation, a "common carrier" status for transmission lines, and heavy government oversight.[30] Fans of Superpower tended to oppose Giant Power, calling it "radical," "fanciful," technically impractical, commercially limited, financially experimental, and "socialistic."[31]

Over time, the question of control doomed both Superpower and Giant Power proposals in front of legislative bodies. The Pennsylvania legislature voted down the Giant Power proposal in 1926, and Congress never held hearings or in any other way formally considered the Superpower plan. The issue of who wielded the greatest authority over electrification lay at the heart of the matter. Progressive politicians opposed excessive private sector control. Private utilities opposed a central government planning agency. Even the Superpower plan, initially embraced by executives of privately owned companies, required government to "possess some control over the operation of local utilities."[32]

The debates over Superpower and Giant Power unfolded in the context of a larger contest over economic control of electric power itself. Advocates of public power pushed for increased government oversight, if not outright ownership, of the electricity infrastructure. The dominance of holding companies increased as private sector ownership was concentrated into fewer and fewer hands. Consumers—domestic as well as industrial—continued to demand more and more power, while some complained that rates, though lower than in the past, were inexcusably high. In fact, during the 1920s, while formal legislative proposals for a planned transmission grid failed at the state and federal levels, the private utility magnates solidified a dominant role in power growth as well as in interconnecting systems.

In the economic contest over interconnected power systems in the 1920s, the holding companies were winning. The private sector generated roughly 95 percent of all the power produced, as evidenced in the graph in figure 3.3.[33]

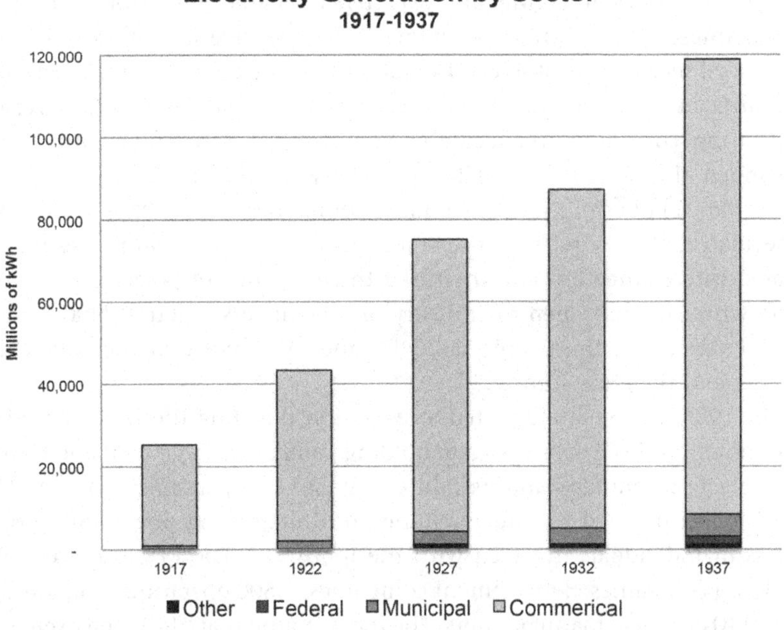

Figure 3.3
Graph comparing electricity generation by commercial, municipal, federal, and other types of power companies, 1917–1937.

Source: Historical Statistics of the United States, Colonial Times to 1970 (Washington, DC: US Government Printing Office, 1975), Series S 44–52, 821.

Through the device of holding companies, an increasingly tight-knit group of shareholders managed an ever-larger portion of the electricity market. After World War I, the public expressed increasing concern about excessive speculation in electricity holding company financial instruments. At the same time, small private and municipal generating plants found it difficult to function economically without joining larger systems, whether through interconnections alone or as part of larger financial organizations. Utilities that operated as part of larger holding companies had access to engineering and financial know-how, otherwise unaffordable for the small independent operator. The success of holding companies during this decade rendered proposals for government oversight of the industry politically impractical. The rise of a putative power trust influenced the development of the grid both politically and economically.[34]

In the first decades of the century, holding companies, though considered suspect by antitrusters, offered benefits to operating companies and consumers alike. Holding companies facilitated access to lower debt, capital, skilled engineering, and experienced management for operating companies, thereby often providing less expensive and more reliable electrical service to customers. In many ways, the benefits resembled the advantages of interconnection—investment in geographically diverse operating companies reduced the financial risk associated with each in much the same way that load diversity improved the efficiency of each plant. But critics saw the holding companies of the 1920s as merely speculative.[35] They rejected the argument that holding companies facilitated interconnection and increased the efficiency of power plants. As one writer offered, men of industry and engineers together "have one simple aim, that the people say 'Oh!' and 'Ah!' and that theirs be the power and the glory forever."[36]

In 1925, the Senate adopted a resolution directing the Federal Trade Commission (FTC) to investigate holding companies in general and General Electric Company and its spinoff, EBASCO, in particular.[37] The 1927 FTC report detailed the stockholding and interlocking directorate positions of individuals involved with the industry.[38] The FTC surveyed 60 holding companies, 3 investment companies, 1,500 operating companies, and 140 electrical manufacturers. The report found that "119 men exercise one-fifth and 57 men one-eighth of the total voting power of the directors in the control of an industry with nearly $7,000,000,000 of investment."[39] This concentrated financial control equated to political influence.[40] The report included extensive details about American Gas & Electric Company (AGE), the holding company that owned half of Windsor station through its subsidiary, Central Power Company. AGE was the first holding company formed by EBASCO in 1906.[41] In 1925, EBASCO held 6.68 percent of the voting power in AGE, which in turn owned seventy-eight subsidiaries. More to the point, the report showed that AGE actively consolidated smaller power companies and at the same time expanded interconnections and improved operating economies across the network. In the case of AGE, each subsidiary issued its own financial instruments and maintained operating autonomy within the larger system. This illustrates that the effects of consolidated ownership in the case of AGE, while clearly offering financial rewards to the privileged few at the top, also offered more generalized benefits to both the producers and consumers of electricity.

The advance of holding companies like AGE did proceed arm in arm with increased interconnections in many instances. For example, by 1929, two hundred utilities produced power in eleven northeastern states. Many of these operated under common ownership, and 45 percent of the companies were interconnected. The advantages of interconnection could be more easily realized when plants operated under the control of a single entity. Often the opportunity to achieve economies of scale and fuel conservation through interconnection served as a justification for the expansion of investor holdings.[42] Holding companies facilitated power exchanges across state lines by avoiding the oversight of state regulatory agencies and by establishing markets for high-cost generating stations.

Interconnections allowed power companies to work around regulatory limitations on regions served, and at the same time, state-level regulations facilitated the collegial interaction among power experts. Before 1907, investor-owned power companies operated through franchise agreements with local governments and frequently competed with each other as well as directly with government-owned companies in the same community. In many cities, congested overhead power lines, frequent excavation of public thoroughfares, heated public disputes about competing utility companies, and graft characterized the process of electrification.[43] Under the leadership of Samuel Insull, and with the support of a variety of Progressive activist groups, the investor-owned utilities sought and obtained state-level regulation of rates in return for monopoly control of regional markets.[44] By 1915, forty-one states had introduced power company regulation by public utility commissions.[45] This had the salutary effect of making it comfortable for utility managers and system operators to share technical information, because their employers very seldom competed for the same customers. In addition, by interconnecting across state lines, utility companies expanded the market for the power they generated beyond their state-designated monopoly region while operating outside of any regulatory oversight. Although interstate commerce was constitutionally the purview of the federal government, there was minimal federal oversight of private sector electrification at this time.

Regulation and monopoly status strengthened the notion that investor-owned power companies offered a public service. Historian Richard Hirsh argues that the utilities, the regulators, and the public struck a bargain—he calls it the "utility consensus"—that guaranteed a profit to the investor-owned companies in return for reliable and equitable service at fair rates for customers.[46] Interestingly, the power experts often described their work in terms of service to the public, regardless of the economic configuration of their employers. The American regulatory structure of the early twentieth century, though later critiqued for multiple reasons, reinforced the collegiality within the fraternity of experts.[47]

Neither fans nor critics managed to resolve whether holding companies in the aggregate benefited consumers and investors in the 1920s or merely increased the wealth of those at the top of the pyramid. Congress finally established pyramiding limits with the enactment of the Securities Exchange Act in 1934 and the Public Utility Holding Company Act in 1935 (PUHCA). Throughout the 1920s, however, autonomous utility managers, like those at AGE, planned and implemented links between systems both within and beyond the boundaries of holding companies and in the absence of a coherent national transmission network plan. Operators and electrical engineers took on the task of determining how to control electric power on these expanding systems.

Frequency Control and a Clock

For many years, when power producers interlinked alternating current systems, maintaining stability was no easy feat. On the earliest networks, two or three companies exchanged electricity on a scheduled basis. During an emergency outage, the affected firm obtained electricity from another operator only until the original equipment was functioning properly again. Systems maintained close to the same frequencies but did not operate interconnected all the time. "Closing the ties" between systems (in other words, allowing electricity to flow between them) required communication between the respective operators first to bring the systems into phase with each other and then to manually switch controls.[48] Once connected, operators maintained close contact to prevent the frequency and load variations on one system from upsetting the stability of the other(s). Utilities kept the ties closed only for the length of time necessary to accommodate the scheduled or emergency power exchange.

By the early 1920s, the information exchange and manual adjustments required to keep electricity moving economically, and without system failures, overwhelmed operators of large networks. The apparatus for managing frequency and load was as important to power networks as the physical power lines and interties. Power companies turned increasingly to automatic measurement and control devices. The foundation technologies included an electric clock and an electric data recorder. By 1931, devices that automatically controlled frequency worked well for system stability, but operators found that this upset load schedules crucial to the economic arrangements between interconnected companies. For another two decades, engineers and operators experimented with techniques for balancing the quality-of-service requirements of the networks with the financial goals of the participants.

Because electricity is dynamic, especially in the form of alternating current, even the simplest interconnected systems require finesse in order to function. On a power system, generators speed up and slow down in response to changing demand. In order to maintain a predetermined frequency, devices called "governors" automatically correct generator speed as demand changes. Originally, power companies employed many different frequencies in North America, but most began to converge on either 60 Hz or 25 Hz by the beginning of the 1900s, and by 1920, technical advances gave the advantage to 60 Hz for almost all applications. In order to share power, operators had to bring their systems into parallel; all connected generators had to be operating at the same frequency and in phase with each other. Once they brought the systems into parallel operations, the men in the control rooms closed the connections.[49] The system operators communicated by telephone, as seen in figure 3.4, and manually switched controls in order to open and close tie lines.

On linked networks, operators cooperated to prevent the frequency and load variations on one system from upsetting the stability of the others. From the earliest days of electrification, nearly all generators had speed governors. On a system with multiple generators, any of the governors might respond to a load change and cause a generator to speed up or slow down randomly.[50] In a single generating station, this might not be a problem, but on a larger system, a governor in one station might respond to a load change taking place far away when a generator in a closer station might be

Figure 3.4
Power plants and equipment: power load dispatcher's office, 1925.

Historical Photo Collection of the Department of Water and Power, City of Los Angeles. © City of Los Angeles Department of Water and Power.

able to effect a more efficient response. For example, on a two-company interconnection, a large industrial customer of one company might activate a production line, suddenly increasing demand, but the response to the change in frequency might take place on the other company's generators. The system operator had to override the automatic responses with manual controls if he wanted to ensure that each generator carried its assigned load.

By the late 1910s, interlinked power systems generally relied on a designated central station, under the watchful eye of a primary load dispatcher, to maintain a stable and functional network.[51] Prior to the introduction of automated systems, this central load dispatcher exercised authoritarian control over all segments of the network. He scheduled the stations much as he would generators within his own station, using the most economical station the most and the least economical the least. On a few networks, load dispatchers simply coordinated their operations without oversight from a central location, relying on collegial goodwill for success.[52]

In *Networks of Power,* Thomas Hughes addressed the load dispatcher's challenge in the 1920s, which he described as a "reverse salient" and "an embracing critical problem."[53] Borrowing from military historians, Hughes defined reverse salients as critical problems that caused a technological system to lag in its development.[54] In military usage, an army has an advancing line of fighters, but a problem may cause one section of the line to lag, and as a result, the progress of the entire line is slowed. This may be described as a reverse salient. If the problem causing the lag is addressed, then the entire advancing line continues together. If not, the direction and rate of progress of the line may change. Similarly, as large technological systems grow, a problem with one operational element may slow development across the system without necessarily bringing it to a full stop. Hughes asserted that failure to successfully address a reverse salient created an opportunity for a new system to emerge. He cited the example of the 1880s battle between DC and AC systems. AC systems prevailed because engineers at the time failed to address a key reverse salient of the DC systems: the limits to long-distance transmission. In the case of growing power systems and the challenges facing load dispatchers, Hughes explained, "Inadequate control, or failure to continuously match supply to demand resulted in instability in the system, which manifested itself as variations in voltage and frequency."[55] He posited that the engineers and operators corrected this reverse salient in the 1920s, but in fact, controlling frequency on expanding networks proved to be an ongoing concern through the ensuing decades.

For utilities, maintaining steady frequency on the power lines increased consumer satisfaction, attracted industrial customers, and minimized wasted power generation. Early technical instruments fell short of delivering absolute frequency control. Operators used meters that measured frequency at an instant. They collected measurements over time to document a change and then manually adjusted governors in response. Temperature variations and wear and tear led to calibration errors and often caused operators to over- or undercorrect the frequency, making the variance worse. In 1916, two problems converged. The demand for better instruments to regulate power system frequency coincided with a century-long search for a successful electric clock.[56]

Long before interconnected power systems captured the public interest, tinkerers, clockmakers, and engineers explored the possibility of creating a reliable electric clock. As Henry Warren, inventor of the Telechron Master Clock, would claim in 1937, "For a hundred years, more or less, inventors yearning for mental exercise have concerned themselves with the problem of using electricity to drive clocks."[57] Experiments in battery-powered timepieces provided the market with a variety of electric clocks through much of the nineteenth century, but these timepieces were fragile, expensive, and no more accurate, in general, than hand-wound clocks and watches. Systems of master regulators connected to secondary clocks, like the one used by Western Union, required the installation of dedicated wires. By the early 1900s, inventors still sought a reasonably priced, sturdy, and reliable clock that would surpass traditional clocks and watches in accuracy, longevity, and low maintenance.[58] Warren, an MIT engineering graduate, joined the fray as a hobbyist shortly after 1900 and patented several battery-powered clocks beginning in 1906. He established a clock-manufacturing business in 1912. By 1915, "the inadequacy of the battery clocks which [he] had been able to design and build impressed [him] so forcibly" that he looked instead to linking a clock to "an existing communication system for the distribution of time."[59]

Modern electric power systems using alternating current presented themselves as an attractive option. Warren noted that power companies had begun to converge on a sixty-cycles-per-second system speed—coincidentally ideal for synchronizing a clock. He quickly developed and, in 1916, patented a self-starting synchronous motor that would spin at the same speed as the generator powering it. This synchronous motor provided the movement for a clock face. Sadly, the accuracy of the clock was limited by the ability of a central station to maintain constant frequency, and by 1916, central station service had not yet achieved this level of control.[60]

Station operators controlled frequency fairly well by the early 1900s, but as loads changed minute by minute in the course of a day, the systems experienced minor frequency variations. Customers complained, whether they were frustrated by lights dimming or by interruptions in industrial processes. For example, textile mills, when connected to a central station for power, depended on a steady frequency to keep machines operating at a constant speed. This was the only way to minimize the fabrication of flawed goods and maximize productivity. Within the central station itself and between interconnected stations, maintaining stable frequency increased operating economy and reduced energy waste. The lack of effective control instruments limited the ability of a station operator to secure economy and customer satisfaction. A successful electric clock driven by a central station required almost perfect frequency control to maintain accurate time, and ironically, central stations required electric clocks to regulate frequency.[61]

Henry Warren faced a conundrum: How could electric clocks relying on frequency generated by a central station provide accurate time if the central station itself could not maintain a steady sixty cycles per second? Warren's solution required two clocks: "one regulated by a pendulum and the other driven by one of [his] self-starting synchronous motors."[62] Warren provided a schematic drawing of his solution in his patent application, as shown in figure 3.5. Warren connected an electric clock (identified as *c* in figure 3.5) to a central station system and a pendulum-driven master clock (identified as *b* in figure 3.5) to the electric clock. Each day, the station operator set the highly accurate pendulum clock to standard time as provided by the Naval Observatory. A special gold hand on the pendulum clock moved with the hands on the electric clock. The station operator could see at an instant if the gold hand moved faster or more slowly than the black hands on the pendulum clock, and this would signal a change in frequency. The operator could then make a frequency adjustment, bring the system back to 60 Hz, and bring the electric clock back to standard time. Warren patented the Telechron clock in 1918. He stated, "The master clock may be considered as a device to maintain a true base line for the frequency."[63]

Warren's innovation offered utilities new economic opportunities. With an accurate electric clock, power companies could sell both electricity and "time."[64] In other words, power companies sold Warren electric clocks to

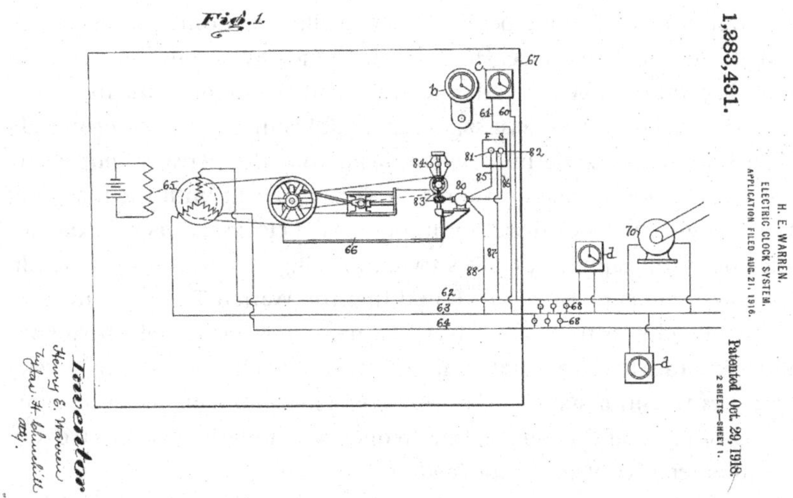

Figure 3.5

Schematic drawing of Warren Electric Clock, 1918.

Source: US Patent No. 1283431, figure 1.

their power customers. The clocks showed accurate and uniform time without winding or recalibration, but only when plugged in. By purchasing and installing these clocks, customers committed to using power all day and all night, creating a permanent load for the local power station. The companies "selling time," as they called it, were then committed to keeping the frequency stable.

Power companies that acquired the Warren system and sold electric clocks found the technology to be financially advantageous. The Warren system provided central stations with a small but continuous and even load, especially useful during off-peak periods when "time" comprised as much as 85 percent of the demand, and resulted in appreciable revenue. Electric clocks also improved public relations for power companies. In the past, to reset their watches and clocks, consumers checked chronometers in a jeweler's window or a Western Union clock for standard time. Until the service was discontinued during the war, they also called telephone operators to request the correct time. With the Warren system, consumers relied on the power company. As one enthusiastic journalist offered, "It is to be hoped that eventually every community that enjoys the use of modern electric service will have this added comfort to be thankful for—a new bond of friendship between the public and the power industry."[65]

Warren predicted in 1919 that the greatest benefit of his Telechron clock for power producers would be its efficacy in allowing utilities to operate in parallel without disturbance. He claimed, "Generally, where many systems are feeding into a large network, the individual variations in the frequency meters are such that it is necessary to make adjustments . . . before an individual station can come into synchronism with the network; but where master clocks are installed it is only necessary to wait until the machines have the right phase relation."[66] Without master clocks, errors in frequency indicators handicapped operators trying to bring systems into parallel. In fact, power company managers found that the Warren Telechron installations could address three objectives: (1) maintaining very tight frequency control in order to sell accurate time, (2) simultaneously controlling several generators within a station in order to achieve automatic and economic loading of each, and (3) regulating frequency among interconnected systems in order to control tie-line loading.[67]

Companies scheduled transfers of power, referred to as "loading," over the tie lines. Robert Brandt, vice president of the New England Power Company, remembered, "The load dispatchers attempted to keep the tie-lines somewhere near on schedule by a process of alternate begging and threatening over the telephones" with their counterparts at interconnected stations.[68] By the late 1920s, system operators using the Warren system found that frequency measurement "now ranks as the most accurate of all industrial electrical measurements," and they could maintain "more accurate speed time than could be hoped for by hand regulation."[69] Warren clocks found widespread use in the power market by the early 1930s.

While numerous technical issues affected the successful parallel operation of multiple plants, the Warren system partially solved the frequency control problem. An operator at a central plant with a Telechron Master Clock provided oversight and correction to all the other plants on the network. In the case of Windsor station and its interconnections, for example, George S. Humphrey, an engineer with West Penn, reported, "After the installation of Warren clocks by the various systems . . . there is now very little difficulty about frequency."[70] As systems grew, however, Nathan Cohn reflected, "This proved to be an arduous task, particularly considering the many other activities for which the operators were responsible."[71] Engineers concurred that to gain the benefits of interconnection, adequate control over the operation of an interconnected group was an absolute necessity. By the early 1920s, utilities operating interconnected, and selling time, sought to automate frequency control.[72]

Precision: Moving beyond a Clock

The Warren clock provided a good method for managing frequency on interconnected systems, but not all utilities sold time, and not all measurements and control could be satisfactorily accomplished solely with the clock. In the early 1920s, utilities turned to Leeds & Northrup Company (L&N), a small company with a very short prior history in the electric power industry. Philadelphia-based L&N offered high-precision instruments manufactured in the United States and, by the postwar years, had garnered the confidence of local utilities. With Philadelphia Electric Company's first request for an automated frequency meter in 1923, L&N became a pioneer in the field of automatic frequency control and later the market leader. By 1948, 90 percent of the interconnected infrastructure in North America used L&N load frequency control equipment.[73]

Morris Leeds organized Morris E. Leeds & Company, later Leeds & Northrup Company, in 1899, hoping to bring affordable precision instrument manufacturing to the United States. Through the nineteenth century, laboratories, classrooms, and industry looked to German manufacturers for highly accurate and reliable measuring instruments. Leeds, a Quaker from Philadelphia, obtained his graduate degree in physics at the University of Berlin and worked for German instrument makers for several years before returning to the United States. Leeds & Northrup established itself as a top-quality company from the start, winning the grand prize at the 1904 Louisiana Purchase Exposition for a uniquely accurate potentiometer, an instrument to measure or control low voltages. Scientific test labs quickly made wide use of the L&N Type K potentiometer. Leeds followed in 1909 with the development of an instrument that could both measure and record very small electric forces, the L&N Recorder, with additional capabilities to activate a controlling mechanism, to transmit data automatically by wire from a remote source, and to make arithmetic computations. Later, the L&N Recorder became a component of automatic control systems. During these early years, L&N targeted research and academic laboratories as its primary market.[74]

In 1911, Leeds took a step unusual for a small company, yet significant for the power industry, and created an Experimental Committee of the board. Through the work of this group, L&N hoped to improve regular apparatus, develop special apparatus on order, test materials, and conduct "experiments required in connection with other work."[75] One year later, Leeds established the Research Department. The Experimental Committee followed industry trends, watched for competition, identified problems with existing apparatus, and defined new opportunities based on reports from the Selling Department. The Research Department developed test data for use by the Selling Department to promote new equipment. In 1916, when L&N decided to expand its market beyond schools and laboratories, the Experimental Committee and Research Department took on even greater importance to the company.[76]

By 1919, L&N still had a very slim presence in the electric power industry. Only one man was "assigned to the development of the Power Plant business and also . . . general scouting in such lines as Ceramics."[77] The total budget for sales to electric plants was just $800. The board saw real potential, however, in the power industry. By midyear, a new committee oversaw the development of power plant equipment. Leeds himself also embraced a vision of automated control that would "perform better than the best human operator taking account of all the factors needed to run a system."[78] The company had patented its first automatic control device in 1917. In 1920, Leeds reorganized the board to increase the focus on the utility market. By 1923, L&N recorders found wide application in power systems, measuring generator and transformer temperatures.[79] An example of a typical recorder is depicted in figure 3.6.

Figure 3.6
Leeds & Northrup Recorder, ca. 1918. This was the prototype for a later automatic frequency control device.
Courtesy of private collection of David L. Cohn.

Through its sales force of fifteen or sixteen people, L&N actively solicited ideas and concerns from its customers. In 1923, for example, L&N learned that utilities needed better regulation of the charts connected to L&N recording devices.[80] Several of the largest power companies in North America—including the Ontario Hydro-electric Power Commission (HEPCO), Philadelphia Electric Company, New York Edison, and Boston Edison—shared this critique.[81] These engineers were otherwise impressed with the L&N product and urged the company to consider linking the chart of the recorder to a Warren synchronous motor for greater accuracy. The sales force asked the company to consider this problem with some urgency.[82]

A direct request from the Philadelphia Electric Company pushed L&N to invent and patent a high-precision recorder specifically to measure and chart frequency on power systems. According to L&N company historians, in 1923, Nevin Funk, chief engineer of Philadelphia Electric Company, told his L&N sales representative that he wanted an instrument that would tell him how much his system varied from 60 Hz. Funk predicted that L&N could sell at least a dozen a year if they worked well. By this time, Philadelphia Electric Company was one of the largest independent, investor-owned utilities in North America. If Funk's company introduced a successful frequency control innovation, other utilities were likely to follow suit. Within the year, the Development Committee reviewed ideas for a recorder that would address Funk's request.[83]

L&N's impedance-bridge frequency recorder (the Wunsch recorder) joined the Warren clock as a second major building block for automated frequency control. Felix Wunsch, an inventor in the L&N Engineering Department, presented his first frequency recorder design in June 1924, and the company submitted a patent application by March 1925.[84] The drawing of Wunsch's invention is depicted in figure 3.7. Wunsch adapted the original Leeds recorder to address frequency measurement. In the patent application, he explained, "My invention relates to a system for measuring or recording the frequency of a fluctuating or alternating current or the speed of a moving system."[85] In retrospect, prominent industry leader Philip Sporn credited this frequency recorder with being the "critical piece" in stabilizing frequency and load control.[86] As a young engineer at AGE, Sporn had conducted some of the early tests of frequency control devices, including L&N apparatus. L&N initially planned for the manufacture of six units for the sales force to show their customers. By 1926, before there was even a patent in place, L&N had sold "at least fifty 60-cycle frequency recorders and three 25-cycle frequency recorders."[87] Figure 3.8 offers a view of the Philadelphia Electric control room with the L&N frequency control apparatus installed. A company brochure published in 1927 touted the benefits of the Wunsch recorder and noted its potential for aiding in interconnections. In 1928, despite unexpected success with 214 recorders sold, the company learned of the product's shortcomings.[88]

At a key 1928 meeting, a young engineer, Leslie Heath, described rival products: a device produced by Westinghouse, to which Heath did not give much attention, and the Warren clock system. He explained that the Warren master clock showed instant deviation from 60 Hz, while the L&N instrument documented change over time.[89] Heath reported that participants in the New England System Operator's Network felt the L&N recorder was still not sufficiently accurate. He declared that there was a "ready market for a more accurate recorder." When two systems with L&N recorders

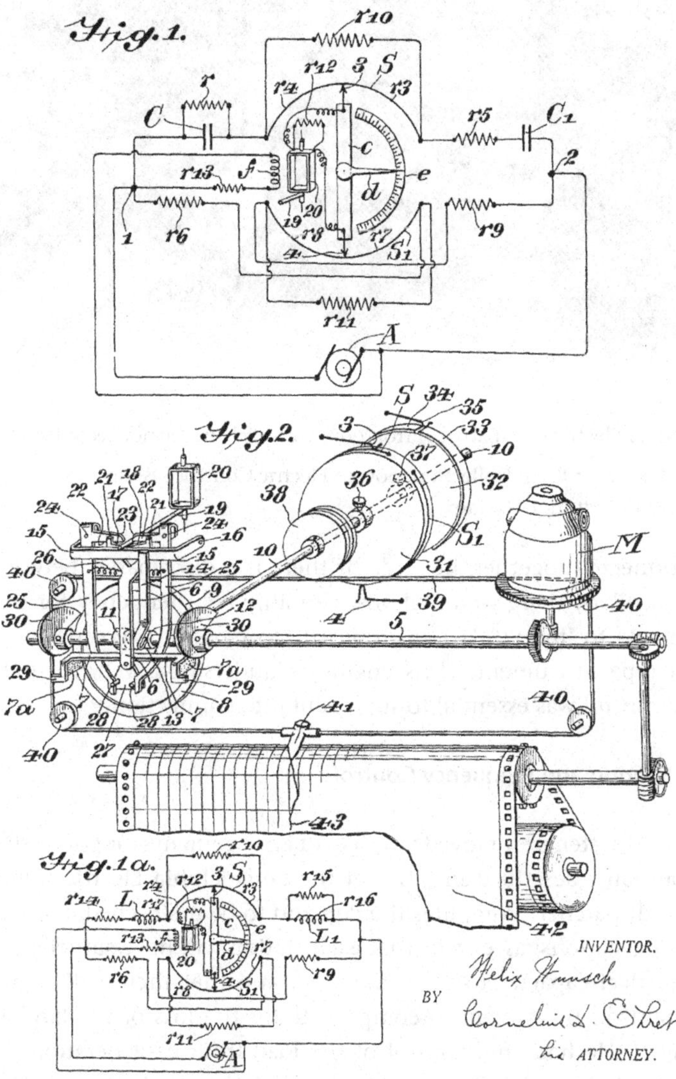

Figure 3.7
Drawing of Wunch's frequency recorder design.

Source: Felix Wunsch, "System of Frequency or Speed Measurement and Control," US Patent 1751538, filed March 27, 1925. US Patent 1,751,538.

Figure 3.8
Philadelphia Electric Company control room with L&N frequency recorder, ca. 1925.

Source: W. Spencer Bloor Collection, now in Electric Control Systems Records, Hagley Library, Wilmington, Delaware.

were connected together, he said, "if there is a discrepancy between the two controllers, there would almost certainly be a tendency for the two instruments to 'fight.'" Heath proposed "to cross our recorder with the Warren type instrument."[90] As engineers understood well, accurate frequency control was essential to successful interconnection.

Power Sharing and Frequency Control

Interlinked systems employed a variety of arrangements for power sharing. As a journalist described in 1919, at the Philadelphia Electric Company, the load dispatcher "computes the demand to be met; schedules it on his generating stations; ascertains that each station will have sufficient steam and electrical capacity to carry the load; and that there will be reserve equipment on the system to compensate for the loss of the largest unit running."[91] He had "full control of the load from its generation (beginning in the boiler room) to the customers' premises."[92] To accomplish this, the dispatchers at Philadelphia and elsewhere often had a dedicated telephone exchange linking them to all connected stations and substations. In California, home to the largest interconnected systems, all operations were "under the supervision of the Load Dispatcher, who has absolute authority of the amount of load each plant or connected company shall deliver."[93] The dispatcher's responsibility was to ensure the maximum of both safety and efficiency in operations.[94] Some networks employed more casual relationships. In the case of a large New England system, according to an engineer, successful operation required "only the complete cooperation of load dispatchers in control of system operation."[95] In this instance, Boston Edison exchanged power with two adjacent utilities, but they did not exchange power with each other. For the large Ohio-based AGE interconnection, no load dispatcher acted as the central dispatcher. Instead, Humphrey described, the various dispatchers "operate on the give and take principle, realizing that the fellow giving help today may be the one who will need it tomorrow."[96]

Utilities owned by a single holding company shared a common interest in profitability and hence cooperated well when operating interconnected. Engineers and utility executives engaged in debate about whether fully autonomous utilities could likewise gain the full benefits of interconnection, especially with respect to energy efficiency. At a special AIEE symposium on interconnection held in 1928, a senior executive with Georgia Power Company claimed, "Unquestionably, the greatest benefits are derived . . . when the interconnected companies while maintaining their independent corporate identity are subsidiaries of one holding company. . . . Full advantage is taken of diversity of time, diversity in rainfall, and in seasonal load."[97] At the same meeting, however, representatives from California and the mid-Atlantic states illustrated that "excellent results are obtained . . . even though the systems are not under one ownership."[98] These autonomous companies relied on well-delineated operating agreements and procedures and claimed they shared power as successfully as the utilities controlled by a single holding company.

The operations of interconnected systems reflected the shared management and divided authority of their respective economic arrangements.[99] In the case of the Pennsylvania–New Jersey Interconnection (PNJ), for example, the parallels between physical management of the network and organizational relationships between the companies were explicit.[100] In this instance, three utilities joined to build the Conowingo Dam on the Susquehanna River in Maryland, and power traveled to Pennsylvania and New Jersey. The PNJ promised to maximize power generated at the hydroelectric dam and minimize reliance on coal-fired steam plants, thus furthering resource conservation. To sustain the profitability of the participants and ensure a fair distribution of responsibilities, the utilities signed an agreement that later became the model for many power pools.[101]

Under the PNJ operating agreement, each member utility designated one person to serve on an operating committee, which then established the policies for operations, exchanges of energy, and forecasts of loads. The chair of the committee rotated according to the order in which the members had signed the agreement. The participating companies retained autonomy in developing their regional power systems and designed and constructed their own power lines but shared planning information.[102] In addition, the committee designated one company to regulate the frequency on the network and another to control the power flow.[103] As described at the 1928 AIEE symposium, the group established a central interconnection headquarters, which "will be in direct communication at all times, through the load dispatchers of the individual systems, with the steam and hydrogenerating stations of the three companies."[104] Despite the careful organization of the operating committee and the effort to protect the autonomy of each participating utility, there were "many delicate problems of load apportionment," and "only the most alert and skilled load dispatching will realize all the possible benefits" of the interconnection.[105]

Other operating groups formed to share management and divide authority on interconnected systems. The typical agreement called for representatives from each company to serve on an operating committee and communicate regularly. In addition to questions of current load distribution, outages, and seasonal changes, operating committees discussed longer-term infrastructure planning. In Chicago and the eastern Midwest, for example, utilities signed a "three-party exchange agreement," coordinated plans for power station construction through a committee of two representatives from each company, and assigned load dispatchers to meet weekly to arrange operating schedules.[106]

Humphrey enumerated the principal operating problems for interconnections: (1) control of frequency, (2) control of voltage, (3) control of power flow, (4) stability, both static and dynamic, (5) disturbances, (6) short-circuit currents, (7) relaying, and (8) dispatching. The engineer described many of these as challenges already met. But the difficulty of controlling the flow of power on his system led the companies to close tie lines only on a scheduled basis rather than operating the entire system interconnected at all times.[107] As a Commonwealth Edison executive explained, the multistate, long-distance "systems can be operated in parallel successfully for such periods of time as may be necessary to meet the needs of operation during emergency transfers of energy."[108] When the long lines experienced surges, however, "there is instability of operation, and lack of continuity is likely to result."[109] A "disastrous situation . . . may develop in a few minutes."[110] All the interconnected utilities in the southeastern states used master clocks and accurate frequency recorders to maintain 60 Hz to achieve continuous parallel operation. Yet, the Georgia Power Company general manager reported, "with the increasing amount of load carried over the lines, the problem of system stability has been encountered," and calculations for load division were "becoming increasingly difficult, and, in fact, practically impossible."[111] The real problem lay with the effect of frequency control at one station on load distribution for the entire system.[112]

In the late 1920s, utilities turned to automatic frequency control on interconnected systems, and this had the inadvertent effect of upsetting scheduled loading of power plants. With automatic control, frequency remained closer to the ideal of 60 Hz, as illustrated in figure 3.9. The graph on the left shows continual frequency variation under manual control, while the graph on the right shows steady automatic control. But as one engineer noted in retrospect, the automatic frequency controller was "imperious," and as it restored system frequency, "it had the adverse effect of disturbing scheduled power exchanges between areas."[113] The chief benefits to interconnection included energy efficiency, economic viability,

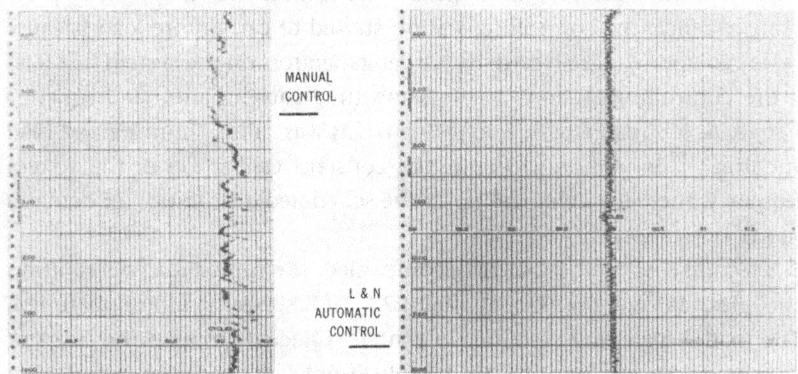

Figure 3.9
Comparison of manual and automatic frequency control on the Pennsylvania Power & Light Company System in 1928.

Source: S. B. Morehouse, "Some Historical Highlights of Operation of Electric Utility Systems on an Interconnected and Coordinated Basis" (paper presented to the Annual Joint Meeting of the Northeast Regional Committee of the Interconnected Systems Group and the System Operation Committee of the Pennsylvania Electric Association, Pittsburgh, Pennsylvania, 1965), 10.

and reliability. Utilities realized these benefits only through careful estimation of demand and determination of which plants should operate at what level and when. Load dispatchers scheduled load distribution to match the anticipated demands with the most efficient generating source. With automated frequency control, operators found their optimization of an interconnected system seriously compromised. As Humphrey explained in 1930, "The factor of maximum economy is rapidly coming to the fore as an essential element in effecting constant frequency [on interconnected systems]. It is possible to maintain constant frequency but at an expense that far outweighs its intrinsic value."[114]

Load changes affected both single generators and interconnected systems in the same way. When a load change occurred, the generator(s) responded by slowing down or speeding up accordingly. As difficult as it was to maintain planned load schedules with manual frequency control, automatic frequency control made it worse.[115] For example, if a large industrial customer greatly increased the amount of electricity in use, the turbines in the generating station slowed down, causing the frequency to drop. An automatic frequency controller detected the dropping frequency and caused the turbines to speed up, all within moments. On an interconnected system with one station managing frequency control, the automatic frequency controller caused the generators in that station to speed up. As a consequence, the controlling station started to carry more load because it was generating more electricity. Other generators on the system lost load to the controlling station, at the same time causing another frequency change. As a British engineer explained, this was called "hunting and load pinching."[116] In addition to requiring constant calibration of the system frequency, these load changes upset the scheduled distribution of demand among the stations.

A second economic problem accompanied interrupted load distribution schedules. As one engineer at the 1928 AIEE special symposium noted, "The load-dispatching system of an interconnected group is the heart of the whole question, and it must be solved not only from the engineering but from the executive standpoint on very broad lines."[117] Another participant suggested, "The engineering of interconnections has been worked out farther and better than the commercial or contract features. . . . There is a need for a simple but equitable form of agreement that does not require a Philadelphia lawyer or a Schenectady engineer to interpret."[118] Utilities participating in an interconnection followed accounting systems designed to equitably distribute the costs and benefits of operating certain stations more and other stations less. If load inadvertently shifted from one company's station to another's outside of the scheduled interchange, a cost accounting problem occurred. How would the accountants address the electricity sales under this circumstance, who would pay, and who would collect? Different owners on a single system had to "sort out, classify, and account for the different classes of power flowing" because this was "the basis on which money was exchanged."[119]

A Field Test and the Load Control Problem

The New England Power Company (NEPCo) took the lead in experiments with automated frequency control. In 1927, NEPCo installed both an L&N automatic frequency controller and a Warren Telechron controller at the Harriman station, a hydroelectric plant interconnected with four different utilities. After testing both devices, the operating engineers determined that each worked well enough to validate the principles of automated control, although neither was yet entirely reliable. Hunting occurred regardless of which device was used.[120] Robert Brandt noted, "If the bulk of the load change, however, should come on one system, then the automatic controllers, while bringing the frequency to normal, would necessarily upset the steady flow of power over the tie-line. From this it appears that it may be necessary to incorporate with straight frequency controllers some sort of tie-line load control."[121] The author lamented that with the growing complexity of systems, "the development of satisfactory apparatus to accomplish all the desired results automatically still appears to be in the distance."[122]

Several customers and NELA pushed Leeds & Northrup to address the related problems of frequency regulation and load control. Philadelphia Electric Company asked for a quote for a combined system in 1928. L&N acknowledged, "Load-frequency control is an important development which promises to result in a considerable amount of business, as, if the System is successful, it will mean application of a Frequency Controller to practically every machine in the Stations in which the System might be installed."[123] The Development Committee determined to devote energies to making a "real success" of the installations at Philadelphia Electric Company, to proceed with marketing of automated control directly to prospective customers, and to investigate and settle the patent situation.[124] But the field of frequency-load control was so young that it was "not possible to determine very accurately at this time the ultimate extent of the field."[125] Nonetheless, Leslie Heath and others took on the task of developing apparatus to simultaneously control frequency and load on interconnected power systems.

Leeds & Northrup faced active competition from much larger and more established electrical manufacturing companies, especially General Electric Company (GE). GE posed "stiff price competition using Warren type equipment" and boasted of technical superiority.[126] As Heath reported in 1929, "Unless we produce substantial advantages over the [Warren] apparatus," L&N would have to meet GE's very low price.[127] Heath outlined the technical considerations of frequency-load control. If every station in a system had a frequency controller, there would be undesirable load shifts. If only one station controlled frequency and the others operated with a fixed load, the controlling station would be subject to "undesirably large load fluctuations."[128] The L&N frequency controller was reliable for keeping frequency stable but could not guarantee correct time. In contrast, GE claimed the Warren system operated in each station at the same time and kept correct time without "undesirably shifting load between stations."[129] Heath concluded that this was theoretically possible, but he doubted its practical success.

The system L&N proposed to pursue combined the L&N frequency controller, a synchronous motor and clock like the Warren system, and separate wires to carry the control signals. In early 1929, Heath applied for a patent for the automatic control apparatus. Heath and several other engineers continued to consider approaches for automatically regulating both frequency and tie-line load. The proposed solution included a separate system of wires (pilot wires) between all participating stations to carry load-controlling signals. By the end of 1929, the team understood that the rapid growth of interconnections would make proprietary pilot wires impractical: "The implication . . . is that practically a national system of pilot wires would be required for the successful carrying out of such a scheme," and the utilities were in no position to promote this.[130] The engineering team approached American Telephone and Telegraph (AT&T), and by 1930, the telephone company not only expressed interest in participating but also engaged in experimental installations at utilities. The engineering team (including Heath and another engineer named Doyle) finally submitted a patent application for this method and system of automated frequency and load control in 1931.[131]

The patent situation posed an additional problem for the engineering team. In 1929, the Development Committee appointed a special committee to address the complicated patent considerations. The art of frequency control appeared "to be very old," with patents dating back to 1896.[132] Practically every element the engineering team wanted to use, other than L&N proprietary apparatus, was subject to prior claims, though they might be invalid. Most significantly, GE, Westinghouse, Allis Chalmers, and AT&T owned all the relevant patents. Heath's team reported that all these companies "appear to be aware of our activities in this field."[133] The inaction of patent owners might imply acquiescence and that patent negotiations would be reasonable. By 1930, L&N believed that their apparatus was ready for a wide market. Executives hoped that within a year, L&N's "own patent situation may be in such shape that we will have something tangible to offer should we attempt an interchange of licenses under the patents involved."[134]

L&N proceeded with collaborators and competitors to develop a working system for frequency and load control. As Nathan Cohn noted in retrospect, there was "relatively little control theory. Simulation as practiced in recent years was not available for control experimentation. It was not, however, especially missed. . . . Experimentation on the best of all simulators, power systems themselves, was feasible, and was practiced."[135] L&N worked with multiple utilities to run field tests of its equipment. As had taken place in 1927, several experimented with both the GE Warren system and the L&N system.[136] The issue of patent infringements clouded this year of experiments. GE's Warren system posed a particularly thorny problem because the L&N engineers found that their solution incorporated Warren's synchronous motor. The L&N patent attorney recommended proceeding with publicity and sales efforts and waiting for notices of infringement to arrive. Echoing a letter from NELA, he stated, "It is of great advantage to yourselves to enter [the field] before it greatly develops."[137] He thought L&N would avoid infringement entirely or secure immunity "because of your friendly relations with others in or about to enter the field."[138]

Taking this advice, L&N moved ahead with direct sales to potential customers. Heath prepared a memo for the company's branch offices describing the experiences at twelve stations that had installed L&N automatic frequency control devices, some of which also used the Warren system. Those stations that controlled frequency from one generator found problems with load distribution. By contrast, a system that shifted responsibility from one station to another depending on operating conditions experienced great improvements in efficiency. On this system, operating losses were reduced to less than one-half of 1 percent for extended periods of time, and station economy rose by 8 percent. Heath reported that a second station with similar flexibility showed promise in the area of biased load frequency control—that is, the generators were adjusted to help a little with load swings without dramatically affecting economy. There were no immediate legal actions from competitors, and Heath encouraged the continued sales of the existing L&N equipment.[139]

Collaboration proved to be fruitful for L&N in the near term. As Philip Sporn recalled, "Results have been achieved by the work of progressive manufacturers, and by co-operation between technicians of many companies, hundreds and even thousands of miles apart."[140] Southern California Edison incorporated the L&N frequency controllers with an in-house time error correction system (similar to the Warren system) to regulate load shifting between steam and hydroelectric plants "as indicated by water storage and flow conditions and economy considerations."[141] At the Carolina Power and Light Company, L&N, together with I. P. Morris (a division of the Baldwin Southwark Company), installed apparatus to automatically load hydroelectric units to achieve optimum water use efficiency. This system also incorporated Westinghouse wattmeters.[142]

AGE conducted an extensive and definitive experiment with L&N and Warren systems for frequency control at three stations. Both collaboration and competition marked this project. Companies in the Pennsylvania-Ohio Interconnection, including West Penn, Central Power, and others, relied on Windsor station for frequency control for the entire system and load control for one area. This resulted in "numerous regulating troubles and arguments with the load dispatching organizations on all sides of it."[143] Both L&N and GE demonstrated their automatic frequency controllers to the system operating committee in 1929. By fall of 1930, with intense observation from NELA and from other utilities, the two instrument manufacturers agreed to lend AGE control equipment for testing on the Pennsylvania-Ohio Interconnection. Heath justified this to L&N by noting that it would be an inexpensive way to learn if the new apparatus worked "under real life conditions."[144] Further, if the company failed to match GE's commitment to this experiment, L&N would lose its prestige. Finally, the trials would certainly illuminate how the L&N apparatus compared to the GE products.[145]

Trials took place beginning in 1931. In frequent handwritten letters, formal memoranda, and committee reports, L&N engineers documented the progress on the AGE system. The experiment also extended to an interconnection with the Commonwealth Edison system. From the beginning, the L&N team closely observed the GE approach: "If we had known before what we do now about . . . what G.E. was supplying we would have changed our layout considerably."[146] The L&N layout is depicted in figure 3.10. At first, the team noted that GE "knows of negotiations on our combined frequency recorder and Warren so a very friendly spirit exists."[147] Later, the team expressed concern that GE had an engineer on-site continuously

Figure 3.10
Installation of L&N equipment for field tests. The frequency time error recorder is on the left; the load limit recorder is on the right.
Source: S. B. Morehouse, "Some Historical Highlights of Operation of Electric Utility Systems on an Interconnected and Coordinated Basis" (paper presented to the Annual Joint Meeting of the Northeast Regional Committee of the Interconnected Systems Group and the System Operation Committee of the Pennsylvania Electric Association, Pittsburgh, Pennsylvania, 1965), 15.

during the tests, while the L&N personnel were there only intermittently. With such close attention, GE was getting very good results. To counteract this "propaganda," one engineer suggested that L&N needed a person to "camp on the job."[148] By the end of 1931, it appeared to L&N that the people at the utility favored GE and gave them all the advantages. It was time for L&N to "stop playing ball."[149] Despite the fact that L&N was "pre-eminent in the field of frequency control, having more installations than all other manufacturers put together," it was necessary to "improve the stability of our frequency control."[150]

Problems between the utility systems plagued the tests as well. Operators at Commonwealth Edison and AGE disagreed about their respective frequency and load control responsibilities. The Pennsylvania and Ohio utilities wanted Commonwealth Edison to regulate frequency and load at the tie line. Commonwealth Edison maintained that due to its excellent internal system regulation, there would be no need to install additional controls at the tie line. Without clear agreement from the operating companies about what was to be controlled, the L&N team lamented that manufacturers could not design satisfactory equipment. Ultimately the executives of the two systems, who both wanted to see unified operations, had to meet to restore a more cooperative spirit.[151]

The results of the yearlong tests indicated that automatic frequency control at multiple stations, rather than a single station, was a step in the right direction. But as Philip Sporn claimed, problems occurred with load shifting where "very rigid contractual relations have been set up between systems."[152] Reporting in *Electrical World*, Sporn and his colleague from American Gas & Electric explained that a setup with "automatic frequency control in each area, supplemented by tie-line control where necessary," would solve the problem.[153] Engineers from all teams began to consider "tie-line bias" to be a workable solution. Tie-line bias is the option of allowing some frequency variation within an established range so that loading stays on schedule.

By this time, power system engineers understood that there were several possible approaches to managing frequency and load control. The methods, listed in Table 3.1, differed by which controls were centralized and which were distributed and by the amount of aid for frequency control permitted at the expense of load control. On a relatively small system, straight (or flat) frequency control allowed a central dispatching location to monitor and correct system frequency. Operators made manual adjustments to address unscheduled load shifting. Small load swings did not create a burden for

Table 3.1
Methods for controlling load and frequency

Methods	Result
Circa 1930	
Central automatic frequency control Manual load adjustment	Effective for smaller systems
Central automatic frequency control Distributed tie-line load control	Large burden on controlling station; intermittent load upsets
Distributed automatic frequency control	Overcorrection and instability
Central automatic frequency control Constant tie-line load control	Fails to aid neighbors
Circa 1936	
Central automatic frequency control Tie-line bias control	Aids neighbors a little bit
Circa 1948	
Distributed automatic frequency control Net interchange tie-line bias control	State of the art

the regulating generator. A slightly more complex system involved one automatically controlled station and tie-line load controllers (also called watt controllers) on the other stations. In this case, the automatically controlled station responded to frequency changes, while the watt controllers responded to load changes at the tie line. This method still allowed intermittent load changes on the interconnection, undermining the efficiencies to be gained from planned load schedules for each station.[154]

Installation of automatic frequency control devices at multiple stations on a system allowed each station to regulate its own frequency. However, every control device tended to respond to changes anywhere on the system, not only to local changes, causing overcorrection and instability. A further iteration of this approach added constant tie-line controls to the system. This control held the load to its scheduled flow at each tie line regardless of system frequency. As a result, individual stations, while retaining their planned operating activities, might fail to help the larger interconnection address major load and frequency changes. The ideal system, yet to be developed in the early 1930s, incorporated frequency regulation and tie-line control at each location. Called "net interchange tie-line bias frequency control," this method allowed for shared management of grid frequency and load distribution with divided authority over each area's scheduled functions, thus maximizing the energy efficiency promised by interconnections.[155]

Utilities and manufacturers shared the concern that centralized frequency and load control had failed to address the diverse needs of interconnected power companies. Following the experiment at AGE, the participating utilities determined that "the principle of having any unit or plant operate as a master frequency station" was not satisfactory.[156] In an unrelated publication, the superintendent of the PNJ explained, "The interconnected system is a synchronized system, but the parts of it are owned by different companies, each of which must look out for its own interests. . . . An ideally regulated interconnected system would economically and promptly distribute load variations over all generating equipment in operation."[157] Robert Brandt summed it up neatly in 1934: "Today we do a great deal and much of it is wrong."[158] He laid out the challenge for the rest of the decade: "What is required is a mechanical device which can not only be made so intelligent that it will be ready to respond, positively and accurately, when there is a real need for response, but, what is fully as important, that it will sit back and do nothing when the logical correction point is elsewhere."[159]

In the early 1930s, many utilities operated in parallel with their neighbors, sharing power on a scheduled basis and during emergencies. When utility operators closed the ties and allowed electricity to move across the network, the systems appeared to function as unified wholes. But in truth, the operators hustled to keep the frequency stable and to avoid untoward load shifts between systems. The fraternity of experts collaborated to build critical control pieces of the expanding power network.[160] With the aid of newly invented automatic frequency control devices, operators found the burden of close observation, overwhelming data collection, and time-sensitive manual control was somewhat lessened. They continued to ensure steady electrical service to their customers. But the load control problems intensified. The frequency and load control problem appeared to be an exceptionally difficult reverse salient within the larger question of who controlled the economic and political future of power systems.

4 Balancing Reliability and Economy, 1930–1940

In the early 1930s, the United States, like much of the rest of the world, was in crisis. National unemployment jumped from an average of 4.7 percent through the 1920s to 8.7 percent in 1930, 15.9 percent in 1931, and more than 20 percent for the next several years. The story of the Great Depression is well studied, but the progress of the power industry, despite the setbacks suffered by the economy, is a different tale altogether. Power companies did cut back production, but at the same time, they continued to expand interconnections. In 1928, the Interconnected Systems Group (ISG), which eventually became the largest network in North America, organized with eleven companies in three states, including the American Gas & Electric Company (AGE).[1] By 1933, forty-four companies interconnected across twelve states.[2] In another five years, ISG and its interconnected systems covered portions of fourteen states, an area extending 450,000 square miles. As the decade progressed, the federal government invested in large-scale electric power infrastructure, much of which relied on interconnections to carry electricity to the intended consumers. With new and very large federal dams, a recovering economy, and preparations under way for a possible second world war, power networks intensified significantly in the late 1930s, as illustrated in figures 4.1 and 4.2.

The depressed economy served as the backdrop to political, economic, regulatory, and technical choices made by public and private sector leaders through the decade. President Franklin Roosevelt excoriated utility executives for financial practices that contributed to the stock market crash of 1929 and the subsequent losses experienced by millions of Americans. At the same time, he promised to develop public power as a yardstick against which to measure the work of privately owned utilities. Through executive action, Roosevelt inserted federal authority into numerous aspects of power development. Congress approved several major pieces of legislation to plan, finance, and manage electric power infrastructure.[3] Just as importantly, in

Figure 4.1
Map showing high-voltage transmission lines and related interconnected systems in 1928.

Source: Report on the Status of Interconnections and Pooling of Electric Utility Systems, Edison Electric Institute, 1962.

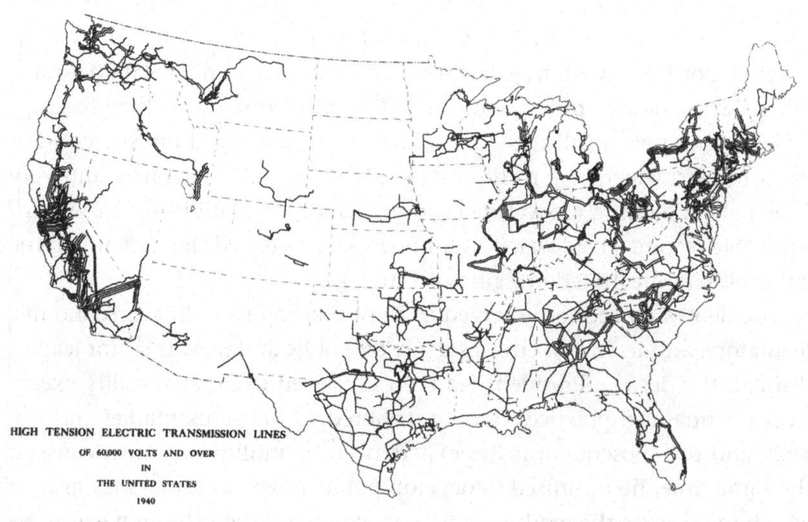

Figure 4.2
Map showing high-voltage transmission lines and related interconnected systems in 1940.

Source: Report on the Status of Interconnections and Pooling of Electric Utility Systems, Edison Electric Institute, 1962.

1935 Congress reshaped the economic structure of the industry through the Public Utility Holding Company Act (PUHCA) and a major amendment to the Federal Water Power Act.[4] The Federal Power Commission (FPC) quickly assumed new regulatory responsibilities with respect to interstate power transactions while holding company executives acted to protect their respective interests.

The power industry had expanded from a handful of stations in the early 1880s to thousands of companies by 1930, yet disunity prevailed. The FPC issued its first, and only partially complete, National Power Survey in 1935.[5] In the opening paragraphs, the chair noted the commission's first task was to create a complete and accurate list of the plants in operation. Such a list did not already exist, despite the fact that numerous public and private entities had completed surveys of US power systems over the prior two decades. Further, the chair stated, "While these studies, which are of a highly technical character, are incomplete they have gone far enough to show that interconnection as it exists today in the United States is not the result of any definitely planned program."[6] The survey described intercompany rivalry, artificial barriers such as state lines, prohibitory laws, and a variety of local tax laws as factors in the haphazard development of interconnections. Out of this disorganized situation—and a pending, yet ultimately unsuccessful, government effort to bring order to the industry—a fraternity of power system experts devised functional solutions to improving the continuous interconnected operations of power systems.

The disaggregated industry shared responsibility for keeping power networks stable yet pursued autonomous economic goals. All power companies sought sufficient revenues to meet their financial obligations, but privately owned utilities additionally pursued profits for investors. Beyond the gross distinction between public and private, individual companies varied greatly in size, type of energy resources used, and variety of customers served. Some assumed a stronger public service profile, while others more clearly sold a commodity. Within a network of interconnected companies, power-sharing arrangements addressed access to primary energy resources, economy of operations, and diversity of demand. Whether agreements were formal or informal, the participants generally attempted to respect each other's distinct economic style. As a result, technical choices tended to follow social arrangements on increasingly complex networks of power.

The New Power Deal: Federal Electrification in the 1930s

Government influence over power systems had both physical and nonphysical manifestations. While federal investment in dams and transmission systems shaped regional networks, the government's expanding regulatory role touched all parts of the industry. Before the 1930s, federal regulation of electric utilities was minimal in the United States. Early in the century, administration officials, politicians, and Congress focused on control of waterpower sites, many of which were on public land or federally regulated waterways. After two decades of debate, Congress finally passed the Federal Water Power Act in 1920, exerting explicit control over the location of hydroelectric dams on federal lands and navigable rivers.[7] Other efforts to insert federal authority over electrification, like the proposed Superpower plan and the 1920s Federal Trade Commission investigation of holding companies, resulted in substantial public attention but little change in the economic activities of utilities.

The Federal Water Power Act of 1920 did rationalize the process of determining who built dams of what size and where across the country. The FPC, as established by the act, exercised oversight over waterpower site development, issued fifty-year licenses to private power developers, and collected annual license fees. The secretaries of war, interior, and agriculture comprised the commission. The FPC, working through these federal agencies, gathered water resource data and determined whether proposed waterpower projects represented the most advantageous use for the region.[8] Notably, within its first three years of operation, the commission issued licenses and permits to double the developed hydroelectric power in the United States.[9] The act also laid out provisions for regulation of rates for power generated at licensed plants and used in interstate commerce. This marked the first federal foray into regulation of power prices in the forty-year history of the industry.

The Federal Water Power Act, however, failed to address a number of ambiguities in utility regulation. Investor-owned power companies were able to sidestep state and federal regulation when selling power across state lines, and this created numerous problems for consumers and government alike. In one prominent 1927 case, *Public Utilities Commission of Rhode Island v. Attleboro Steam and Electric Company*, the Narragansett Electric Light Company, located in Rhode Island, sought to increase rates for power sold to the Attleboro Steam and Electric Company, located in Massachusetts. The Rhode Island Public Utilities Commission granted the rate increase, finding that the old rate was harmful to both the Rhode Island utility and its in-state customers. The Attleboro Steam and Electric Company appealed this increase to the Rhode Island Supreme Court, which found that the higher rate would become a burden on interstate commerce. The US Supreme Court concurred: "The rate is therefore not subject to regulation by either of the two States in the guise of protection to their respective local interests;

but, if such regulation is required it can only be attained by the exercise of the power vested in Congress."[10] While this established that interstate rate regulation was entirely a federal matter, no statute allowed the FPC, or any other federal agency, to address the situation.[11] In 1930, Congress amended the Federal Water Power Act to increase the independence of the FPC and reformulate the membership from three administration executives to five individuals approved by the Senate. Over the next few years, this change led to a reformulated commission but barely increased federal authority over electrification.[12]

Debate over how to regulate interstate trades of power, how to bring holding companies under control, and whether to expand overall federal involvement in electrification continued into the 1930s. The financial crash of 1929 undermined the solvency of numerous utility holding companies, and many public figures blamed the Depression on the excesses of the "power trust." Before and during his presidency, Franklin Roosevelt identified Samuel Insull of Commonwealth Edison as a particular target in diatribes against the private power industry. By 1928, Insull controlled interlocking utility holding companies worth $3 billion. In 1932, his financial empire had collapsed, bringing major financial losses to thousands of investors, and Insull stood accused of fraud.[13] According to Roosevelt, "The Insull failure has done more to open the eyes of the American public to the truth than anything that has happened. It shows us that the development of these financial monstrosities was such as to compel inevitable and ultimate ruin."[14] Despite the later acquittal of Insull, Roosevelt made power industry reform a target of his activities through the 1930s.[15]

Campaigning for president in Portland, Oregon, in 1932, Roosevelt invoked electrification as a means of addressing the welfare of the people. He suggested that the public sector could institute equitable distribution of electricity and fair rates through federal regulation of interstate utility holding companies and federal development of power sites on major rivers. Federal river power systems in each of the four quarters of the country could be "forever a national yardstick to prevent extortion against the public and to encourage the wider use of that servant of the people—electric power."[16] Roosevelt stopped far short of proposing a fully public electric system, stating that "as a broad general rule the development of utilities should remain, with certain exceptions, a function for private initiative and private capital."[17]

Government expenditures nonetheless resulted in major additions to growing regional power networks. Roosevelt sent his Tennessee Valley Authority (TVA) message to Congress within one month of his 1933 inauguration, and the bill to create the agency was introduced the next day. By mid-May, despite significant utility opposition, Roosevelt had signed the authorizing legislation and, with Congress, had officially initiated central government participation in power generation and transmission on a large scale. Congress chartered the TVA to build dams, generate and transmit power, install flood control and navigation improvements, institute regional economic planning, and provide jobs in the Tennessee River Valley. During 1933, Roosevelt authorized construction of the Grand Coulee and Bonneville Dams on the Columbia River, representing both a significant federal investment in hydroelectric power and the addition of more than two million kilowatts of installed generating capacity in the Pacific Northwest. By 1939, federal investment in electric power facilities totaled more than one billion dollars. During the 1930s, private sector power generation increased by nearly 46 percent, but federal power generation increased by a factor of twenty.[18] Although private utilities controlled the majority of the industry, as illustrated in figure 4.3, the federal government was busily shaping the landscape of electrification across the country.[19]

In addition to funding for large-scale dam projects, numerous federal agencies invested in the construction of new transmission and distribution facilities across the country. During Roosevelt's first one hundred days in office, Congress created the Public Works Administration (PWA), which spent $50 million in loans and grants for power projects within two years. The authorizing legislation included "transmission of electrical energy" within the list of projects eligible for funding.[20] Between 1933 and 1935, the PWA financed thirty-one power and light projects in twenty states, the majority of which served cities with populations under twenty-five thousand.[21] Roosevelt established the Rural Electrification Administration (REA) by executive order in 1935, and Congress passed the Rural Electrification Act in 1936 to accelerate electrification of agricultural regions. The REA provided a revolving loan fund to rural power cooperatives for building distribution lines. Congress authorized the REA to spend $50 million in its first year of activity and $40 million in each of the ten succeeding years. In 1937, Congress created the Bonneville Power Administration specifically to transmit power from federally funded dams in the Columbia River Valley to customers, with priority given to publicly owned utilities. As a result of these actions, the federal government established regional systems that demonstrated the elasticity of consumer rates, contested the dominance of integrated private utility systems, and provided a competitive advantage to public utilities in certain markets.[22]

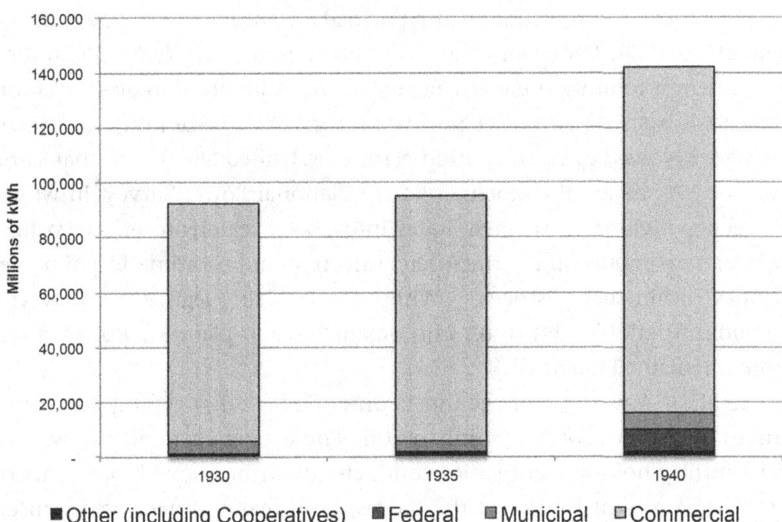

Figure 4.3
Comparison of electricity generated by private power companies and the federal government, 1930–1940.

Source: Historical Statistics of the United States, Colonial Times to 1970 (Washington, DC: US Government Printing Office, 1975), Series S 44–52, 821.

On the consumer side, federal agencies further contributed to electrification in the 1930s. New Deal–era programs targeting rural and low-income Americans added to the demand for larger generating facilities, longer transmission lines, and more widespread distribution networks.[23] Programs that encouraged residential customers to switch to electricity increased the customer base for utilities. With more consumers, utilities expanded the network of power lines, including long-distance transmission lines and interconnections. The mid-decade National Housing Acts provided home rehabilitation loans and appliance purchase loans intended to modernize and electrify dwellings. The Federal Housing Administration established wiring standards for buildings qualifying for mortgage insurance, further pushing for electrification. The Reconstruction Finance Corporation provided a loan to the municipal power company of Los Angeles for construction of its major transmission line from Boulder/Hoover Dam to Southern California.[24] Numerous other agencies encouraged consumers to electrify and modernize through a variety of loan and credit programs. The federal role, while undefined as a comprehensive policy, influenced power growth across the board.[25]

The Roosevelt administration also pursued a planning role in power development. In 1934, Roosevelt created the National Power Policy Committee in an attempt to unify the many federal agencies involved in electrification and establish an actual power policy.[26] Yet in 1935, fifteen different agencies were involved in electrification across the United States.[27] In that same year, the FPC issued the aforementioned National Power Survey, in which the agency divided the United States into power regions to effect "voluntarily or by compulsion . . . the interconnection and coordination of power facilities within such districts."[28] Without authorizing legislation, however, the administration's adventures into organizing and planning power development remained fractured.

The PUHCA was one of the most contentious and sweeping power initiatives of the Roosevelt administration. The act restructured the way in which utility holding companies conducted their business. Proponents of the PUHCA sought to correct the harms, both real and perceived, caused by unregulated holding company expansion in the prior two-and-a-half decades. Opponents sought to preserve a status quo that facilitated operating efficiencies while enriching utility investors. As finally enacted, the PUHCA required utilities involved in interstate power or securities transactions to register with the Securities and Exchange Commission (SEC). In addition, holding companies could not purchase or sell securities without SEC approval. Further, the act called for the simplification of holding companies. A holding company could petition to remain whole if its subordinate operating utilities were interconnected or capable of being interconnected. Otherwise, the holding company had to dissolve. This favored the retention and expansion of interconnections. Holding companies prepared their own reorganization plans but had to acquire SEC approval to stay in business. In the years following enactment of the PUHCA until the SEC completed its reviews in 1955, the number of holding companies dropped from 214 to 25, and the subsidiary assets under holding company control dropped from 922 gas and electric utilities and more than 1,000 nonutility companies to 171 utilities and 137 nonutilities.[29] Those that retained their holding company status operated entirely within one state or demonstrated that their interstate subsidiaries were contiguous.[30]

Congress also expanded the regulatory powers of the FPC by amending the Federal Water Power Act in 1935. The amendments gave the FPC the authority to regulate interstate sales of power and to require interconnections between utilities when demanded by the public interest. By the end of the decade, the FPC performed its work under six different acts of Congress, and its authority extended to flood-control activities related to hydroelectric power and regulation of interstate natural gas transmission.[31] With regard to electrification, the FPC described its duties in conservationist terms:

The rounding out of the Commission's responsibility for proper utilization of the country's power resources reflects a broadening of the concept of conservation. In terms of this broadened concept, conservation includes:

1. Encouragement of the orderly development of waterpower resources under conditions precluding their exploitation for excessive profits;
2. Sound coordination of all power-generating facilities to assure the most economical supply of power to distribution centers;
3. Uniform original cost accounting to eliminate hidden cost barriers to consumption, and the development of comparative cost and price standards as a quasi competitive stimulus to better management.[32]

The agency defined a conception of conservation that reflected Progressive Era notions of wise use of water resources, with protections for the long term. This view included widespread power consumption as a positive attribute.

The private utilities unsuccessfully litigated against New Deal legislation. Power companies filed ninety-two suits opposing PWA allotments, but in 1938, the US Supreme Court affirmed the federal role in financing of public power facilities. Utilities also filed fifty-eight suits against the SEC regarding the PUHCA and thirty-four additional suits against the TVA. Wendell Willkie, president of Commonwealth and Southern Corporation, one of the larger private utility holding companies in the country, was an outspoken critic of Roosevelt's policies. Commonwealth and Southern's subsidiary utilities provided electricity to 4.3 million customers in four southern states, all potentially affected by TVA projects. Willkie argued that the electric business "was the child of a single parent—private enterprise."[33] Ultimately, Willkie sold the affected utilities to the TVA. Willkie challenged Roosevelt on many elements of the New Deal power program and ran against him, unsuccessfully, in the 1940 presidential election. Despite this full-out assault on the federal government's efforts to shape electrification of the United States, the New Deal legislation survived.[34]

The Army Corps of Engineers, the Bureau of Reclamation, and the Tennessee Valley Authority all participated in dam construction during the 1930s. Major projects included the Boulder/Hoover Dam on the border of Arizona and Nevada in 1933, the Bonneville Dam in the Pacific Northwest in 1938, and six TVA dams in the Tennessee River Valley. Grand Coulee, also in the Pacific Northwest, went into service in 1941. As figure 4.4 illustrates, installed generating capacity owned by the federal government increased dramatically in 1936. The chart begins with 1920, the first year in which federal census records provide consistent data. The chart ends in 1942, indicating the federally installed capacity available as the country entered World War II. Each jump in capacity can be linked to the opening of large hydroelectric plants at new federal dams.

The financial viability of big dams depended on access to big markets. As the FPC noted in 1929, "Water-power sites without a market are of no practical value, and although within the past decade great strides have been made in the art of transmitting power for long distances at low cost the limit of economic transmission is now about 300 miles under normal conditions."[35] In the case of the Boulder/Hoover Dam, the Los Angeles area offered the primary market for power.[36] At a distance of 266 miles, the power line linking the dam to its largest customer, the Los Angeles Department of Water and Power, neared the limit of cost-effective transmission. In other regions of the country, both the Tennessee Valley Authority and the Bonneville Power Administration explicitly addressed interconnection as a method for transmitting power from federal dams to a widespread consumer base.

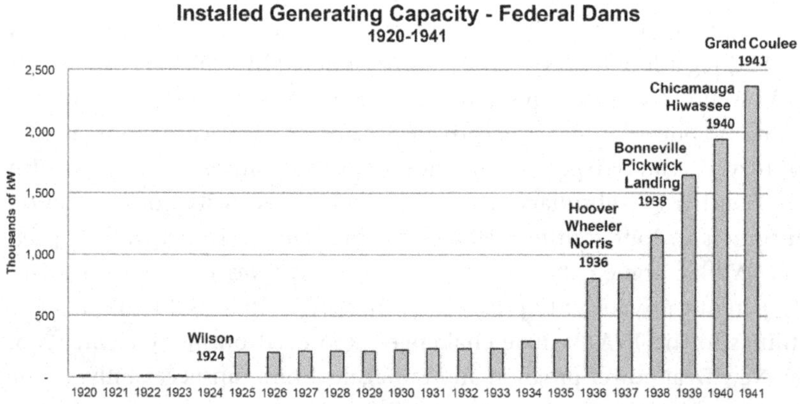

Figure 4.4
Total installed hydroelectric generating capacity, 1920–1942. Years in which the largest federal hydroelectric plants opened are noted.

Sources: Historical Statistics of the United States, Colonial Times to 1970 (Washington, DC: US Government Printing Office, 1975), Series S 74–85, 824; Bureau of Reclamation website (https://www.usbr.gov/); Tennessee Valley Authority website (https://www.tva.gov/); US Army Corps of Engineers website (http://www.usace.army.mil/).

The big federal dams of the New Deal introduced large amounts of power into regional networks and magnified the need for effective frequency and load control in interconnections. The federal commitment to dam construction stemmed in part from conservationist goals of developing and managing watersheds to maximize energy production. Load control was essential to achieving those goals. With the big dams came bulk power transfers—that is, the sharing of very large amounts of power across tie lines. These power transfers also increased the potential for greater interruptions of frequency and scheduled power exchanges. As a letter from Leeds & Northrup (L&N) to the US Department of the Interior explained, "From the time the first unit comes on the bus, until the ultimate in generating and tie-line capacity is installed, one of the principal operating problems [for a big dam] will be the directing and maintaining of a scheduled power transfer on certain tie-lines or a combination of them, or maintaining the system frequency."[37]

The new power deal of the 1930s changed the landscape of electrification in multiple ways. In one sense, the effects were minor; investor-owned utilities continued to dominate the market for electric power and evaded federal control of system planning. On the other hand, through legislation and investment, the president, Congress, and several federal agencies expanded the power infrastructure, encouraged consumption, and introduced economic limits to the actions of holding companies. The actions by Congress to more closely regulate holding companies created incentives for power companies to interconnect. The PUHCA gave an advantage to holding companies with contiguous subsidiaries when they linked those subsidiaries into power pools. Likewise, the FPC attempted to assess the status of the power industry and offer plans for growth. The commission could even compel interconnections under certain circumstances. Federal dams, federally funded transmission lines, and federal loans to rural cooperatives added capacity and reach to growing power networks.

For system operators, new federal dams and larger interconnected networks introduced significant demand for effective power control techniques. For instrument manufacturers, this represented new market opportunities, including an expanding demand for machines that gathered and analyzed data as well as machines that automatically controlled systems. Leeds & Northrup, among others, followed the 1931 frequency control test on the Pennsylvania-Ohio Interconnection with aggressive pursuit of contracts for control instruments on the Boulder/Hoover Dam project. The test had indicated the limits of automatic frequency control, so L&N continued to experiment with innovative approaches to power control. Within a few years, L&N had new patent applications in process and a foot in the door with the Bureau of Reclamation.

Instruments of Measurement and Control

In its 1935 survey, the FPC astutely observed, "The control of power is a social as well as an engineering and economic problem."[38] While the FPC referred to the potential for a social catastrophe should power systems fail, system operators understood that their technical duties had social and financial dimensions. As L&N engineer S. B. Morehouse remembered, in 1926, West Penn Power Company closed ties with Duquesne Power and Light Company, and "parallel operation . . . was achieved."[39] Although this solved many technical problems, it brought out "people problems," because Duquesne sold time, and West Penn did not.[40] Duquesne regularly opened and reclosed the tie line in order to keep the clocks on its system within sixty seconds of accurate time, which caused constant interruptions to operation of the network as a whole. System operators had to ensure not only physical stability on the network but also adherence to the agreements between companies to generate certain quantities of power at certain times and to honor the related financial arrangements. Successful operation of an interconnected system guaranteed the customer a steady supply of electricity and promised the power company—whether public or private—the most economical operating conditions. To meet the social and financial obligations of his employer, the system operator sought techniques that balanced reliability and economy.

The physical challenges of interconnection were by no means limited to frequency and load control. For example, to design and operate an intertie, engineers also addressed "voltage, reactance, resistance, losses, and charging current of the line, and of wattless generating capacity at the receiving end of the line."[41] All these conditions, in addition to frequency and demand, changed constantly throughout the day and the year. As power pools grew larger, the volume of information about operating conditions grew as well. A system operator had to attend to reports on every detail, from frequency to voltage to patterns of demand, in order to achieve both reliability and economy. As Nathan Cohn said in retrospect, "It is a self-evident maxim that what you cannot measure, directly or inferentially, you cannot control, or at least you ought not try to control."[42] Advances in automated data collection and analysis, coupled with automatic control techniques, presented new opportunities for the leading apparatus manufacturers. Ironically, each new improvement in data gathering, calculating, analyzing, or applying to system control seemed to be accompanied by new, or previously hidden, challenges.

Power systems offered an early market for machine data gathering and computing. From the early 1900s, instruments measured load, output, frequency, and voltage on generators and power lines. Load dispatchers (later referred to as system operators) developed a high level of skill for manually calculating system behavior and estimating future demand based on this data. By the early 1920s, some companies used telemeters to gather data from distant locations and deliver the data automatically to the load dispatcher. Telemetering involved the transmission of "instrument readings or their equivalents from one point to another" for operations management, not for billing purposes.[43] Philadelphia Electric Company used a central telemetering system to collect information about both frequency and demand across its networked system. The chart in figure 4.5 illustrates the telemetered data recorded during the company's peak load on February 10, 1926, at 6 p.m. As Morehouse explained, "Very few system operators had this comprehensive information on the system they were dispatching."[44] The first telemetering devices were costly and impractical to place in more than one location on a system. But Robert Brandt, of the New England Power Company, reflected that without telemetering, it would have been impossible to address loading along with automatic frequency control.[45]

Interest in telemetering reached a watershed in 1928, when articles about the equipment and its use started to appear regularly in the technical literature.[46] Technical writers covered new telemetering apparatus, integration of telemetering with system control devices, and use of telemetering on power systems. Engineers designed "totalizers" to enhance telemeters. A totalizing machine automatically added up the multiple bits of data delivered to a central station by telemeters. In 1933, L&N published a bulletin describing proprietary telemetering and totalizing equipment.[47] The company explained that the advantageous distribution of load on an interconnected system required the dispatcher to have direct and immediate knowledge of all station and load conditions at all times. Telephone communications were inadequate for the task. In 1934, an engineer from the Washington, DC, area cautioned that interconnection was "justifiable when it makes economies possible," but this could not be achieved without the highly accurate data made available through telemetering and totalizing.[48] As a New York utility operator affirmed, telemetering made it possible to achieve energy efficiency on the network.[49] By the end of the 1930s, telemetering and totalizing apparatus found wide application on interconnected power

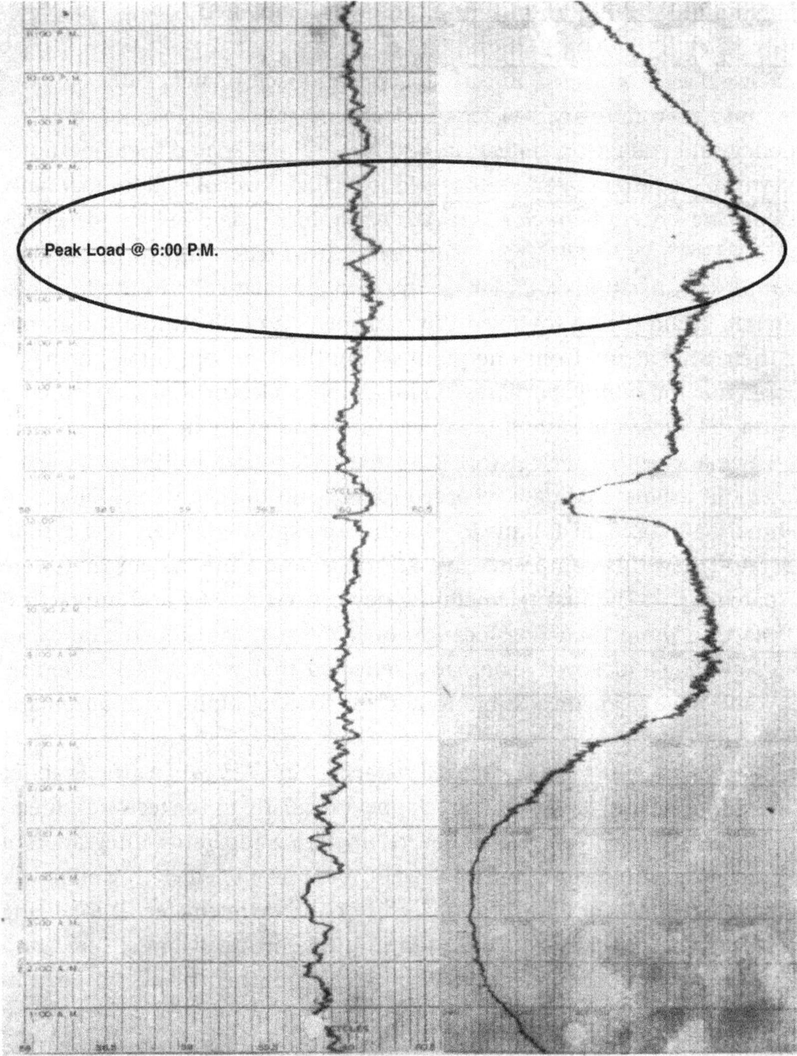

Figure 4.5

Philadelphia Electric Company's frequency and total system generation on February 10, 1926. Chart shows frequency (on left) and demand (on right). Oval indicating peak demand at 6:00 p.m. added by author.

Source: S. B. Morehouse, "Some Historical Highlights in the Operation of Electric Utility Systems on an Interconnected and Coordinated Basis" (paper presented to the Annual Joint Meeting of the Northeast Regional Committee of the Interconnected Systems Group and the System Operation Committee of the Pennsylvania Electric Association, Pittsburgh, Pennsylvania, 1965), 8.

systems in the United States and abroad. These measuring and calculating tools helped system operators bring power control on networks into closer alignment with the social and economic relationships among their employers.

With the advent of extensive data collection facilitated by telemeters, engineers began to contemplate a fuller analysis of system operations beyond the algebraic sums performed by totalizing equipment.[50] As early as 1924, investigators at the research lab of American Telephone and Telegraph Company, working with Western Electric Company, Inc., experimented with the use of an instrument that would automatically measure, record, and analyze electrical frequency.[51] This instrument offered the industry a major benefit: speed of calculation. In 1929 and 1930, engineers from MIT and General Electric (GE), working in collaboration, published the results of their lab experiments with a device they titled the Network Analyzer, designed to compute data collected from an AC power system. This project acknowledged the challenge of completing a daunting number of calculations in order to determine the behavior of a power system under various conditions of operation: "The chief function of the Network Analyzer is to serve as an experimental substitute for the lengthy and generally impractical calculations of . . . electric power networks."[52] While the apparatus described in 1930 served a research function in an MIT lab, the design engineers gave thoughtful consideration to its application to real interconnected power systems. This early computer illustrated the benefits of system modeling for the electric power industry.[53]

Operators of complex interconnections hoped to better understand how their network functioned in order to both optimize energy efficiency in daily operations and plan for expansion. The analyzer corresponded directly to an actual power system, with elements representing components such as load and transmission lines.[54] It modeled a system with up to "8 generating stations, 60 lines and cables or other connecting elements, 40 loads, 4 ratio-changing transformers for closing loops, and any desirable number of synchronous condensers."[55] The designers expressed confidence that the analyzer could adequately represent the key engineering problems on power networks. As is evident in figure 4.6, the Network Analyzer took up quite a bit of space and required hands-on attention from the individuals preparing calculations.[56] Over the next several years, MIT and GE improved the Network Analyzer, under the leadership of Vannevar Bush, and sold the first device for commercial use in 1937.[57]

Figure 4.6
The MIT Network Analyzer in operation on a power system study.

Source: The Solution of Commercial Power-System Problems on the M.I.T. Network Analyzer (Cambridge, MA: Technology Press, 1931). Copyright Massachusetts Institute of Technology, used with permission.

While this analyzer modeled steady-state operations easily, operators completed additional manual calculations to represent changes on the network, such as major load increases, failures, or the addition of a new generator. GE introduced a "new" Network Analyzer in 1938. The following year, the federal government approached MIT about expanding the device to model a possible federal power grid. During World War II, the Network Analyzer provided important data for determining the effects of emergency interconnections in various parts of the country. Over the ensuing years, engineers used the Network Analyzer to model system behavior, to understand the limits of long-distance and interconnected transmission, and ultimately to improve the control of shared power on the grid.[58]

In the early 1930s, engineers and system operators relied on less-sophisticated tools for analyzing and controlling power networks. In the more advanced power control rooms, like the one at Windsor station, operators used data from frequency recorders to evaluate their automatic control field test and augmented this with phone conversations, meetings, and negotiations about what worked and what did not.[59] The utility engineers reached a consensus about the limitations of the automatic frequency control equipment from different manufacturers. The automatic instruments did regulate frequency very well, but they overburdened the central controlling station and upset planned load distribution across the larger network. With automatic frequency control, the operators improved reliability on their networks but upset social and economic arrangements. The equipment manufacturers took note and began to design, patent, and test a variety of tools that combined automatic frequency control with automatic load control.

Big Dams and Better Controls

In an effort to gain a larger foothold in the control instrument market, the technical and sales teams at L&N continued to collaborate as they had in prior decades. This company's approach was emblematic of the ongoing competition among manufacturers supplying the power industry. In 1935, L&N engineers filed a patent application for a system that allowed one station to provide frequency control while others held tie-line load within "capacity or contractual agreements" yet assisted somewhat with frequency.[60] Called "tie-line bias frequency control," this approach achieved the desired shared management and divided authority reflected in operating arrangements on a network. The word "bias" referred to a setting that determined how much assistance the controller allowed before restoring scheduled power trades. The marketing team at Leeds & Northrup urged expedited development of high-speed frequency control for this very active field. Other companies posed aggressive competition; from Australia to Detroit, General Electric, Westinghouse, and a number of smaller companies vied for the business of measuring and controlling power exchanges on networks.[61]

L&N maintained close ties with client utilities, described active installations in marketing campaigns, and sought and won contracts for control equipment at many major projects built during the decade. The company's engineers tracked the competition closely and gathered testimonials from utilities using L&N equipment. Leslie Heath, now sales manager, made a point of sharing installation results with his sales team in the field. For example, in a 1934 memo, he forwarded to the sales team information about recorders and meters on the Pennsylvania Railroad system, the Philadelphia Electric system, and the Philadelphia Electric interchange with the Aldred Company, another large regional utility.

He later provided details of a New York Edison test of totalizing equipment.[62] The company focused a great deal on the Boulder/Hoover Dam project, under construction at the time. The final agreements regarding power generation at Boulder/Hoover Dam called for each of the power companies with contracts for electricity to operate their own hydroelectric facilities at the dam.[63]

In 1934, Nathan Cohn, then a young engineer tasked with selling L&N apparatus on the West Coast, faced direct competition from GE for the Los Angeles Department of Water and Power (LADWP) contract for control devices. This city-owned utility claimed 15 percent of the power to be produced at Boulder/Hoover Dam. Cohn focused initially on persuading the Bureau of Reclamation to write technical specifications that encompassed L&N's control equipment.[64] L&N engineers spent time in Denver in April 1934 successfully lobbying the bureau in person. Back in Los Angeles, GE had reportedly told LADWP that GE equipment was favored over L&N apparatus following the 1931 field test at Windsor station. According to Cohn, the GE approach to sales was to "find a new listener on the top-deck, and simply point out that they were low bidders by $1200.00 and should therefore get the award." Cohn told his supervisors at L&N that GE failed to meet the specifications for the job.[65] Cohn provided LADWP and the neighboring private utilities with letters from Commonwealth Edison Company in Chicago, the United Electric Light and Power Company in New York, the New England Power Engineering & Service Corporation in Boston, the Ohio Power Company of Canton (formerly the Central Power Company), and the West Penn Power Company of Pittsburgh—all field test participants—extolling the performance of L&N automatic frequency control equipment. In contradiction to GE's message to LADWP, the testimonial from Ohio included confirmation that the utilities in that area had dropped GE control equipment in favor of L&N apparatus: "GE is no longer in use while L&N is."[66]

Other factors worked in L&N's favor. Cohn took engineers from LADWP to visit a power station that had installed both a GE Warren clock and L&N frequency control equipment. The operator there described trouble with the Warren clock and explained in detail how the L&N equipment performed. Cohn also provided the city with charts from the 1931 field test that compared GE and L&N equipment, while GE failed to provide any testimonials or hard data from other installations. In addition, the private utilities with contracts for power from Boulder/Hoover Dam—Southern California Edison, Southern Sierras Power Company, and Los Angeles Gas & Electric Corporation—all agreed to purchase L&N

controllers. LADWP was already considering purchase of the Los Angeles Gas & Electric Corporation.[67] Presumably future interconnection between these utilities would be simplified if all used the same control equipment.

Cohn's approach worked. On March 20, 1935, he reported back to his supervisors that LADWP and the three private companies had "definitely approved our frequency and load control equipment."[68] He noted that not only had the company beaten GE on these contracts, but the L&N apparatus would replace a Westinghouse system currently in use in Los Angeles. Leeds & Northrup installed instruments in the control room at Boulder/Hoover Dam, visible on the back wall in figure 4.7 and opposite the system operator in figure 4.8. Each of the Los Angeles–area utilities installed its own control facility at Boulder/Hoover Dam. Following this project, L&N continued to pursue work with the large federal dam installations.

Figure 4.7
Main control room for the operation of units for public agencies at elevation 743 in the central section of the power plant, October 21, 1936. The carrier current board is in the foreground, the miniature remote control board for the main units to the right and for the station service units to the left, and recording instruments in the background.

Courtesy of the Bureau of Reclamation.

Figure 4.8

Supervisory control board furnished by the city of Los Angeles, located in the main control room, elevation 743 in the central portion of the powerhouse, October 21, 1936. Courtesy of the Bureau of Reclamation.

Through these years, system operators across the industry employed tie-line bias frequency control techniques, but not always with success. The most commonly used approach called for one power station to regulate frequency for the entire system and for tie lines to maintain scheduled loading, with minor and gradual adjustments of load as needed to help with frequency control. Automatic load controllers triggered removal of this aid after a certain period of elapsed time regardless of whether or not system frequency had stabilized. This approach allowed each autonomous company to help out the network just a little bit without significantly altering planned load schedules. The technique caused ongoing problems for certain interconnections. By 1937, L&N had introduced this approach in three locations: in the Northeast, in the Carolinas, and on a small system in the eastern Midwest. As Cohn recalled, "[In 1937] it fell to my lot to place into service . . . these new 'tie-line bias frequency controllers,' at the Twin Branch Station of Indiana and Michigan Electric Company. Getting satisfactory operation was, however, not easy, and in fact we didn't achieve it."[69] He described a controller that would "cooperate nicely, act to assist the remote areas and permit its own tie-line to go off schedule."[70] But as soon as the aid to the remote area was withdrawn, there was a further upset in frequency, hunting between areas, and "a totally unneighborly kind of operation."[71] The connected companies preferred to honor their power exchange agreements as closely as possible, thus failing to fully aid the controlling station with frequency regulation.[72] While the two other L&N installations seemed to work as intended, the bias controller at Twin Branch acted to aggravate frequency deviations rather than returning the system to a stable frequency.

Jack Girard, the operations chief at Indiana and Michigan Electric Company, exhibited a strong commitment to making interconnections work. Girard "had a clear concept of the potential benefits of interconnected operation, and in preparing new contracts with neighbors he stipulated their use of tie-line bias control to optimize reliability and economy."[73] Cohn and his associates determined that the flaw in the bias control approach used on Girard's network was the scheduled withdrawal of frequency regulation. The withdrawal limited the natural response of each generator to frequency changes, overburdened the controlling station, and caused further upsets to system stability. Cohn rebuilt the controller to allow the tie lines to restore the scheduled loads essentially on their "own accord," and it worked: "And of course we did most of the experimental work over the midnight shift."[74] He later remarked, "It was another example of using the system as our test simulator."[75]

As a result of this engineering on the fly, "sustained" tie-line bias frequency control became the new industry standard and remained such until the late 1940s. Notably, this technical approach incorporated a shared response to trouble but autonomous regulation to bring the system back into synchrony and scheduled loading. Sustained control, however, did not fully distribute frequency and load control authority to discrete participants on an interconnected network. According to Cohn, Indiana and Michigan also tested the very first known instance of a fully distributed approach. In 1938, Indiana and Michigan negotiated an interconnection with Public Service Company, thus joining a pool of five power companies. Cohn and Girard met regularly to "analyze operations and explore new and expanded needs." They "sent many restaurant tablecloths to the laundry laden with exploratory and tutorial sketches."[76] As a result of these discussions, Cohn proposed and Girard agreed to try a further variation on distributed control called "net interchange" tie-line bias frequency control. In this novel approach, subareas regulated frequency and load internally, and the larger pool controlled only for the net of power that crossed the tie lines between the smaller networks. This worked well for the five-company pool but remained an experimental approach for many years.

By the end of the decade, Leeds & Northrup dominated the market for load and frequency control apparatus on major interconnected systems. In a 1939 bid for equipping the Grand Coulee Dam with attached materials dated 1940, the company enumerated and described installations on nearly every major federal dam in the country and most of the largest interconnected systems. The bid opened with a strong statement: "Our experience with the regulating problems of over fifty power systems on which we have over seventy control installations prompts us to make this broad presentation."[77] L&N provided the Bureau of Reclamation with descriptions of installations at Boulder/Hoover Dam in Nevada, Norris and Wilson Dams in the south, and Bonneville Dam on the Columbia River. It detailed controls used by the Tennessee Valley Authority and in several other locations, including New England, central Texas, and the Pacific Northwest.

True to form, the bid included a further denunciation of the GE system, except when the Telechron clock appeared as a component of the larger L&N system.[78] Over the years, this piece of technology reflected the shifting alliances between manufacturers, the need for cooperation between rivals, and the organic cobbling together of solutions that characterized development of the grid. In the 1920s, GE's Warren Telechron clock had posed a potential patent challenge for L&N; in the early 1930s, L&N and GE had collaborated on tests of their equipment on the AGE system; and in the mid-1930s, GE and L&N had competed directly for major installations of control instruments on federal dam projects.

Part of the pitch to the bureau covered the potential for Grand Coulee's role in future interconnections and the need to resolve problems as they arose. L&N made the point that, as had been the case at the TVA, "it was not possible to foresee all of their present interconnections and types of regulation which the various plants have since been called upon to handle. . . . [L&N] wished to insure that the control which [the bureau] will install will operate satisfactorily with similar equipment" in the Pacific Northwest.[79] The bid offered a variety of equipment options but stressed the benefits and flexibility of sustained tie-line bias control. L&N argued, "Similar changing requirements are being handled automatically on other systems, and it has been definitely proven, repeatedly, that the present extent of interconnections could not exist without automatic control to coordinate the regulation of the various areas."[80] In essence, L&N promised the tools and capabilities to meet unforeseen demands and operating conditions and to engineer solutions as the interconnections in the Columbia River basin grew.

The proposal also described the Midwest Power System, which stretched from Chicago to the Gulf of Mexico and from Pittsburgh to Texas. Of all the systems, this one perfectly exemplified the need for L&N's premiere package of automated controls that allowed both cooperation and independence among the participating utilities: "You can appreciate that this Midwest interconnection involves many different operating companies, owned by different financial interests, no one of which has authority to dictate the policy to be followed."[81] Part of the Interconnected Systems Group, this huge system operated through a Test Committee that decided what apparatus to employ and how to control the flow of power. In an attached letter, the chairman of the Test Committee, who also worked for Commonwealth Edison, noted, "It required considerable effort to have these [L&N] controllers installed. They were installed from an appreciation of their merits in the attainment of the high standard of service to which this interconnection subscribes."[82] Using tie-line bias frequency control, utilities within the interconnection maintained authority over their autonomous operating areas while sharing responsibility for stability and energy efficiency across the network. In the end, Leeds & Northrup won the contract to install controls at Grand Coulee Dam.

With growth came the exposure of limitations. By the early 1940s, most interconnected power systems in North America employed some form of partially distributed and automated frequency and load control. But many of these systems still relied on a central station to govern overall network frequency, and as more utilities interconnected, the burdens on this station increased. On some systems, the configuration of the control network could be described as a "cascading of controllers."[83] The load controller on the tie line farthest from a central frequency-controlling station acted to hold its own load steady and contribute a bit to frequency control. At the next tie line closer to the central station, the controller adjusted load both for overall frequency control and for its own as well as the farther station's scheduled load. Thus each controller closer to the central station carried a larger burden of adjustment. This limited the potential of an interconnection to expand indefinitely and created operating problems for controllers at the heart of a network. The intense demand for electricity during World War II and the resulting increase in the scale of interconnections exaggerated these problems. Innovations during and immediately after the war led to nearly complete distribution of automated control responsibility among power plants cooperating on a network, mediated by analog computers.[84]

5 Power Transformations on the Home Front, 1935–1950

The long-distance transmission lines of 1940 appear as a skeleton of today's grid. As illustrated in figure 5.1, connections between power systems stretched east, west, and north from the country's coastlines. Yet diversity dominated the power industry—not just in terms of the makeup of individual companies but also in terms of how they used interconnections. The industry and its thousands of operating parts were still more disaggregated than linked. It was rare for a power pool to operate interconnected all the time. Further, the physical properties of electricity, the variety of organizational approaches to interconnection, and the state of technical knowledge challenged the ability of operators to maintain system stability and preserve economic relationships when sharing power. World War II significantly accelerated the process of interconnection, as illustrated in figure 5.2, and advanced both the technology and social capacity for operating across networked systems.

World War II provided a laboratory for close collaboration between government and industry and rapid expansion of interconnections. War industries required huge quantities of power in specific locations in order to meet defense demands. The national government closely controlled the use of raw materials in order to supply war needs first. By building interconnections rather than new generating facilities, utilities were able to meet those needs more quickly and with fewer capital and material resources. In addition, the interconnections facilitated maximum use of existing plants. Engineers continued to experiment with techniques for managing power on the networks and made advances in both operating practices and automatic control apparatus. By the end of the war, the power industry was well situated to meet pent-up consumer demand for electricity and, soon thereafter, new demand from a growing industrial sector. From the late 1940s to the 1950s, the utilities accelerated construction of new and larger power plants as well as more interlinked transmission lines. At the same time, the private power sector, preferring autonomous operation and independent decision making, withdrew from close collaboration with the federal government.

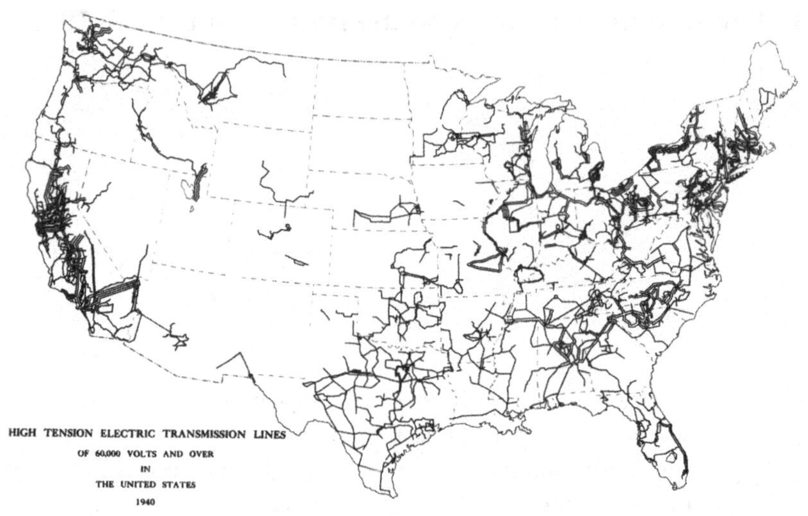

Figure 5.1

Map showing high-voltage transmission lines and related interconnections in 1940.

Source: Report on the Status of Interconnections and Pooling of Electric Utility Systems, Edison Electric Institute, 1962.

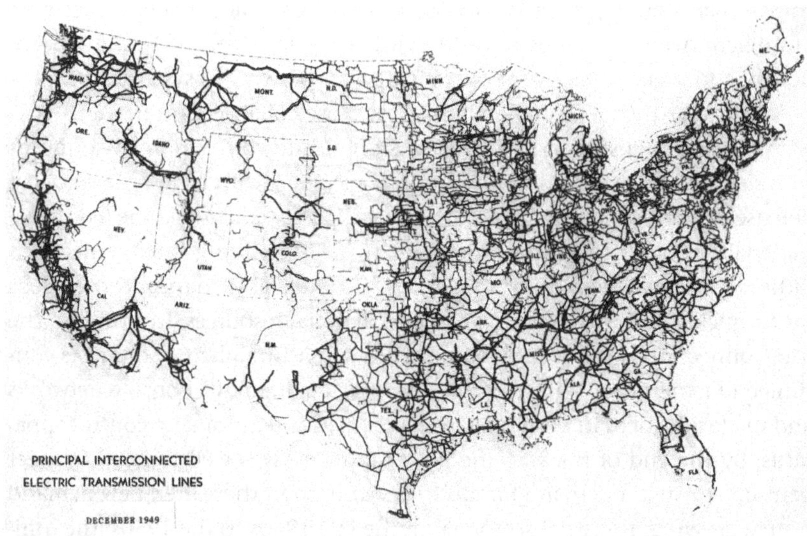

Figure 5.2

Map showing high-voltage transmission lines and related interconnections in 1949.

Source: Report on the Status of Interconnections and Pooling of Electric Utility Systems, Edison Electric Institute, 1962.

The Pieces of the System

Despite the economic depression, electrification had continued throughout the 1930s. Power production fell in 1931 and 1932 but picked up again in 1933 and continued to increase throughout the decade with only a slight dip in 1938.[1] As demand for electricity grew, the electric power industry invested in new facilities in the same piecemeal fashion that marked the prior fifty years. By 1940, thousands of companies produced electricity, and more than half were publicly owned, as shown in figure 5.3. Of these, some were so small that they served as few as ten customers, while the largest systems served millions. Private power companies dominated the market, as illustrated in figure 5.4. They produced the vast majority of electricity, despite the fact that they comprised only one-fourth of the 4,457 utilities operating in the United States at that time.[2] By the mid-1930s, nine large holding companies controlled half the power production.[3] In this sense, there was a semblance of unity on the private sector side. After passage of the Public Utility Holding Company Act of 1935 (PUHCA), however, holding companies had to justify the integrated nature of their businesses to the Securities and Exchange Commission (SEC) or sell off noncontiguous subsidiaries and reduce their overall influence.[4]

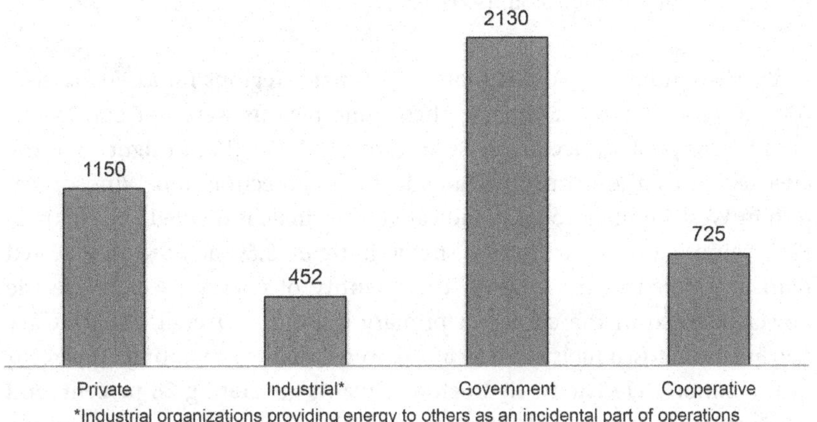

Figure 5.3

Number of electric utilities operating in 1941 by category of ownership.

Source: Federal Power Commission, *Directory of Electric Utilities in the United States, 1941* (Washington, DC: Federal Power Commission, 1941), iv.

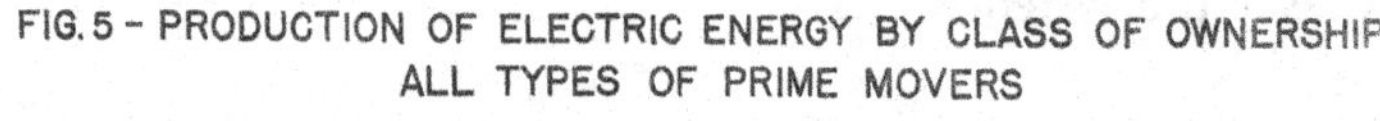

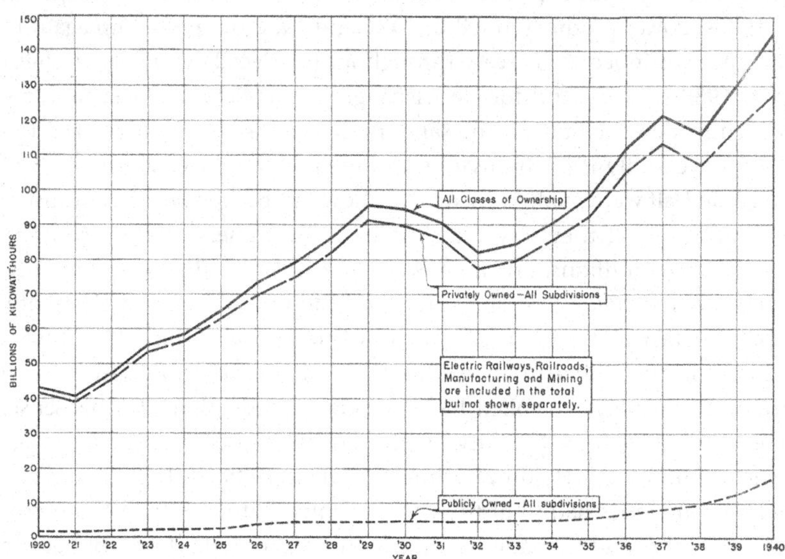

Figure 5.4

Comparison of electric energy production by class of owners. *Top line,* all classes of ownership; *middle line,* privately owned utilities; *bottom line,* publicly owned utilities.

Source: Federal Power Commission, *Electric Power Statistics, 1920–1940* (Washington, DC: Federal Power Commission, 1941), ix.

By 1940, the FPC had designated nine major regions for assessing electric power activities. In general, these nine regions were not contiguous with power pool service areas. As an example, note that in figure 5.5, the area served by the Pennsylvania-Ohio Interconnection and Windsor station in West Virginia, roughly indicated by a circle, is divided among three FPC regions along state lines. As shown in figure 5.5, each region reflected marked differences not only in the quantity of energy used for electric power but also in the choice of primary resources. Overall, the industry depended on fossil fuels, with steam-powered plants generating 70 percent of the nation's electricity, hydroelectric plants generating 28 percent, and combustion engines generating 2 percent.[5] Contrary to national trends, the Pacific and mountain states (Washington, Oregon, California, Montana, Idaho, Wyoming, Colorado, New Mexico, Arizona, Utah, and Nevada) relied

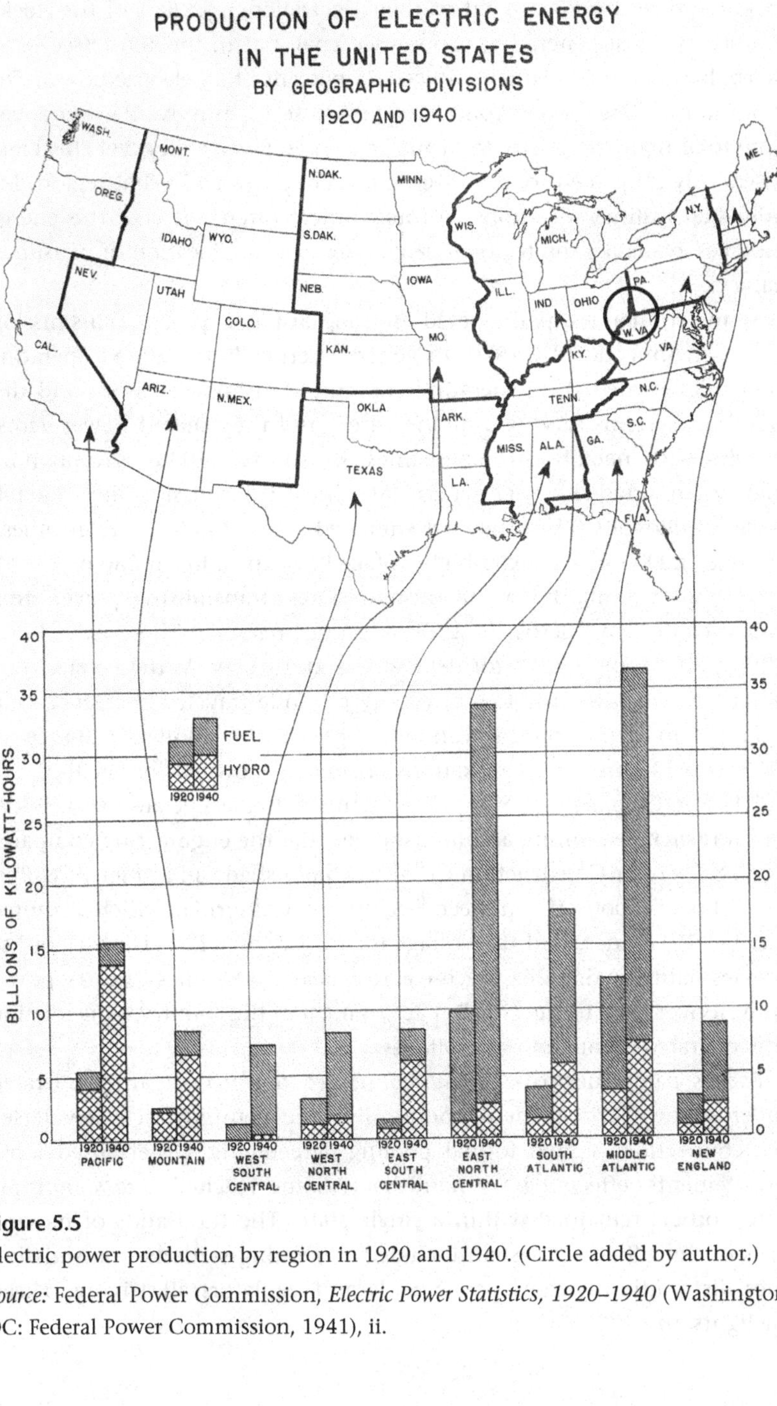

Figure 5.5

Electric power production by region in 1920 and 1940. (Circle added by author.)

Source: Federal Power Commission, *Electric Power Statistics, 1920–1940* (Washington, DC: Federal Power Commission, 1941), ii.

on waterpower for 80 percent of their electric energy.[6] East of the Rocky Mountains, most generating plants used coal, but in the Tennessee River Valley, hydroelectric plants produced 75 percent of all electric power. The mid-Atlantic states—New York, New Jersey, and Pennsylvania—produced more total hydroelectricity than the Tennessee River Valley, yet this comprised only 20 percent of the electric power generated in that region. For individual utilities, decisions to interconnect often reflected the energy resources available, the regional geography, and the location of consumer markets.

Many companies participated in long-distance power transmission and interconnection by 1940. In general, vertically integrated operating companies owned the generating stations, transmission lines, and distributions systems serving a defined area, and they shared power across interties with neighboring companies. By the late 1930s, several hundred systems traded power across state lines. Transmission lines operating at 33 kilovolts (kV) or greater stretched across 145,000 circuit miles.[7] Of these, 3,000 miles of extra-high-voltage lines operating at 230 or 287 kV were located primarily in the western states, transmitting power from hydroelectric dams across long distances to markets.[8] These extra-high-voltage lines were notable for their scale and capacity. As the voltage on a power line increases, the loss of energy per mile traveled decreases, thus making it more economical to transmit power over a longer distance. In 1913, the 240-mile Big Creek transmission line owned by Pacific Light and Power Company (PL&P) in Southern California operated at 150 kV.[9] This was such a stunning accomplishment that the engineering company, Stone & Webster Construction Company, published "an attractively illustrated book" about the project.[10] Southern California, which acquired PL&P in 1917, increased the voltage to 220 kV in 1923.[11] The city of Los Angeles built the first 287 kV line across nearly 300 miles to carry power from Hoover Dam to the city. But across most of the country, transmission lines operated at much lower voltages.

Each separate network of interconnected power companies operated under its own arrangements. Some, like the Pennsylvania–New Jersey Interconnection, signed formal pooling agreements. Others used loose arrangements, effected by a handshake. Some reached across multiple states; others remained within a single state. The thousands of moving parts of the US power system included generating stations, transmission lines, distribution systems, power pools, and regulators, all striving to keep the lights on.

Preparations for War

In the years preceding the United States' involvement in World War II, President Roosevelt and his senior administrators attempted to bring order and comprehensibility to the nation's power system.[12] These initiatives reached beyond the new laws that reshaped the financial structure of the industry to address planning for future power development. In 1933, Roosevelt ordered the Federal Power Commission (FPC) to conduct a national survey "of the water resources of the United States as they relate to the conservation, development, control, and utilization of waterpower; of the relation of waterpower to other industries and to interstate and foreign commerce; and of the transmission of electrical energy in the United States and its distribution to consumers."[13] When published by the FPC in 1935, the first volume of the survey specifically addressed "problems of unemployment, industrial recovery, and national defense."[14] Already, the FPC anticipated the potential of military engagement to challenge the nation's power production capacity. The energy shortages of World War I had not been forgotten.

The survey made several provocative points for both government planners and private utility owners. The studies completed by this date indicated that eastern and midwestern industrial centers of the country threatened to run short of power once manufacturing picked up its former pace. This could be "disastrous in case the United States should become involved in war."[15] Further, the survey urged that "careful planning under Federal supervision of new power plants and facilities for transmission is required to promote the safety and welfare of the Nation."[16] The survey projected a one- to two-year time frame to plan and construct new steam-powered generating plants and two to seven years for a large hydroelectric plant. The FPC offered to take the lead by providing a chart for future power development, both public and private.

Perhaps less threatening to private interests but just as significantly, the survey emphasized the gap between existing interconnection practices and the potential for saving money and increasing utilization of existing facilities through more extensive and highly coordinated integration of power systems. For example, in describing the Interconnected Systems Group (ISG) area, the survey stated that the participating operating companies "rely upon each other to a certain extent . . . but are not operated as a closely coordinated system."[17] In examining the potential power demand and available capacity in seven of the nine districts across the country, the survey concluded that interconnection and coordination could obviate the need for much of the new capacity otherwise required. Up to 1935, the survey acknowledged, there had been no formal plan for integrating power systems. The FPC seemed to suggest that the full benefits of interconnection—in terms of money saved, power produced, and national interests secured—could only be achieved under official federal leadership.

The FPC findings urged a larger role for the federal government in planning future electrification and in directing the operating practices of interconnected systems. Notably, these survey results coincided with the passage of the PUHCA. In addition to passing laws that governed the financial structure of power companies, the Roosevelt administration was also attempting to infiltrate the private sector's physical and technical arrangements. While the FPC gained new powers under the PUHCA, the private utilities did not accede authority for planning and integrating systems. Instead, they continued business as usual at the operating level while girding for negotiations before the FPC, the SEC, and in many cases, federal judges over their financial relationships. The many agency heads within the Roosevelt administration likewise continued to operate within their own domains as they pertained to electrification. Roosevelt failed to consolidate authority over electrification under a single federal agency.[18] Both government officials and private utility executives did, however, begin to share a focus on war readiness in the last years of the decade.

Early in 1938, the National Association of Railroad and Utility Commissioners issued a warning that power-generating capacity should be increased well in advance of engagement in another world war. By this time, President Roosevelt had asked the FPC and the War Department to survey the country's generation and transmission capacity. The resulting report indicated that wartime power needs in the 1940s could be anticipated to be four times greater than peacetime power needs. The FPC proposed beefing up interconnections to achieve greater capacity, while the War Department proposed scrapping widespread networks and instead expanding regional generating facilities in order to have a surplus for armaments production.[19]

In September of 1938, Roosevelt appointed the National Defense Power Committee (NDPC) to plan a more formal preparation for war. Composed of Secretary of the Interior Harold Ickes, Frederic Delano of the National Resources Committee, chairs of both the FPC and the SEC, and senior representatives of the War Department, the NDPC addressed who should build a stronger transmission system with greater generating capacity, how it should be funded, and who should operate it.[20] At the same time, the Public Works Administration received an allotment of $200,000 to research the feasibility of constructing a national power transmission network. Meanwhile, private utility executives argued that sufficient electricity already existed to satisfy war production needs.[21] Two years of dispute marked this attempt by the federal government to plan and build a national power network.[22]

Questions of control plagued government efforts to organize for a possible war. The FPC, the NDPC, and Interior Secretary Harold Ickes separately pursued wartime power plans. In 1939, Roosevelt combined the NDPC with the now dormant National Power Policy Committee into the new National Power Policy and Defense Committee (NPPDC), with Ickes as chair.[23] By 1940, administration infighting had doomed the NPPDC to a lame duck role. In the vacuum, the president asked the FPC to coordinate more closely with the War Department.[24] The FPC initiated a survey of manufacturing industries and their projected demand for power to complement monthly reports from the utilities about capacity.[25] Leland Olds, chairman of the FPC, explained to the industries that the FPC hoped to translate defense order dollars into kilowatts in order to answer, at short notice, "any question concerning the power situation in any of the 48 areas" the commission studied.[26] Notably, federal administrators and utility executives alike looked to interconnections as a solution for preparing the country for war. They quibbled, however, over who should plan, finance, build, own, and operate the networks.

While the utilities resisted direct federal control of their activities, they participated in planning activities. By 1939, the FPC, in cooperation with private utilities, had mapped out power supply areas for defense work, where to interconnect and exchange power, and how to protect against hostile acts. With modeling and testing provided by private utilities, the FPC "investigated the feasibility of a system of high capacity transmission interconnections . . . with a view to the more economical use of existing capacity and greater assurance against interruption of service in any one of the important centers of defense production."[27] The FPC also surveyed twenty-five geographically defined power supply areas essential to industrial defense production to assess the economic feasibility of interconnection and coordination.

Private power company executives expressed confidence in their ability to match the anticipated rise in demand for electricity as the country geared up for war. Throughout 1940, industry leaders assured the public that the "government can count upon the availability of an adequate power supply for national defense without the need for expenditures or other special measures on its own account."[28] Utility leaders claimed, "Private power resources are equal to the demands of national defense and require no assistance from . . . government funds."[29] Industry had a "tremendous undisclosed" surplus of power that could be "brought in through interconnection."[30] But the holding company executives were not tone deaf. Because they were still in the throes of untangling financial relationships under the 1935 Public Utility Holding Company Act, utility owners took pains to clarify that physical integration would not necessarily equate to corporate integration.[31]

Some analysts reflected that the prewar industry was unstable and both physically and financially unprepared when "confronted with new, and in some regions unprecedented, need for its highly essential product."[32] This proved true in 1941. A regional drought and accelerated defense production in several southeastern states led the FPC, on June 27, to urge citizens and utilities to limit nondefense uses of electricity.[33] On the same date, the commission issued seven orders for utilities in southeastern states to interconnect. This type of federal intrusion into private sector power activities had not been practiced since the end of World War I. Policy experts expected the federal government to flex even greater oversight both during and after the war to effect regional grids.[34] Reluctant to depend entirely on the intelligence and goodwill of the private sector, the Roosevelt administration took additional steps to ensure some federal control over the electric power system.

Leland Olds called a series of meetings with utility executives beginning in June 1941 to assess the power situation. Olds held meetings in the Northeast, Atlanta, Chicago, Portland, and Denver. Describing a power emergency, Olds stated, "I cannot overemphasize the gravity of the present power situation."[35] Following the FPC meetings, the US Office of Production Management established a special unit, headed by James A. Krug, chief power engineer with the Tennessee Valley Authority (TVA), to "handle all defense power problems."[36] Krug announced plans to create giant power pools in the Southeast and the New York–New England areas to supply major aluminum and magnesium production efforts. "The day of emergency is here," Krug explained. "Reserves long provided for such contingencies must now be called upon to the limit of their capacity."[37]

Roosevelt consolidated his administration's authority over power systems throughout 1941 and 1942. The TVA deployed engineers to New York to assist with interconnection between several very large private operations including Consolidated Edison, the Niagara Hudson Power Corporation, and the Pennsylvania–New Jersey Interconnection. The SEC acted in November 1941 to facilitate financing for private utilities seeking to interconnect. While the federal government had failed to exert this degree of authority over power system planning during the 1930s, the threat of joining in a global war spurred both the public and private sectors to collaborate more closely. Following the country's formal entry into the war in December 1941, Roosevelt took further steps to orchestrate power for defense manufacturing.[38] In January, he created the War Production Board (WPB), which subsumed the Office of Production Management and exerted greater influence over private industry activities. In April, he conferred additional authorities on the WPB, allowing it to allocate equipment for power development, determine supply and demand for war and civilian purposes, take over planning in districts with limited power supply, and work out arrangements to ensure electricity would be available for war industries. The FPC would "make suggestions" to the WPB while continuing with its ordinary responsibilities but only order interconnections after conferring with the WPB.[39]

During 1942, the FPC waived one of its key interconnection policies and allowed intrastate companies to join interstate pools without falling under federal regulatory control. In numerous orders approving emergency interstate interconnections, the FPC provided an exemption that would cease to be in effect ninety days after the end of the war. This policy shift proved significant for investor-owned utilities in Texas. Until the 1940s, several major investor-owned utilities in Texas refrained from building interties across the state line, thereby avoiding any federal regulation of power transactions. Following the FPC policy change, several Texas utilities joined the Southwest Interconnected Power Systems group in order to supply power to key defense manufacturing plants in Arkansas and farther east. Most withdrew from the interconnection after 1945, and a completely autonomous grid still serves the majority of Texas today. The FPC also began closer management of power sales across the country's borders with Canada and Mexico. In some cases, the commission permitted companies on the Niagara River to divert more water for power than previously allowed. In others, power sales were either curtailed or expanded in order to accommodate war production activities on either side of the border. By December 1942, federal war agencies had already commandeered significant authority over power systems.[40]

Pooling beyond Mere "Puddles"

Participants in both government and utility planning circles quickly understood that interconnection provided the only rapid route to essential power resources, offering multiple advantages for defense preparation. With interconnections, utilities avoided new plant construction by sending excess power from areas with low industrial activity, like New York City, to regions with intense war production, like upstate New York. Large power pools took advantage of the diversity of types of generators, needs of customers, time zones, and regional rainfall differences. Further, by avoiding new plant construction, the WPB diverted material resources and manpower to the construction needs of defense industries. Finally, interconnection allowed power pools to prepare for direct attacks, sabotage, or other causes of outages that might cripple defense producers. As Philip Sporn reported to the American Society of Civil Engineers, "If a plant is put entirely out of commission, the only immediate replacement can come from interconnections."[41] While private utilities had created interconnected systems during peacetime to achieve operating economies, conserve natural resources, and increase reliability, these were mere "puddles" when compared with the power pools effected for national defense.[42]

Pooling took place both at the behest of the federal government and through private utility initiative. Between 1941 and 1944, the FPC ordered forty-five interconnections, the majority during the first two years, many in the Northeast, the TVA area, Texas, and the Pacific Northwest. The southeastern United States served as the testing ground for accelerating integration. Over the prior decade, area utilities had built extensive interconnections but did not generally practice regional coordination. In response to the drought of 1941, the power branch of the Office of Production Management issued Limitation Order L-16, which directed interconnected utilities to maximize flow to areas serving defense industries in the Tennessee River Valley. This resulted in a 40 percent increase in power availability in the drought-stricken region. Evidently, not all consumers appreciated the focus on military use of electricity over domestic use. One utility in Tuscaloosa published ads, illustrated in figure 5.6, to explain to customers the need for cutbacks even when there appeared to be no shortage of local power. Interconnections followed in the northeastern corridor, in Texas, and in several midwestern states in 1942.[43]

Both public and private power companies and consultants offered plans for wartime power production. For example, Arkansas Power & Light Company learned in May 1941 that Arkansas was under consideration for a large aluminum-producing facility. The utility organized the Southwest Power Pool (later integrated into the Southwest Interconnected Power Systems group) with eleven neighboring companies in eight nearby states and proposed interconnection plans to the WPB, in direct competition with a similar proposal from the Rural Electrification Administration. Ultimately, the WPB approved a compromise plan that incorporated government and

Figure 5.6

Alabama Power Company explains power curtailments to customers.

These advertisements appeared in *The Tuscaloosa News*, November 5, 1941, and November 26, 1941.

private entities. To conserve copper that would otherwise be used in transmission lines, however, the War Production Board curtailed some interconnections.[44] Ebasco Services, Inc., formerly part of Electric Bond and Share Company, proposed to the US government a comprehensive plan to link isolated and municipal plants into existing power pools and increase capacity for war production.[45] The 1942 study illustrated the amount of additional power available from existing plants through improved load factor and greater economy in operations. Similarly, in 1942, the Nebraska Power Company and Kansas Gas and Electric Company built a 270-mile transmission line to aid in war production, increase operating economies in a thirty-one-state power pool, and protect against interruptions. The public and private power sectors, despite expressions of distrust, worked with each other and federal agencies to increase available electricity.[46]

The interconnections built through federal and private sector cooperation proved critical to meeting defense power demands. All told, through interconnections and careful planning of operations, the government and private utilities together assured that hundreds of billions of kilowatt-hours of electricity traveled across roughly two hundred thousand miles of power lines to both defense and domestic users. With only a 25 percent increase of installed capacity from 1940 to 1945, the nation's power system generated nearly 60 percent more electricity during the war years.[47] No peacetime era matched this phenomenal record. Only through close coordination of operations and rapid installation of new interconnections could the industry achieve such unusual gains in power production without new power plants.[48] Different regions told their own stories.

The Northwest Power Pool reported an impressive record of power delivery by late 1944. Six utilities created the pool in 1942, and WPB Order L-94 mandated interconnections with Bonneville and Grand Coulee Dams.[49] The pool provided power for new defense industries along the Columbia River, including the Hanford nuclear facility (top secret at that time). The diagram in figure 5.7 provides a schematic representation of this network of public and private utilities. The pool linked generating plants across five states—Washington, Oregon, Montana, Idaho, and Utah—and capitalized on the combination of hydroelectric power, steam power, differing time zones, and diverse customers to produce 4.5 million horsepower of capacity around the clock. This equaled the quantity needed "to build one 10,000 ton Liberty ship every day, or turn out (censored) [*sic*] Flying Fortresses a day, or to produce 275,000 pounds of aluminum every 24 hours."[50]

By the end of the war, the Southwest Interconnected Power Systems pool reached from Nebraska to southern Texas and from Tennessee to New Mexico, an area covering roughly eight hundred thousand square miles. In this region, existing generating plants produced more power than was needed for domestic purposes and enough for expanding war industries, but not in the right places. As utility executive C. S. Lynch explained in a journal article, "The additional loads for war purposes located in the entire area are about equal to the capacity available in the territory as a whole. . . . When reduced to areas this is not true."[51] Power companies were challenged not

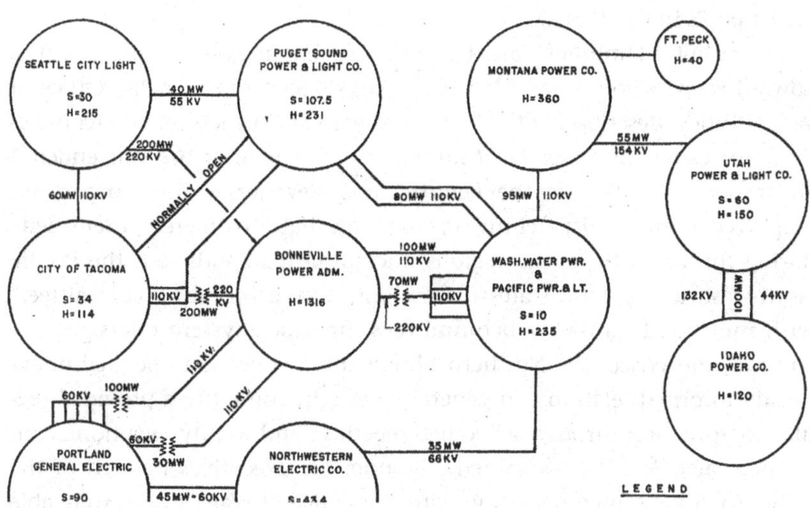

Figure 5.7

Circle diagram of Northwest Power Pool.

Source: W. C. Heston, "Kilowatt-Hours Pooled for War," *Electrical West* 92, no. 3 (1944), 52.

only to move sufficient electricity to points of major demand, like the new aluminum processing plant in Arkansas, but also to carry peacetime loads. Lynch continued: "To do this, power must be moved from one area to another and proper accounting made of it."[52] Texas utilities joined this pool to supply the Arkansas aluminum plant following the FPC emergency orders promising exemption from federal regulation. This pool also periodically joined the East Central and Southeast pools to form a "superpool with over 21,000,000 kw [kilowatts] of resources."[53]

Pioneering Control Schemes into Obsolescence Every Few Days

Through the efforts of the War Production Board, the FPC, the SEC, and private utility executives, several new power pools organized and tackled the challenges of operating interconnected, and at maximum capacity, for defense production. Regional differences, historical experiences, and particular technical preferences informed how each pool functioned. In the aggregate, however, the utility industry and the engineering community gained significant insights into how to manage and control power exchanges among multiple entities across great distances. The lessons learned helped the industry respond to the postwar industrial boom and the advent of giant power pools in the 1950s.

Operating techniques varied across different regions of the country. Edward Falck, who followed James Krug as director of the WPB's Office of War Utilities, described both the importance and variety of interconnections to readers of *Power Plant Engineering* shortly after the war ended.[54] Older power pools, like the Pennsylvania–New Jersey Interconnection, employed a central dispatching organization that was "merely intensified" during the war.[55] In several regions, including the Southwest, the Pacific Northwest, and central states in the East, a "top operating committee," with multiple localized subcommittees, provided system oversight. By contrast, the Wisconsin–Northern Illinois interconnection operated under mostly informal relations. In general, oversight committees provided regular coordination through scheduled meetings and weekly telephone conference calls. As Falck explained, "Responsible executives in each of the utilities working in co-operation with the Office of War Utilities were able to carry out the necessary co-ordination throughout the operating departments of their respective systems."[56] Actual power dispatching, however, took place in the control rooms of the participating utilities.[57]

Historically, companies sharing power on an interconnected system might have differing operating objectives that guided the choices made by system operators. Choices regarding frequency and load control reflected whether each company placed the highest value on system stability, financial agreements, resource management, or operating economy. During the war, however, priorities shifted. Interconnected systems had to address three new issues. First, the war industries demanded more power in certain locations, but the utilities were constrained to rely on interconnections rather than new generating capacity to meet those demands. Second, as a result, the power pools were larger and more complex than they had ever been before. Third, the utilities attempted to keep tie lines closed all the time in order to take maximum advantage of generating capacity, while in the past it was rare to operate this way. In the larger and more complex systems that shared power nearly full time, the familiar challenges of frequency and load control became more nuanced.

System operators understood all too well the need for careful control of exchanges to achieve maximum delivery of power for wartime purposes: "Control of some kind is necessary in order that the power transferred from one system to another over the interconnecting tie-lines may conform to plans."[58] By the waning months of the war, automatic frequency control was "almost universal, and automatic tie-line control [was] becoming more and more extensively used."[59] As one Leeds & Northrup (L&N) engineer remembered, "Control schemes were pioneered into obsolescence every few days."[60] New control problems plagued interconnected systems. For example, operators noted an increase in unscheduled—also called "inadvertent"—exchanges of electricity across tie lines between systems. These unplanned power trades resulted in uneconomic operation and frequency regulation problems. Power experts tested new control strategies and advanced technologies on the very systems that worked around the clock to deliver electricity. Operators approached successful resolution of the frequency and load control problem but failed to eliminate inadvertent exchanges. In fact, power experts wrestled with inadvertent exchanges well past the end of the war and through to the end of the century.[61]

In the Southwest Interconnected Power Systems group, the scheme for managing both scheduled and inadvertent exchanges presaged automatic control techniques adopted by the entire industry after the war. The pool functioned as several discrete control areas, each with a central dispatcher responsible for that area's operations. The area dispatcher handled local load changes, allowing the pool as a whole to control frequency and load on a net interchange basis. This radical approach was also tested on the Indiana and Michigan Electric Company system in the early 1940s.[62] Using data from extensive field measurements, the technical staff analyzed trends and established generation plans for both intra-area and interarea power dispatch. Because "estimating loads, making units available, and scheduling interchange, are manual operations which require extensive communication," each area was equipped with telemetering apparatus and reliable communication equipment, as L&N engineer S. B. Morehouse reported.[63] With telemetering data that provided accurate information about actual system conditions, "more capacity was gradually assigned to regulation and coordinated results improved correspondingly."[64] With this approach, the Southwest Interconnected Systems group made greater and more economic use of available generating and transmission capacity, achieved continuity of service without loss of facilities, experienced less inadvertent interchange between areas, improved system frequency, improved economy at individual stations, and enjoyed excellent cooperation among participants through a true "good neighbor policy."[65]

In a similar manner, the Northwest Power Pool organized as a single system with multiple operating units. In addition to the top operating committee that included one representative from each pool member, a coordinating committee of four engineers conducted studies and supervised pool operations out of a central office. Power plants in this region supplied electricity to critical aluminum plants along the Columbia River as well as the Manhattan Project's Hanford nuclear facility. The pool moved electricity over thirteen thousand miles of interconnecting lines from regions with little war load to these high-energy-demand war industries.[66] Operators accomplished this feat by keeping ties closed at all times, which allowed the electricity to flow continuously between utilities.[67] As an engineer reflected at the time, "Controlling the energy in a Goliath such as this super power pool requires high operating skill and expert dispatching along with careful scheduling of power exchanges."[68]

The system operators agreed to manage local load changes locally to minimize frequency upsets across the entire pool. The coordinating committee used historical data and network analyzers to simulate operating conditions and transmission line faults before scheduling interchanges. Automatic telemetering and tie-line load and frequency control apparatus aided the process. Grand Coulee usually served as the frequency controller for the entire system. As a result, "variations in frequency [were] less for the pool as a whole than formerly under separate system operation."[69] The committee scheduled interchanges on an hourly basis with a goal of zero net deviation from the schedule between operating units. The Northwest Power Pool advanced the art and science of frequency and tie-line load control on a net-interchange basis.

The lessons of new pool-operating experiences included numerous unsolved challenges. While the systems achieved phenomenal improvements, Morehouse claimed that "interconnected operations of many power systems throughout the country [were] still in the pioneering stage."[70] For example, the problem of inadvertent interchange, also known as the unscheduled exchange of power, while partially resolved, continued to plague engineers and system operators. Both automatic control settings and manual control decisions could cause unscheduled power shifts. Engineers accurately predicted that larger power pools that included hundreds of units would experience these problems on a magnified basis. By the end of 1945, system experts defined a need for further study, modeling, and practice in the field of automated control. Nonetheless, the advances brought about during wartime hastened the development of net interchange tie-line bias frequency control, the operating standard of the industry for the coming decades. As utility executive Robert Brandt described, "No scheme has been devised yet which will permit tie-lines to be scheduled for loads as near to their operating limit as this one and this, of course, was extremely important during the war."[71]

Cooperation, coordination, and dedication to the defense cause proved to be the hallmarks of successful electrification during World War II. The private sector took pride in utility achievements. The year of peak demand during the war, 1944, "justified all the predictions which the electric utility industry has consistently made as to its ability to supply the war load of the United States," according to the Edison Electric Institute.[72] Cooperative action allowed utilities to pool regional capacity and carry extra loads.[73] Falck recognized that for the private sector, "the emphasis . . . was to assure the maximum co-ordinated use of existing facilities and to minimize the amount of additional generating capacity required."[74] Falck explicitly noted that utilities arranged operations on their own to achieve policy directives set forth by the WPB. At the same time, according to another electrical engineer, government agencies understood the "vital importance of power" and stepped in with "drastic action to bring about full co-ordination of the utility."[75] Engineers working for both utilities and the government "knew perfectly well that operating economies could be effected by cooperation and interchange between systems, and the impetus of a national emergency promptly overcame the long-standing resistance to integrated operation."[76]

Poised for Expansion

The concerted effort to integrate generating and transmission systems created an extraordinary situation for power producers at the end of the war. Either they had overbuilt for a postwar economic bust or they were underprepared for a postwar boom. Two issues had dogged utility executives as they addressed wartime electricity demands. First, although they had initially proposed expanding generating facilities by ten million kilowatts, industry leaders feared finding the industry in a state of overcapacity as the war ended. The War Production Board limited planned expansion of generating plants to 5.5 million kilowatts in 1942, but this did not completely eliminate the potential problem. Second, the strategy of relying on interconnections to increase available power for war production led directly to a drop in generating reserves. By 1945, the industry operated with the lowest reserve capacity in decades, creating a high risk for blackouts and brownouts in the face of sudden high demand. In the end, greater capacity and extended interconnections allowed utilities to meet an unprecedented and unexpected surge in domestic demand in the late 1940s.[77]

Early in the war years, utility executives waffled on the question of overbuilding. As the 1942 Ebasco Services, Inc., report noted, "Utilities . . . usually tried to prepare for all expectable load . . . and a large number of power developments, aggregating millions of kilowatts, have been projected. . . . There is the possibility or even probability that large excess capability will be available with the dropping off of war load."[78] In 1944, American Gas & Electric Company chief engineer and senior executive Philip Sporn acknowledged the possibility that "installed electric capacity immediately after the war may be anywhere from three to seven years ahead of the long-term trend."[79] Regional differences affected the engineers' assessments of capacity growth. In the Pacific Northwest, the additional capacity created through interconnection was "a war drama with peacetime implications that can have a far reaching influence on the industrial future of this region."[80] The Edison Electric Institute went so far as to congratulate the industry for approaching the end of the war "with no great overcapacity of generating plant."[81]

While utility operators feared the financial burdens imposed by excess capacity, they equally dreaded operating with insufficient reserves. Power producers historically relied on reserves to meet peak loads, to address incidents of unexpectedly high demand, to aid in emergencies, to compensate for unplanned generating unit shutdowns, and to allow scheduled shutdowns for repairs and maintenance. Before the war, US utilities boasted a 26 percent reserve capacity. Because the War Department held expansion of generating capacity to a minimum and the industry maximized the use of installed capacity through interconnections, operating reserves fell steadily. By 1943, the reserve had dropped to 13.2 percent, and after the war, it hit a dangerous low of less than 5 percent. Industry specialists preferred a reserve of at least 15 percent to ensure adequate capacity to address all eventualities. Utility managers responded by carefully managing system operations and taking full advantage of interconnections while awaiting new and larger power plants.[82]

Advances in managing interconnections placed utilities in an enviable position to meet postwar demands with great flexibility. Experiences with large power pools in the Southeast, the Southwest, the Pacific Northwest, and the Midwest demonstrated that utility operators could control and deliver power on a timely basis to specific regions. The use of biased tie line and frequency control apparatus and, in some cases, the further use of early net-interchange procedures allowed operators to manage power flow with a minimum of system interruption. Neither the federal government nor the utility industry knew what to anticipate heading into the late 1940s. With power pools, control technologies, and war experiences, however, the industry proved itself ready for the tasks of reconversion and, later, domestic industrial and economic expansion.

A brief postwar depression marked economic activity in the United States, but electrification picked up even before new industrial growth took place. Industrial production dropped in 1946 and did not reach wartime levels until 1948. Production fell again in 1949 and then began steady growth in the 1950s. By contrast, the pace of US power production slowed in the opening months of 1946, picked up speed by August, and exceeded the prior year's activity by December. Power production then exploded in 1947 and grew at an average rate of 44 percent every five years for the next two decades. Figure 5.8 illustrates the percent change in power production compared to industrial activity and population growth in the United States during the postwar years.

The growth in power production took place independent of the rise and fall of industrial activity and the steady but slower growth in total population. The extra capacity for power production created during the war, especially through interconnections and extended operation of existing plants, met immediate postwar demand. Mostly residential customers,

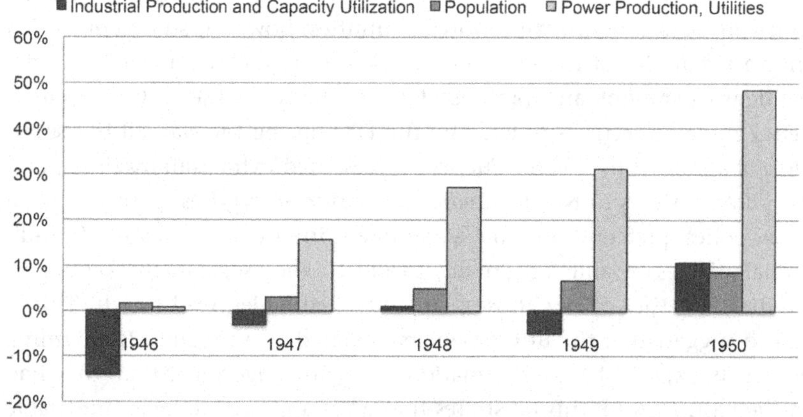

Figure 5.8
United States percentage change in industrial production, population, and power production, 1945–1950.

Sources: Federal Reserve System Industrial Production and Capacity Utilization Data Download Program (https://www.federalreserve.gov/datadownload/); *Historical Statistics of the United States, Colonial Times to 1970* (Washington, DC: US Government Printing Office, 1975), Series A 608, 8, and Series S 32–43, 820.

but some commercial customers as well, sought electricity for new appliances and for extended use of lighting and motors. As economic activity accelerated, the power industry added momentum to electricity growth through active marketing. In the 1940s, more than two hundred companies adopted Reddy Kilowatt, an advertising slogan developed during the prior decade to advance the cause of electrification. For the utilities, bigger loads justified the construction of bigger power plants, and bigger power plants in turn encouraged greater interconnection.[83]

Following the war, engineers credited interconnections with allowing the United States to maintain industrial supremacy. Further, they credited the technical demands of the larger power pools with solidifying advances in power control technologies. As the chairman of the War Production Board reflected, "We never once had to slow down production because of any lack of electric power."[84] The power industry prided itself with meeting both war production needs and civilian power demands during the war years. The utilities formed giant power pools in the 1950s to capitalize on interconnections, share techniques for managing the grid, avoid government oversight, and retain autonomy over their own operations and profits.

The war experiences illustrated to both government and utilities the efficacy of centrally directed planning and highly coordinated power production and distribution. Investor-owned utilities, however, still commanded the vast majority of the electricity market and quickly returned to independent operations and profit seeking. Notably, of the forty-five emergency interconnections ordered by the FPC during the war, all but seven were abandoned by 1947.[85] The war represented extraordinary times, and in ordinary times in North America, capitalist enterprises, even regulated monopolies, preferred minimal government interference. As a result, individual utilities expanded regionally to address their separate market opportunities, whether or not those plans matched widespread trends. This is not to suggest that the utility owners operated in a vacuum. If anything, electricity experts shared information even more vigorously than they had in the past. Instead, this illustrates that what appeared to be an inevitable direction of growth was not pursued in a unified manner.

6 Nuances of Control in an Increasingly Interconnected World, 1945–1965

For the power industry, the two decades following World War II were "the golden age of electric utilities" and "the good old days."[1] Pent up demand immediately after the war and industrial expansion in the 1950s created an ideal climate for power system growth. With no significant political or economic pressure to reverse the trend toward integration, discrete networks that had expanded rapidly during the war soon stretched across the continent. In 1950, Nathan Cohn claimed, "A grid approaching country-wide extent is not too fantastic a contemplation for the future."[2] At the end of the decade, a utility engineer serving on the Interconnected Systems Group (ISG) Test Committee forecasted the likelihood of a coast-to-coast network.[3] As illustrated in figure 6.1, the skeleton of high-voltage power lines at the end of the war seemed to predict the future grid. In 1960, transmission lines with voltages of sixty thousand or greater crossed the country like a dense web, as aptly displayed in figure 6.2.[4] By 1965, a full 97 percent of the generating capacity in the United States was interconnected in five large systems, and two years later, interties linked all but Texas and Quebec into a single network.[5] Indeed, the United States' first fully realized National Power Survey, issued at the end of 1964, predicted functional interconnections crossing the entire United States by 1980.

The practical concerns of operating a network of widely diverse power systems on such a large scale remained daunting. For example, in the central and eastern parts of the country, more than two hundred smaller electricity pools crossing thirty-nine states and two provinces shared power in a single network. Within this network, however, many ties were "sufficient for emergencies only."[6] Each new expansion of power pools introduced new complexities to the process. As older technologies for controlling power became obsolete, utilities relied heavily on system operators and

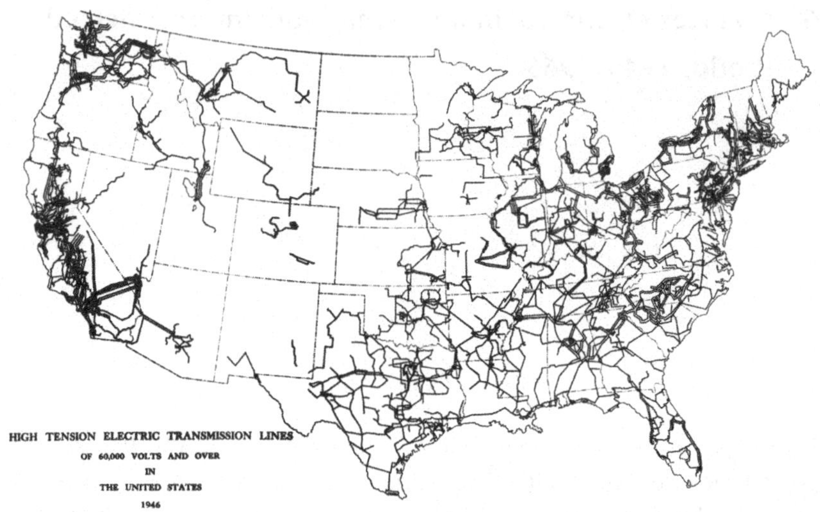

Figure 6.1

Map showing high-voltage transmission lines and related networks in 1946.

Source: Report on the Status of Interconnections and Pooling of Electric Utility Systems, Edison Electric Institute, 1962.

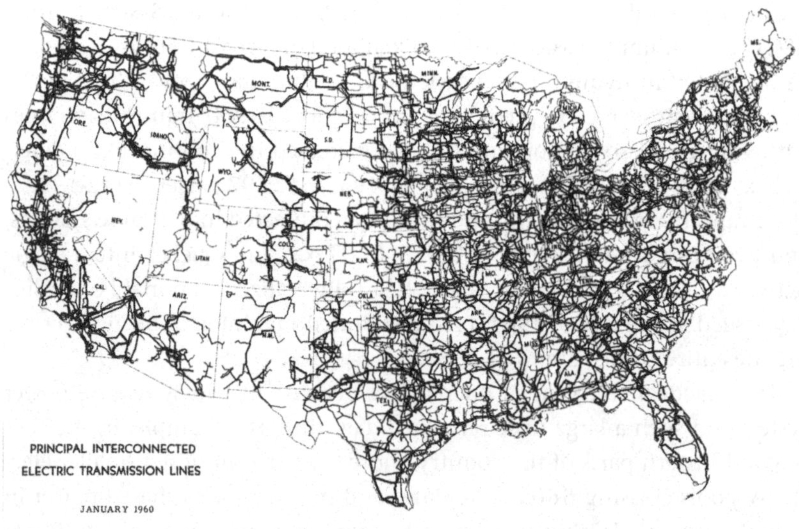

Figure 6.2

Map showing high-voltage transmission lines and related networks in 1960.

Source: Report on the Status of Interconnections and Pooling of Electric Utility Systems, Edison Electric Institute, 1962.

manufacturers to collaborate on new techniques. While there were no major blackouts, the nature of the failures taking place, particularly in the 1950s, indicated that interconnected power systems harbored underlying weaknesses. The fraternity of experts developed new and better techniques for managing electricity interchanges. They adopted more sophisticated instruments, including digital computers, for analyzing and modeling system behavior. And they formed an organization through which they prepared for nationwide interconnections.

In the earliest years of electrification, the engineers, inventors, managers, and dispatchers who designed and operated power systems established a network of experts through both formal and informal means of communication. Through journals, meetings, professional associations, and visits to each other's facilities, the experts developed a body of shared knowledge about the practice of generating and transmitting power. This continued through the century, and the avenues for communication expanded. Universities established programs dedicated to power engineering. Professional subgroups formed within entities like the American Institute of Electrical Engineers and the Edison Electric Institute. As autonomous power companies established interconnected systems, the individual operators developed working relationships through which they shared not only information critical to their particular network but also information of general interest to power systems operations.

Power experts negotiated a fine line between commodity and service as they advanced techniques for controlling power. When the technical challenges threatened the overall stability of interconnected systems, the experts argued vehemently about alternative approaches but generally coalesced around solutions that preserved the service aspect of electrification. The matter of a control instrument setting called the "bias setting" serves as an example of this. When stability was not at stake, operators considered disputed options and made individual choices that met the economic and operating goals of their respective employers. In the case of selecting cutting-edge tools to improve economy loading, for example, utilities did not converge on a single solution and made a slow transition from the more familiar analog approaches to the very new digital approaches. In this way, both the nature of technical challenges and the process of resolution reflected the social operating arrangements that characterized North American power systems.

The Federal Power Commission (FPC), with the support of Congress, initiated a formal consideration of coast-to-coast interconnection in the early 1960s. Under its authorizing legislation, the FPC was responsible for promoting and encouraging interconnection and coordination. To this end, the agency made several aborted attempts to carry out comprehensive planning, beginning with the 1935 National Power Survey. Economic depression, war, and then postwar reconversion and expansion directed the agency's attention elsewhere. In light of new generating technologies, advances in long-distance transmission capacity, steady growth in demand for power, and the trend to integrate across power systems, the FPC took up the survey idea again in 1962. The FPC stated that the project would assess the industry status, project future needs, and prepare plans to meet those needs. This foray into power planning triggered a mixed response across the industry. Representatives of investor-owned utilities expressed suspicion about the government's intentions, while advocates for public power encouraged the project. In the end, stakeholders from all sectors collaborated with the FPC in developing the survey, though not everyone agreed with every finding.

During the 1950s, control engineers and system operators refined techniques for automatic frequency and load control to bring about greater system stability. Together they delineated both problems and solutions through a process of argument, collaboration, and voluntary adoption. At the same time, utilities pursued cohesion across multiple power pools through development of the North American Power Systems Interconnection Committee (NAPSIC). It is notable that this took place under schemes of state and federal regulation established by the end of the 1930s and unchanged until the passage of the Public Utilities Regulatory Policy Act of 1978, the 1964 National Power Survey notwithstanding. State regulators attended to the rates charged and areas served by utilities within state boundary lines. Federal regulators addressed rates charged for interstate trades. Utilities themselves governed the stability of the interconnected power lines on a voluntary basis. The industry approached a national grid by institutionalizing certain systems of shared management and divided authority, while the government prepared by offering up a national survey that suggested, but did not determine, future paths for growth.

Postwar Patterns of Growth

Increases in the size of transmission lines and the scale of interconnections accompanied the economic expansion experienced by the United States between 1945 and 1965. In 1940, slightly fewer than three thousand miles of the highest-voltage transmission lines crossed the United States.[7] That number increased nearly ninefold by 1965.[8] Before World War II, the majority of transmission lines operated at 161 kV or below, with some in the western states operating at 230 and 287 kV. A surge in construction of higher-voltage lines began in the early 1950s, initially to carry power in the West Virginia–Ohio area and on the Bonneville Power System. FPC maps, reproduced in figure 6.3, indicate the extent to which utilities closed the gaps in parallel operations during the postwar years. It is telling to compare the 1939 map and the 1962 map. In 1939, many small interconnected systems served various regions of the country, and large areas appeared to have no electric service. By 1962, a handful of very large networks seemed to serve the entire country. The shaded areas represent general regions in which systems operated interconnected, but parts of those regions were not integrated into the networks. Nonetheless, by the early 1960s, most utilities could exchange power, at least in an emergency.[9]

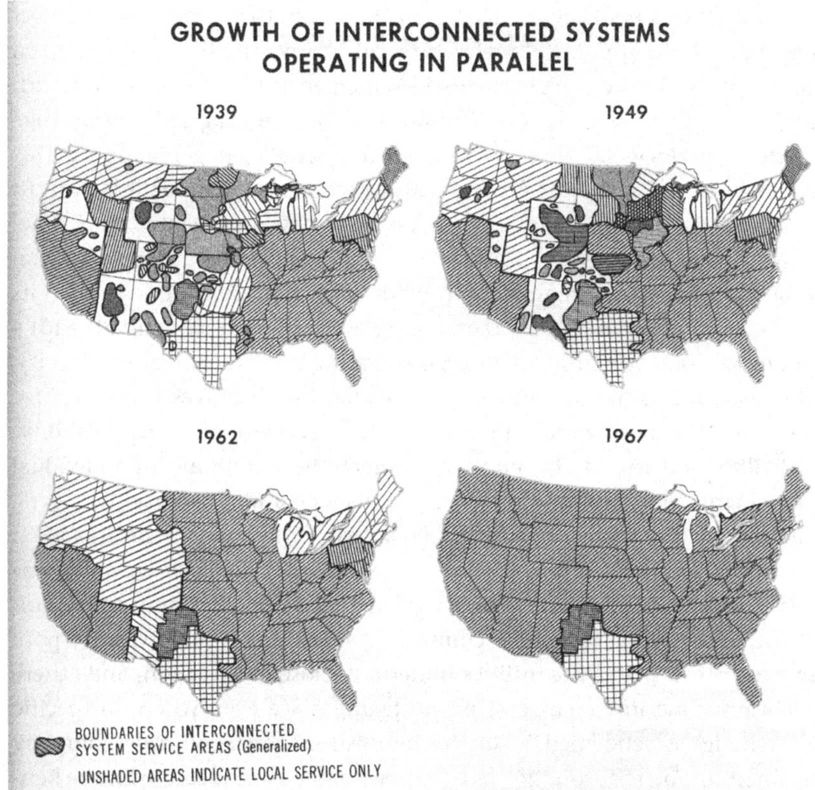

Figure 6.3
Areas of the United States served by interconnected power systems in 1939, 1949, 1962, and 1967.

Source: Federal Power Commission, *Prevention of Power Failures: An Analysis and Recommendations Pertaining to the Northeast Failure and the Reliability of the U.S. Power Systems, A Report to the President by the Federal Power Commission* (Washington, DC: US Government Printing Office, 1967), figure 11, 35.

By 1964, nearly two hundred major systems or pools aggregated power from the vast majority of the 3,600 entities producing electricity in the United States.[10] These networks served the Pacific Northwest; the Pacific Southwest; Texas and part of New Mexico; the central states stretching from the Rocky Mountains in the West to the southeastern seaboard in the East; and the mid-Atlantic states, New England, and eastern Canada in the Northeast. A few smaller grids served portions of the United States as well. While electricity reached most areas of the country through interconnections, numerous independent small power producers dotted the landscape, and vast stretches had no electricity at all.

Utilities shared key objectives in operating interconnected, but the development of each pool occurred along distinct lines that continued to reflect regional priorities, political preferences, and corporate differences. First, in all the pools, utilities agreed to provide emergency assistance. Second, with the exception of parts of Arizona, most utilities also agreed to exchange economy energy. In other words, utilities scheduled generation across the pool to use the most economical power sources first. Third, in most of the giant pools, the utilities shared backup power available at any time, called "spinning reserve," thus reducing the additional generating infrastructure each system required to provide sufficient spinning reserve on its own.[11] Fourth, but less consistently, the pools provided participants with a forum for joint planning of future generating and transmission capacity. Engineers and utility operators identified all these objectives as essential for obtaining the most energy-efficient operations and for ensuring reliability.

Utilities organized the pools to protect the autonomy of individual participants while sharing the energy efficiency and reliability objectives. The Edison Electric Institute described a variety of arrangements in a 1962 report on interconnections.[12] The Southwest Power Pool, for example, was a "voluntary organization with a rigid network for maximum utilization of the resources of all for the common good."[13] In Florida, with no pool agreements in place, the utilities undertook close cooperation, and "there is always some interchange of power taking place even when no specific interchange is scheduled."[14] In the Illinois-Missouri Pool, each company retained complete autonomy, but members agreed to cooperate for the benefit of the entire pool. The Northwest Power Pool was a voluntary organization in which utilities made plans independently but in harmony with each other.

The planning feature of the pools offers the clearest expression of the way in which these organizations allowed for both autonomy and unity. Whether corporations, municipal utilities, rural cooperatives, or government agencies, each participating entity built its own physical facilities. In some extraordinary instances, groups of utilities invested together in infrastructure for a common purpose. For example, in 1952, fifteen utilities in Ohio formed the Ohio Valley Electric Corporation (OVEC) to provide transmission lines and generating plants for an Atomic Energy Commission uranium enrichment plant. The OVEC power plant at Portsmouth, Ohio, is pictured in figure 6.4. In general, however, power providers invested in their own new infrastructure, raised their own capital, and obtained returns through their own regulated rates. Within ISG, the country's largest network, there were multiple smaller pools, some of which had central coordinating offices. Utilities in each subarea engaged in integrated planning for increased capacity. For example, the Illinois-Missouri Pool had no central pool coordinator, yet the member companies participated in quarterly load review and planning and then staggered construction. In the Iowa Pool, "usually the more deficient company installs the next unit required."[15] All the utilities in ISG, however, retained distinct corporate autonomy.

Figure 6.4
X-533 on November 17, 1955, W Switchhouse Extension looking northeast.

Reprinted with permission, courtesy of Portsmouth Virtual Museum, http://www.portsvirtualmuseum.org.

While the mechanisms differed, the overall coordination of system expansions had the effect of creating a seemingly coherent network of power companies. With a wide variety of pools and interconnection agreements, the grid began to look more and more like a single entity, much like the image conveyed in the 1967 map in figure 6.3. The operating committees that provided oversight of daily operations further engendered a sense of integration and cohesion, yet individualism reigned. Some pools used automatic load and frequency control; others did not. Some pools operated only on scheduled and emergency interchange; others kept power flowing at all times. Some operators fretted over inadvertent power exchanges, which caused problems for both frequency control and accounting. Others did not address inadvertent flow. The emerging nationwide grid could best be described as a hodgepodge of public and private companies operating in locally determined fashions yet maintaining a consistent flow of power to customers.

Economy Dispatch and the "Look Within"

System operators and control engineers engaged daily in the effort to improve the efficiency of power networks. As Cohn put it, "This era was highlighted by the adoption in many areas of 'a look within' . . . for economy."[16] In particular, the "look within" referred to the allocation of load among generators, taking into consideration line losses, in order to achieve the lowest cost per kilowatt-hour delivered to the customer. An engineer with Philadelphia Electric Company argued, "The most important tangible advantage is the fuel saving that can be realized by more exact economic load allocation."[17] Economic load sharing had long been one of the primary goals of building interconnected systems. During the 1950s, control engineers and operators expressed "increased interest and emphasis on the overall problem of operating systems at optimum economy."[18] While the fraternity of experts shared the common goal of achieving greater energy efficiency, the practices differed from power plant to power plant.[19]

Within increasingly complex interconnections, the calculation of economical load distribution likewise became much more difficult. In effect, operators faced a "big data" problem. The quantities of data available to operators, and the speed with which it was generated, continually exceeded the capacity of available tools—whether human calculators or the high-tech instruments of the day—to process and analyze the information. Before 1940, system experts documented the operating histories of each generator, calculated the incremental costs, and predicted demand to establish load distribution schedules among the different plants on an interconnection. Each steam-powered generator on a system, for example, burned fuel at a different unit rate depending on its age, size, load, and operating efficiency. As operators planned, they took into account these variations. Companies sharing power in a pool used curves or charts to determine the order in which generators across the network would be loaded. At companies using telemeters, operators collected reams of data about demand and power flow. In 1939, engineers working for Consolidated Edison Company in New York wrote, however, "In order to prepare the schedules required to follow changing system conditions for the vast number of probable combinations to be run on a system of any size, a force too large to be practical would be needed."[20] These engineers designed an incremental cost slide rule to aid the process.

The full cost of generation includes both the incremental cost of each new kilowatt-hour produced at the plant and the value of energy lost as electrons move across transmission lines. In practice, operators did not include transmission losses in economy loading calculations before the 1940s. For the scale of power trades undertaken during the 1920s and 1930s, distributing load based only on incremental cost proved satisfactory. As the power industry accelerated production to meet wartime demands, however, operators realized that their techniques for determining economic loading left money and resources on the table. The practice of preparing economy loading schedules in advance based only on incremental cost calculations no longer met the need to maximize production as cost-effectively as possible.

Engineers introduced numerous innovations for economy loading in the early 1940s. In 1942, E. E. George, an engineer with Ebasco Services, Inc., published an equation that successfully integrated transmission line losses into economy loading calculations.[21] Operators and engineers soon experimented with computers and calculating machines to speed the analysis of data for economy loading. Initially, they tested both analog and digital machines for solving the combined transmission loss and incremental cost calculations. George worked with Purdue professor J. B. Ward to configure a network analyzer to mimic multiple combinations of generators, transmission lines, and loads. In 1949, George and Ward reported on the ability of the network analyzer to complete the loss calculations more efficiently than they could by hand. The fraternity of experts were divided on the question of whether analog or digital tools offered superior results. George and Ward believed the network analyzer, an analog machine, was superior because "practically every power system has engineers familiar with the use of the network analyzer."[22] By contrast, a utility engineer argued, "It is possible the numerical solution may very well be carried out with greater facility by means of automatic digital calculators."[23]

In the early 1950s, equipment manufacturers raced to produce an effective working computer for economy loading. Often manufacturers worked closely with client utilities. General Electric (GE) and American Gas & Electric Company, part of ISG, experimented with punched-card machines and then a small digital machine called the "Transmission Loss Penalty Factor Computer."[24] E. D. Early, an engineer with Southern Systems, also part of ISG, introduced an analog solution titled the "Early Bird."[25] An early installation of the Early Bird is depicted in figure 6.5. Both projects reported annual savings of $200,000. Westinghouse, Goodyear Aircraft Company, IBM, and numerous others, including Leeds & Northrup Company (L&N), pursued the market.[26] As companies introduced more sophisticated computing technologies, they also sought to automate the economy load control activity.[27] In 1955, the journal *Business Week* reported on this application of computing to electrification, and *Electrical World* devoted an entire issue to the topic.[28] In general, engineers described digital solutions in laudatory terms, but utilities acquired the more familiar analog apparatus for their power plants.

The search for effective automatic economy dispatch also resulted in new opportunities for manufacturers to compete in the control apparatus market. By the early 1950s, L&N dominated the narrow category of automatic frequency and load control apparatus. Yet the company had to fight to gain equal footing for this new application. L&N regularly went head-to-head with Westinghouse to win major contracts for load control equipment, including economy loading computers. In sales approaches, L&N focused on the company's ability to assemble a thorough and sophisticated load control system that would ultimately save utilities money. Westinghouse underbid L&N every time. In the case of a Cleveland Electric Illuminating Company project, for example, "We [L&N] could not demonstrate a dollar savings at the coal pile to offset our price disadvantage." In this instance, Westinghouse underbid L&N by almost 60 percent.[29]

Over the ensuing ten years, power experts debated the merits of digital and analog solutions to economy loading. Digital machines processed larger quantities of data at faster speeds and produced highly accurate results. Engineers could easily modify programs for new system configurations, although they had to expend time and care to program the digital machines correctly. These computers tended to be significantly more expensive than analog machines, but a utility could rent time or

Figure 6.5
Installation of the "Early Bird" at Southern Power Company's Coordination Office in June 1954.

Source: "Transcripts from Leeds & Northrup Company's Presentation to American Electric Power Service Corporation," 1959, W. Spencer Bloor Collection, now in Electric Control Systems Records, Hagley Library, Wilmington, Delaware.

divide costs among multiple departments. On the other hand, analog computers delivered sufficiently accurate results to address the control problems at hand during the 1950s and early 1960s. For example, when processing the same data, L&N analog computer results differed from IBM digital computer results by less than 1 percent.[30] While slower, analog machines operated even when partially disabled. And an analog computer processed several equations in parallel, while a digital computer only processed data serially. As a professor from Northwestern University explained, "The digital machines are designed to do many arithmetical computations at high speed, while the analogue in some way forms a model of the actual system to be studied."[31] In an industry that adopted computing technologies early on, the transition from analog to digital for this application proved problematic. In 1974, consulting engineer John Undrill explained that interacting with the network analyzer "gave the user an excellent qualitative 'feel' for the behavior of the system," while this was lost as "digital computation took over the work of power system sharing systems."[32]

Some in the technical fraternity clung to the analog systems. Analog computers responded to electric impulses in essentially the same way power systems responded. This was an added value for young engineers attempting to grasp how electric power networks functioned. The "newbies" modeled system behavior with the computers and felt confident that they understood how electricity flowed. Advocates persuasively argued that the analog computers saved utilities real money. On the other hand, the engineers favoring digital computers rightly argued that these new machines offered much more rapid and accurate calculations of much more complex equations. While the model built into a digital machine might not be a perfect analogy of an electrical network, the speed and power of the computer itself allowed rapid system response. In addition, as the technology advanced, digital computers combined system operating metrics with other types of data important to utilities, including information about finances and personnel. This offered utilities much more detailed productivity management. By the late 1950s, the "look within" for economy included a glance ahead to the digital age.[33]

In this instance, the fraternity of experts disputed techniques and apparatus, but consensus was not necessary across the industry. Individual networks selected different methods for economy load allocation within their own areas, and this had no ill effect on the neighboring network. Often the choice of digital or analog computing reflected the approach to power transactions and billing systems already in use. Because systems looked *within* for economy, there was no pressure for uniformity of economy control apparatus *between* systems.[34]

As the competition between analog and digital solutions to the economy loading problem played out in the professional press and in the marketplace, the fraternity of experts tackled the question much as they addressed similar control problems in the past. They debated and defined the challenge through professional meetings and publications. They collaborated across multiple boundaries. Utilities worked together and with manufacturers and academics. Individuals and manufacturers competed overtly and covertly. Engineers critiqued each other's work in formal publications and raced to gain the confidence of customers through private meetings and presentations. Finally, individual power companies chose control techniques that suited their autonomous needs while sustaining the flow of information, ideas, and power across their interconnected systems. As a whole, the industry moved from one set of solutions to another over decades without oversight from a central authority and, in this instance, even without voluntary standards. The economy loading problem entailed coordination between individuals and organizations but allowed for fully independent choices. Other control problems required closer adherence to commonly accepted solutions lest the larger network be placed at risk.

Argument and Collaboration: The Question of Bias

Within a network of systems, disparate utilities were bound together by an obligation to maintain stability with near neighbors as well as distant operators. The mechanisms for cooperation included enlightened self-interest, loose agreements, contractual commitments, and shared knowledge through technical societies. An episode of repeated operating trouble on a large network, for example, triggered two years of heated debate over a seemingly minute technical detail. The controversy over a fractional setting on a single piece of control apparatus emerged in the postwar years. The industry converged on a standard that not only allowed companies to maintain system stability in the mid-1950s but also served the industry well for decades to follow. Notably, the adoption of the standard was entirely at the discretion of each system operator.

By the 1940s, utilities industry-wide used sustained tie-line bias frequency control to regulate frequency across interconnected systems.[35] This approach allowed generators to respond briefly on their own to frequency changes before control apparatus effected a system-wide correction. Through the 1930s, engineers and system operators experimented with arrangements that required one station to control frequency for an entire network while bias controllers on the tie lines kept power trades close to the predetermined schedules. On larger and more complex networks, this approach overburdened the frequency-controlling station. In 1938, the Indiana and Michigan interconnected system tested net interchange tie-line bias frequency control, which required each subarea to correct for frequency and load locally while the controllers on the tie lines between subareas corrected for the net of exchanges across the entire system.[36] During the war, the Northwest Power Pool and the Southwest Interconnected Power Systems Group both employed variations of net interchange tie-line bias frequency control.[37]

In the late 1940s, L&N engineer Nathan Cohn tested fully distributed net interchange tie-line bias frequency control with a newly formed power pool, the United Pool, which included companies in Iowa, Illinois, Kansas, and Missouri.[38] This test eliminated master frequency control from a single station and also employed the concept that each subarea controlled frequency and load internally.[39] Fully distributed net interchange tie-line bias control allowed "generation of the system as a whole" to "'breathe' with varying demand."[40] Cohn reflected, "Operations were fully in accord with expectations."[41] Robert Brandt of the New England Power Company similarly determined that fully distributed control resulted in more stable and economical system operations across a complex network.[42] Cohn published the mathematical reasoning underpinning this new approach to load and frequency control in 1950.[43] His formula for distributed net interchange tie-line bias frequency control is depicted below.[44] Cohn and Brandt, through their work with collaborating power systems, established a functional approach to power control that was widely adopted in the 1950s and is in continuous use in the twenty-first century.

$$E = (T_1 - T_0) - 10B(F_1 - F_0)$$

On the left-hand side of the equation, E represents Area Control Error (ACE), a term that means the difference between the actual power interchange and the scheduled power interchange.[45] For system operators, the goal is to keep E at zero, or nearly zero, thus meeting the agreed-upon power sharing between companies.[46] On the right-hand side of the equation, $(T_1 - T_0)$ represents the difference between actual power flow over the tie line and scheduled power flow over the tie line. The term $10B(F_1 - F_0)$ represents the difference between the actual frequency and the intended frequency, multiplied by ten times the frequency bias. The frequency bias, B, "is a conversion factor which is intended to represent the energy deficit producing 0.1 hertz frequency change in a control area. As a practical matter, this represents the energy of inertia of the rotating masses in the control area, and how much energy a 0.1 hertz change impacts their rotating mass's energy, i.e. how much additional energy a 1/10 change in frequency will draw from the rotating masses. The 1/10 accounts for the 10-multiplier in the equation."[47] The contemporary version of this equation is

$$\textit{Area Control Error} = (T_a - T_s) - 10B(F_a - F_s)$$

The success of distributed net interchange tie-line bias frequency control depended in part on the selection of bias settings.[48] The bias is the setting on the controller that determines how much help the system will provide to correct a frequency error before returning to scheduled loading.[49] The bias setting is usually a percentage of a system's "natural characteristic"—that is, the response of system governors to frequency changes in the absence of any other automatic or manual controls. Generators speed up and slow down naturally in response to changes in demand. If the demand goes up, the generator slows down; if the demand goes down, the generator speeds up. This is much like a horse pulling a load: added weight causes the horse to slow down, and a lightened load allows the horse to speed up easily. After some length of time, depending on a variety of characteristics, the system will return to the established frequency. The elapsed time is called the "prevailing natural generation governing characteristic," or the "natural governing response," or simply the "natural characteristic."[50] The decision of how high or low to set the bias, and therefore how much to aid the system, fell to the system operators. In the early 1950s, operators began to care deeply about fractional differences in bias settings. The eminent GE engineer Charles Concordia reflected, "When the frequency changed by a tenth of a cycle (or hertz) it was sufficient cause to establish a committee to study the problem for a year and a half."[51]

Early in the years of experiments with automatic apparatus on interconnections, operators and engineers had learned that overly tight automatic frequency control resulted in unwanted and uneconomical load shifts. The biased control had the benefit of minimizing the burden on single machines or single areas to handle all frequency changes on the interconnected system while maintaining scheduled load transfers and restoring the system to desired frequency quickly. Operators typically set the bias at a percent of the natural characteristic of the area under control. Some set the bias at precisely 1 percent of the natural characteristic; others chose settings above and below. For many years, the bias setting gained modest attention, and the effects on the systems during a disturbance were barely noticed. By the early 1950s, however, operators and engineers did begin to notice. As Cohn put it, with newly expanded interconnected systems, "large blocks of load or generation can now be suddenly lost, causing significantly large changes in frequency, but with the system still holding together."[52] These larger disturbances revealed, for the first time, the effect of bias settings, and the differing preferences of each autonomous system operator, on network operations.

The problems caused by inappropriate bias settings emerged during trouble on a segment of the ISG network. At its founding in 1933, the operating executives of the utilities that became ISG met to address "the tremendous frequency control problem their system operating people had to contend with."[53] The ISG Test Committee, formed at that initial meeting, served as the heart of the system's efforts to standardize operations across a rapidly expanding interconnection. ISG pioneered early methods of automated control and by the early 1950s provided recommended operating standards to more than seventy participating utilities.[54]

The operating standards issued by ISG reflected nearly two decades of experience in voluntary collaboration among engineers and operators across the interconnection. The Test Committee regularly performed trials of different types of controls and settings on interties, communicated through typed newsletters and meeting reports, and encouraged voluntary participation of member utilities in tests, surveys, and reporting. In 1951, the Test Committee established that "bias settings are to be increased on all systems . . . to reflect approximately 1 percent of system load per 0.1 cycle frequency departure."[55] Because utilities participated in the interconnection on a voluntary basis, this standard was styled as one on a list of six "recommendations as to operation of the interconnection systems."[56] In other words, no individual utility was under an obligation to meet the standard, but all were encouraged to do so.[57]

Despite the widespread usage of tie-line bias frequency control across North America, individual power pools elected to use different bias settings. Like ISG, most used 1 percent as the internal standard. As Cohn reported in 1950, however, "Operators are sometimes of the opinion that operating with a very small bias will minimize local generation changes."[58] The operators believed this lower setting improved the economies of the local system. As the scale of interconnections grew, disagreements about bias settings intensified. By 1954, a meeting on the subject left members of ISG with "battle scars."[59] One year later, engineers were "well aware of the controversy which is raging throughout the power industry with regard to what bias should be used and how this bias should be calculated."[60] The ISG experience both exemplified the challenge facing utilities and unified the power system fraternity around a solution.

Large-scale problems began on the ISG network in 1955. On February 2, the Tennessee Valley Authority (TVA) Shawnee power plant experienced trouble, causing part of the Illinois-Missouri power pool to disconnect and resulting in interruption of the scheduled generation in multiple systems across Iowa, Missouri, Illinois, and Kentucky, all part of the Northwest Region of ISG. Evidently, a fault on a generator at the Shawnee plant initiated the trouble, and a combination of operator decisions, bias settings, and techniques for tie-line control led to the small cascade of problems. Two weeks later, representatives from the largest affected utilities met to address how to operate during this and other types of emergencies. Participants agreed on the facts and calculations but differed considerably as to the desired operation under each of the various emergencies.[61]

There were six additional disturbances on the ISG interconnection over the next three months, including one in March on the OVEC system. Following this event, Howard Stites, an electrical engineer with Central Illinois Public Service Company, reported to the ISG Test Committee that the trouble lasted only four minutes and, in this particular case, "demonstrated that the frequency bias [of 1 percent] was satisfactory."[62] Stites reported on all seven disturbances at the May meeting of the ISG Test Committee, resulting in a decision to research the problem more thoroughly. ISG held a "discussion on increasing bias obligation" during a full system meeting at the end of that month.[63]

Separately, several of the affected utilities turned to the instrument manufacturers to help sort out the issues surrounding the faults and outages occurring on these larger interconnections. By this time, well over 90 percent of the interconnected utilities in North America used L&N tie-line frequency and load control apparatus on their systems. Three of the ISG utilities—Union Electric Company, Central Illinois Public Service Company, and Illinois Power Company—invited Cohn of L&N to attend an April 8 meeting to discuss the February trouble. Cohn used the opportunity to explain that bias causes "each area to do its share of frequency regulation" and also matches "the area's . . . governing characteristic."[64] Addressing the February trouble, he noted that the bias settings were lower than the natural characteristic, resulting in improper control responses and exacerbating the unfolding problems. Cohn and his colleague, W. Spencer Bloor, recommended that the utilities adjust bias settings upward.[65]

Continued trouble on different parts of the vast ISG network led operators to actively address the growing problems. Following a June 21 disturbance, again on the OVEC system, the Northwest Regional Committee asked the Test Committee to undertake a more detailed study of the bias setting question. Regional Committee Chair E. S. Miller suggested, "It may go a long way towards resolving the percent bias argument."[66] Study of the OVEC disturbance began in July, with a major goal of determining "the advisability of increasing our bias percent."[67] The Test Committee membership at that time included representatives from the TVA and from nine utilities based in eight states. The Test Committee regularly reported findings and recommendations to the full Interconnected Systems Group, composed of seventy-five utilities and including some of the largest and most influential electric companies in North America. A study by the Test Committee clearly held interest for the affected members but also potentially influenced practice across the continent and perhaps around the world.[68]

The Test Committee assessment began with a survey of all ISG member utilities, not only those participating in the OVEC system. In August, the Test Committee sent a memo to the utilities asking for data and considered the results during the following months. On October 25, 1955, the Test Committee issued a report that contended, "The present frequency bias of 1 percent per .1 cycle of deviation is inadequate, unrealistic and detrimental to Interconnected System operation."[69]

The Test Committee continued to probe the question of bias settings. The October report summarized findings based on survey responses, but committee members recognized that those responses had been incomplete and inadequate. At a December meeting, the committee decided to hold a special gathering early in the next year for "conducting additional research on the bias subject" and to formulate a plan for bringing the issue to full ISG membership in the spring.[70] The "plan of attack . . . was to invite a representative from each of four (4) industries, which industries are associated with the problems of system regulation, to meet with the Test Committee."[71] Chairman L. V. Leonard extended invitations to representatives from L&N Company, General Electric Company, Westinghouse Electric Corporation, and Woodward Governor Company. Leonard also exchanged correspondence separately with Cohn of L&N, describing at length the makeup of ISG, the functioning of its four Regional Committees and the Test Committee, and the evolution of the decision to tackle the bias setting controversy.[72]

All four manufacturing companies had been competitors in the automated control field, and the L&N team did not hesitate to exploit the coming meeting for marketing purposes. The correspondence from Leonard provided L&N with added insight into the utilities' concerns. Bloor met with John Donaldson, a TVA engineer and client, just a week before the planned Test Committee gathering and discussed the TVA records regarding the 1955 failures. In the course of this meeting, Donaldson noted that only recently had they "been able to see this regulating effect since 1) the system didn't used to be strong enough to hold together and 2) only recently have they experienced such large load losses."[73] Both Bloor and Donaldson, like others in the industry, connected the recent industry expansion and the growth in the size of electricity loads to the mounting evidence that bias settings mattered a great deal to system stability. Before departing, Bloor ensured that Donaldson had a full set of L&N publications on the topic of frequency and tie-line control to read before the upcoming special Test Committee meeting.

The Test Committee convened in mid-February, and the representatives from the four manufacturers made their presentations. Speaking for L&N, Cohn discussed the theoretical fundamentals of system regulation, the basic concepts of automatic control and operation, the priority of customer service, and the status of present control techniques. He addressed a number of questions about bias control, including the effect of different bias settings on tie lines after a fault occurs on a system. He outlined various scenarios and likely outcomes based on settings below 1 percent, at

1 percent, and above 1 percent. This set the stage for Cohn to argue to the Test Committee that the best outcomes occurred when the bias setting was equal to or greater than 1 percent of the system's natural characteristic. Following the presentations, the utility members met in a private session to discuss the talks and decide how to proceed. The Test Committee evidently found the L&N presentation sufficiently compelling to ask Cohn to address the full ISG committee later in the spring.[74]

The elevation of the bias setting issue to the full ISG committee prompted industry-wide interest in the topic. On March 1, Cohn accepted the invitation to meet with ISG at their Des Moines meeting at the end of April. One day later, the American Institute of Electrical Engineers (AIEE) Committee on System Engineering asked Cohn to turn the material into an AIEE paper to be presented at their summer general meeting in June. Cohn agreed to submit the paper to AIEE. In the meantime, he received a letter from Russ Purdy, senior executive with Commonwealth Edison and vice chairman of the AIEE Committee on System Engineering. Purdy underscored the lack of unanimity among system operators regarding increasing the bias setting: "I can state flatly that there is still a wide divergence of opinion within the group."[75] Purdy included himself among the skeptics, citing the lack of uniformity in apparatus used, the inadequacy of system telemetering across interconnections, and the capacity shortages that might inhibit generator responses under increased bias setting control. In his reply, Cohn emphasized that system operators had the final say in how interconnections should work: "We are always glad to discuss the theoretical aspects of this problem, but in the final analysis it is the operating people themselves who will want to determine what operating practices they will use."[76] Cohn met the submittal deadline for the paper and distributed it widely for comment in advance of the June AIEE meeting.[77]

The April ISG meeting proved educational for members of that interconnection. More than 120 individuals representing the ISG membership gathered in Des Moines on April 26 and 27. The schedule included two blocks of time, totaling nearly three hours, for Cohn's presentation. This was preceded by a report from the Test Committee. Earlier in the year, the Test Committee had distributed new surveys to the participant utilities and compiled more complete data regarding regional response to the disturbance of June 21, 1955. In the new report, the Test Committee concluded, "The data submitted . . . does indicate that the present 1 percent bias is too low."[78]

In his own talk, Cohn repeated his injunction against using theory as gospel, perhaps responding to the critique from Purdy: "I just can't emphasize too strongly that the bias setting problem is your problem and my analysis here is simply a theoretical one of how I think bias operates and does its regulation job."[79] Cohn walked the attendees through the underlying theory of bias control and the expected effects of bias settings above, below, and equal to the natural characteristic. He explained that the setting should be determined by each system based on the degree to which the system can and will aid a neighboring system with a load change. In the most general terms, when the bias setting was below 1 percent, the area under control provided less help, the frequency moved farther away from normal, and it took longer to return to stability. With the bias setting above 1 percent, the area under control provided extra help and the frequency moved toward normal, though to a lesser degree than the movement away under a low bias setting. Cohn again placed the responsibility squarely with the operators for determining their own settings: "For the function of responding to remote load change, the bias setting is very important. . . . Simply ask your self the question: What do I want the bias regulators to do on the contribution function of responding to a remote load change?"[80] In other words, how much did the operator want to help his neighbor at his own expense?

Cohn validated the autonomy of system operators as decision makers on a significant technical issue, at the same time expanding the theoretical basis for decision making. He clearly leaned toward a bias setting no lower than 1 percent that enhanced mutual aid but he stopped short of advocating this as a standard. Both the chairman of ISG and the chairman of the Test Committee reported to Cohn that his presentation "was exactly what we wanted" and that "everyone has a much better understanding of the function of load regulating equipment than they . . . had heretofore."[81] During the meeting, ISG adopted the recommendation that "each system set its bias equal to its natural system characteristic."[82] Nonetheless, the final decision on bias settings rested with each system operator managing his own discrete subnetwork on the interconnection.

The presentation at the June AIEE meeting elevated the issue of bias setting to international status and aggregated a solid professional community behind the changes Cohn advocated. Cohn's presentation at this meeting matched the April ISG talk, with greater attention given to mathematical models of his theory. Fifteen well-known and well-respected engineers, most from very large operating utilities in the United States and Canada, provided comment on Cohn's paper in advance of the meeting. One academic noted, "There have been many more, and unanimously very favorable discussion of this paper than of any other AIEE paper which I can remember."[83] Most concurred that the bias setting question, though seemingly minute, was timely and of great significance to the industry. As Purdy noted, in the past the bias setting was reached by "observation of system reaction in time of trouble by some calculation and considerable arbitration, which often included a sizable factor of ignorance," with the result that "the 1.0 percent bias became something of a standard."[84] Purdy emphasized the importance of considering practical limitations as well as theory in determining a setting. Some favored Cohn's approach, while others found reasons to dissent.[85]

In addressing the comments, Cohn clarified that the controversy rested on two potentially incompatible priorities: rapid restoration of system stability and economy. He noted that five commenters "call attention to the conclusion that bias ratios greater than one have correspondingly less effect on system parameters than bias ratios that are less than one."[86] Thus if rapid restoration of stability was the top priority, a bias setting greater than one had a significant advantage. On the other hand, another commenter noted "that regulation would be minimized, and hence economy improved, when the bias ratio equals one," and if ties are fully loaded, systems "would prefer to have the bias ratio smaller than one" to realize greater economy.[87] In reminding operators that they held the final authority for determining bias setting, Cohn also pushed utilities to determine whether their first responsibility was to meeting stability obligations or economy goals. In the coming years, more sophisticated apparatus made this balancing act easier to achieve.

The bias setting discussion found its way into widely distributed publications following the June meeting. The journal *Electrical West* featured highlights of the conference in the July 1956 issue, including coverage of Cohn's talk. In January 1957, AIEE decided to elevate the paper from conference status and include it, along with discussions and closure, in the printed transactions. In addition, L&N produced reprints of both the ISG and AIEE talks for distribution to clients around the globe.[88]

In the ensuing years, utilities converged on a standard bias setting equal to or greater than the natural characteristic. Notably, like the question of frequency, the standard for bias setting remained voluntary. Cohn's presentations had solidified support for bias settings that favored providing aid from one utility to another while rapidly restoring system stability rather than offering reduced aid in the interest of greater economy. In an internal memorandum at one utility, a Test Committee member reported, "While there was much controversy last year on the suggestion of bias equal to system characteristic, I was very pleased to find that there is now apparently complete agreement after a year's operation under that plan."[89] One engineer wrote in later years that the mid-1950s shift to the higher bias setting led to improved system frequency, closer adherence to tie-line schedules, and a reduced regulating burden.[90] In May 1957, the ISG Test Committee issued updated operating recommendations that stated, "Each individual operating company should set the bias setting of its tie-line load controller equal to or as close as possible to its natural system characteristic as estimated to apply to its system peak load (for the current year). . . . In no case should the bias be set at a value of less than 1 percent of estimated system peak load (for the current year) per .1 cycle change."[91] Operating recommendations issued in 1960 repeated this wording, as did the very first set of operating recommendations issued by NAPSIC in 1963. The newly created NAPSIC included utilities across the entire continent, suggesting that the standard for bias setting had gained universal acceptance in the power industry.[92] The approach of ISG reflected the approach of the industry as a whole: "All changes are, of course, subject to approval by the main group, and they are always only recommendations to the members of the Group because the Group is a voluntary group."[93]

The bias setting debate resulted in a voluntary standard that served the industry well for the rest of the century and beyond. In 1950, the ISG Test Committee urged system operators to set their bias at *approximately* 1 percent per 0.1 cycle change. System operators made decisions about bias settings that served the interests of their own employers, often to the detriment of the larger network. Following heated debate and then consensus, ISG issued the new standard: at no time should system operators set the bias at *less than* 1 percent. By the early 1960s, this tiny fractional change was universally accepted across North America, resulting in more stable network operations to the benefit of power customers.[94]

Contemplating the Grid

By the late 1950s, individuals in both the public and private sectors contemplated the very real possibility of interconnection from coast to coast. Experts working for power companies addressed the grand challenge of a national grid much as they investigated the rather less exciting issues posed by bias settings and economy loading devices. They collaborated across organizational boundaries and developed a process that embraced the common interests of all interconnected entities while protecting the autonomy of each. Through the creation of NAPSIC, the power companies established a forum for setting voluntary standards applicable to only the issues that mattered to grid stability.

At almost the same time, officials in the federal government initiated a project to assess the status of American power systems and delineate a path for growth that would position the country well for anticipated demand. Through the National Power Survey, the FPC hoped to offer a blueprint for interconnections, at a minimum, and if possible, a mechanism for increasing federal oversight of the industry. When finally issued, however, the survey reflected the heavy involvement of private utilities in its development and suggested, rather than directed, pathways to full coordination across the country. Just as the systems of shared management and divided authority characterized the approach of industry experts to completing a national grid, so patterns of tension between the public and private sectors characterized this iteration of federal intervention in power systems growth. By the end of 1964, both public and private entities shared a vision of national interconnections, acknowledged the likely shape of the grid, and reached yet another accommodation around the status quo for moving forward.

The power company and government initiatives focused on very different matters. The priority concerns of the power system experts grew directly out of their experiences with technical challenges like determining the ideal bias setting. Because so many different entities controlled the flow of power through multiple styles of operating agreements and with a variety of business goals, individual system operators were concerned with system stability. By contrast, the National Power Survey addressed much broader issues, including whether the industry could produce sufficient power to meet future demand, what electricity would cost, and where the primary energy resources would come from. As a result, the system operators formed NAPSIC with very narrow responsibilities, while the FPC offered the survey with the broadest possible implications.

The challenge of interconnecting across the entire continent appeared quite thorny to individuals within the largest and oldest power pools. Regional differences from the Pacific Northwest to the New England coast presented a variety of issues the fraternity of experts had not fully addressed. Within ISG, operators took the lead in discussing how to manage the next likely step in the growth of the power grid. In late 1959, members of the ISG Test Committee began to anticipate "the possibility in future years of a coast-to-coast network." At the next Test Committee meeting, in early 1960, the participants again considered that "ties with other Interconnections are possible to the east and west in the future."[95] This led to concern that the operations of non-ISG interconnected groups were inferior to those within the giant power pool. The next step involved inviting "key operating people outside of our Group to the St. Louis meeting," breaking with the past tradition of keeping these meetings closed. Thus began the process of forming an organization, potentially international in scope, to bring about stable operations should a coast-to-coast grid be realized.

This initial effort by the Test Committee to anticipate the ramifications of broader system interconnections paid off within two years, when operators from across the United States and Canada met to discuss coordination. At the instigation of ISG and in conjunction with the Group's annual meeting, system representatives from Philadelphia to Los Angeles and from Texas to Oregon convened in Omaha on April 25, 1962, "to discuss the Future Operations of Systems."[96] At the meeting, representatives reported that with the recent or impending closures of ties between systems, "all systems in the United States except the Texas New Mexico Area, plus systems in Canada," would be operating in parallel.[97] Meeting participants expressed interest in developing "the most desirable operating organization to effect the parallel operation of all systems."[98] They formed a temporary Interconnection Coordination Committee Working Group, chaired by a representative of ISG, and agreed to hold follow-up meetings in June. In addition, those in attendance agreed to limit discussion to two broad areas: first, the formation of an "informal voluntary association of operating personnel," and second, several technical questions and certain accounting topics.[99]

The working group quickly developed a plan. Following meetings in June and August, the group outlined the framework of a new organization that would provide coordination to all the participants in the rapidly emerging grid. The new entity would be informal, voluntary, and broadly representative. Ten operating areas or pools comprised NAPSIC: the Northwest Power Pool, Pacific Southwest Interconnected Systems, Rocky Mountain Power Pool, New Mexico Power Pool, Canada–United States Eastern Interconnection (CANUSE), Pennsylvania–New Jersey–Maryland Interconnection (PJM), and the four regions of ISG. Two representatives from each area or pool would serve on the committee. The working group addressed basic organizational details regarding officers, subcommittees, and a schedule of meetings. More significantly, the committee listed the primary operating matters to be addressed by NAPSIC: frequency and time error standards, bias settings, time error correction procedures, methods for handling unintentional energy exchanges, and response to emergencies.[100]

NAPSIC convened for the first time in January 1963, just months after the initial planning meetings. NAPSIC's first chairman, W. S. Kleinbach from the PJM Interconnection Office, identified two priorities for the member utilities: economy and coordination. He explained, "There is much to be done nationally and internationally in the area of economic integration of power resources and in coordinating day-to-day operations including load-frequency and time control of all interconnected systems."[101] With that, NAPSIC set to work, adopting the organizational guidelines outlined by the Working Group as well as recommendations for tie-line bias settings (identical to those used by ISG) and action in emergencies. The attendees also participated in a degree of coordinated planning, sharing details about upcoming interconnections and tie-line closures, discussing interactions with the FPC, and reviewing a variety of other issues, including nuclear attack warning systems and the Edison Electric Institute Task Force on National Defense.

Although the private, investor-owned utility sector took the lead in organizing NAPSIC, the committee included government-owned power companies as well. The narrow interests of NAPSIC—frequency control and time-error correction, for example—cut across the boundaries of ownership and affected every power company on an interconnection. For utilities to build links across the continent, all participants had to agree to common operating goals that ensured power to customers on demand. For this reason, NAPSIC established a locus of authority for reliability but did not address the freighted debates about ownership.

NAPSIC continued to provide the industry with recommended operating guidelines and reliability standards for the ensuing seventeen years. Beginning with the 1965 Northeast blackout, numerous landmark events triggered reconsideration of the nation's power system. In 1967, the utilities and the US Bureau of Reclamation closed the ties between eastern and western systems, briefly establishing a true national grid. In 1968, at the urging of the federal government, the utilities formed a second organization, the National Electric Reliability Council (NERC), to address regional reliability planning and coordination. Another major blackout, this time in New York in 1977, led to renewed public concern about the reliability of America's electricity networks. The energy crises of the 1970s framed the establishment of the Public Utility Regulatory Policies Act of 1978, a law that opened the door to technical innovation, deregulation of electricity markets, and corporate restructuring in the following decades. Throughout these years, engineers and system operators turned to NAPSIC to vet and tweak voluntary standards. In 1980, NERC finally subsumed NAPSIC into a single organization, under the NERC name, in order to merge the reliability operating, planning, and standards activities.[102]

With the creation of NAPSIC, the power industry had finally established an entity that had eluded large utilities, politicians, and engineers for decades. The independent interconnected systems created NAPSIC without fanfare, publicity, political endorsement, or regulatory demand. NAPSIC served as a clearinghouse for stability issues for all the power companies, both public and private, that operated interconnected across the continent. In addition, at the outset, NAPSIC provided a forum for a level of national grid operation and some discussion of autonomous system plans unprecedented in the industry's history. Yet through its very organizational structure, NAPSIC preserved the independence of government agencies, privately owned utilities, municipal companies, and rural cooperatives and respected the wide variety of systems developing across the continent. NAPSIC was the embodiment of shared responsibility and divided authority.

Federal Planning

Just as the system operators and engineers organized a national committee to address national interconnection, leadership in the federal government also took steps to plan for a grid. In 1962, the FPC proposed to tie together the nation's power systems by 1980. With an investment of $380,000, the FPC planned to survey the industry and determine where demand was growing fastest, which areas were importers or exporters of power, and what the capabilities and plans of existing companies might be. As Joseph Swidler, chair of the commission, pointed out, "Development of the nation's power system has come 'in a spotty kind of way.'"[103] The commission hoped a thorough study and projection through 1980 would allow every utility—whether large or small, public or private—to make its own plans "keyed . . . to a national scale."[104] The project offered high hopes for a coherent strategy for power system growth across the country.

When completed, the survey delineated two primary objectives: lower rates and higher-quality services for the American consumer. The survey set a target price for 1980 that would be 27 percent lower than the average 1962 price.[105] In addition, the survey proposed to reduce coal dependency by forty-five million tons per year through more efficient plant design. American consumers stood to save a potential $11 billion if government agencies and utilities followed the ideas outlined in the final report. The survey sketched out a plan for massive infrastructure expansion to meet anticipated demand in every sector of the country. One map, replicated in figure 6.6, projected a fully interconnected system that facilitated massive movement of electricity from east to west and north to south to take advantage of time zone and seasonal differences. These exchanges could result in reduced rates for regions currently burdened with high-cost electricity.[106]

Figure 6.6

Projected power exchanges in 1980.

Source: National Power Survey: A Report by the Federal Power Commission, 1964 (Washington, DC: US Government Printing Office, 1964), figure 115, 213.

The National Power Survey of 1964 enshrined the confidence and optimism of the power industry. The report of the survey opened with expressions of enthusiasm for a "new era of low-cost power," new technologies, larger machines, competitive nuclear power, economical extra-high-voltage power lines, and interconnections across broad geographic areas.[107] In short, the power industry had the techniques, the know-how, and the public support to perhaps triple capacity for the anticipated demands of the relatively near future. Not only did the survey predict growth in both demand and supply, but it also practically guaranteed enormous savings through advances in technology and practices that resulted in maximum efficiency. In addition, the survey proposed that a national grid promised both equity of service and conservation of resources for the American public.[108]

Swidler embarked on a politically fraught mission when he first introduced the power survey project in 1962. Private utilities, rural cooperatives, and municipal power companies all maintained aggressive lobbying activities to mitigate federal actions that might infringe on their respective market sectors. Swidler attempted to bring together public and private support for the survey by noting that the United States is "probably the only civilized country in the world that does not have a coordinated national electric system."[109] By this date, the central governments in many countries, including Britain, France, and Germany, had assumed control of their respective transmission networks, if not their entire electrification systems. Swidler downplayed a 1961 effort by investor-owned utilities to produce an interconnection plan as "merely a consolidation of local and regional plans . . . limited by local interests and restrictions."[110] Private industry responded by noting that previous federal proposals for national grid systems had been a "prelude to nationalization of all utility companies."[111] A *Wall Street Journal* editor further suggested that the utilities already had all the data the FPC sought, and the survey, therefore, was just "a Federal move in the direction of a nationalized power industry."[112] Swidler counteracted these objections by appointing a diverse advisory committee in March 1962. He reported to President John F. Kennedy, "The industry leaders, representing both private and public power, all agree that the national power survey is in the public interest."[113]

Over the next two years, a broad array of engineers, utility managers, fuels specialists, lawyers, and federal administrators collaborated to produce a report. Through twelve special advisory committees, individuals from the public and private sectors hammered out detailed reports on an array of topics ranging from power requirements, to legal matters, to distribution details, to regional differences. During this time, the FPC weathered an array of controversies that repeatedly brought opposing views into the public spotlight. Whether the rural cooperatives challenged the FPC's intent to support private construction of extra-high-voltage power lines or the FPC took the industry to task over a lack of coordinated research efforts, the debates over the federal role in power development continued.[114]

As the industry anticipated the release of the report in late 1964, the news media focused on the potential for additional controversy. Gene Smith, a reporter for the *New York Times* who covered power issues, compared the report release to "D-Day" for US power companies. While Swidler reassured investor-owned utilities that the survey would not be an attack on their independence, industry leaders expressed concern that the incoming President Johnson had a pro–public power record in Congress. As if to shore up their position, the utilities announced numerous new pooling agreements in the final weeks before Swidler presented the completed survey.[115]

With the release of the National Power Survey report in December 1964, controversy still revolved around the FPC's intentions, although widespread support for the plans soon emerged. In his initial coverage, Smith noted a generally positive response from a limited sampling of the industry: "Battle lines between public and private power seemed to be broken."[116]

The *Washington Post,* however, reported that private utilities were "voicing a mixed reaction."[117] On the public power side, skepticism prevailed. Clyde Ellis, representing the National Rural Electric Cooperative Association, claimed cooperation could be achieved "only through drastic change in the negative attitudes of the commercial power industry."[118] Alex Radin, speaking for the American Public Power Association, cautioned that local communities might lose the freedom to choose municipal service over investor-owned companies.[119] The *Wall Street Journal* was similarly caustic, suggesting the FPC's intentions were thinly veiled: "The industry's record of past achievement hardly adds up to a solid case for sweeping new Federal powers to guide its future."[120] It questioned whether regulators were "simply as preoccupied as ever with another kind of power."[121] Even the executive advisory committee for the survey acknowledged that there were differences of opinion between individuals involved in the project and the published findings.[122]

As the different interest groups took time to study the survey report, they began to express support. In presenting the report, Swidler reiterated that the survey was not intended as a blueprint for development, that it would not trigger a legislative overhaul, and that the national grid was not "anything we'd build"—that is, not a federal government project.[123] Within days, a utility spokesman concurred that the survey "did not advocate a national Government-controlled grid. Instead it suggested a partnership within the industry."[124] The utilities found the survey proposals to be in line with their own plans for 1980. Swidler assured small utilities, municipal companies, and rural cooperatives that the FPC had no intention of increasing their regulatory burden. By expanding power interconnections, he claimed, "we're trying to get them in on the benefits of pooling, technology, and the like."[125] In January 1965, the FPC sent its annual report to Congress and highlighted the survey as a major accomplishment of the prior year. Smith reported, "Strangely enough, the F.P.C.'s assessment of the industry seemed to cut across the traditional battlelines between public and private power and to call for continuing the cooperating between the Government's ratemaking body and the investor-owned segment as a strong means of achieving the goals of the survey."[126]

While sniping continued periodically both among lobbying groups and within the press, the National Power Survey appeared to represent a consensus across the public and private sectors in favor of greater interconnection, larger power plants, closer coordination and planning, and a long-term goal of benefitting consumers. While public power advocates doubted the sincerity of private sector commitments to lowering rates, they embraced the notion of "a national power pool jointly owned by public and private utilities."[127] The private utilities, likewise, saw the survey as vindication of the plans they already had in place for extra-high-voltage lines and giant power pools. Even better, with widespread participation in the preparation of the survey report, the engineering and utility operations community found their ideas about efficiency, economy, and conservation happily endorsed on a national scale.

In early 1965, the federal government celebrated a completed National Power Survey, the power companies worked together on stability issues through NAPSIC, and the public discourse favored a nationwide grid. None of these actors anticipated the event that would soon undermine the reputation of the power industry and call plans for future growth into question. On November 9, 1965, the country's first major blackout threw the northeastern states and part of Canada into darkness and heightened public awareness of interconnected power systems. Deemed a major disaster, the Northeast blackout led to questions about the size and shape of power systems, who should control them, and whether the federal government needed more authority over the utilities. By this date, the industry had weathered two world wars and a major economic depression. But the strength of America's power system was also its weakness. Shared responsibility and divided authority allowed utilities to respond autonomously to a small failure, and this resulted in a big problem.

7 Drifting "Lazily" into Synchrony

From Blackout to Grid, 1965–1967

On the morning of February 7, 1967, four tie lines connected 94 percent of the electric power systems across the continent into a single network. For the first time in history, a kilowatt-hour of electric energy generated on one coast could illuminate a lightbulb on the other.[1] Yet Americans met this moment with a collective shrug. This might have been the result of a less popular event that had occurred just over a year earlier. On November 9, 1965, the lights went out for thirty million North Americans in the worst power outage in history. This was a disaster for the electric power industry, which had enjoyed more than eighty years of celebrated, though not uncontroversial, growth and had approached expansion and interconnection with the enthusiasm and dedication of missionaries. The 1967 East-West Intertie, and its quiet completion, is emblematic of what changed—and what did not—as a result of the 1965 blackout.[2]

When the 1965 Northeast blackout occurred, the presence of transmission networks that reached almost coast to coast caught many by surprise. Promoted as technologies of resilience, conservation, and national security, interconnections developed with very little public scrutiny to this point. As postblackout surveys showed, the average American barely knew that transmission grids existed. But failure catapulted interconnections and their attendant issues into the limelight. Just as engineers verged on tying the continent together into one giant machine, the benefit of the whole approach was called into question.

The 1965 blackout exposed the dark side of interconnection: those linked in success were also linked in failure. Until this event, most power system experts had not contemplated a cascading breakdown of such enormity.[3] A half-century of studying, evaluating, inventing, adjusting, and operating interconnected power systems instilled confidence in engineers, operators, and managers. From the perspective of the experts, the technology was sound, the practices tried and true. Although there had been critics, most observers considered power pooling essentially beneficial in all respects. The interlinked transmission lines carried electricity to consumers, promised efficiency and economy, allowed more thoughtful use of energy resources, facilitated lower prices, and increased the profit margins of the investor-owned utilities. The blackout pushed utility executives and politicians alike to reconsider the feasibility of building networks in bits and pieces.

In the United States, utilities were moving quickly toward coast-to-coast power transmission in 1965. For national grid advocates, interconnection represented economic strength, international political power, technological know-how, and engineering derring-do. It also exemplified the very American value of equity in service by bringing power to every corner of the country. At the same time, as a collection of federal installations, private transmission lines, and international links, the networks personified capitalist democracy at work. Investor-owned utilities time and again resisted government oversight of interconnections in favor of self-determination. Objections came from public power advocates, who questioned private sector control of a public service. The cascading failure took place in the operating area known as the Canada–United States Eastern Interconnection (CANUSE), in which thirty investor- and government-owned power companies shared electricity. It began with a perfectly functioning yet problematic relay setting, but it manifested through varying approaches to automation and decision making across the participating utilities. For engineers and system operators, the organic growth of the grid up to this point had been both challenging and manageable. But a massive blackout suggested that greater uniformity, and perhaps more authoritarian oversight, would ensure a more reliable future.

Cascading failure reflected the advantages and disadvantages inherent in systems of shared management and divided authority. With so many individuals and entities responsible for maintaining stable operations, the initial reaction to the blackout inevitably involved finger pointing. The follow-up studies revealed great variety in installations, inspection programs, information flow, and monitoring techniques. Further, utilities adhered to settings, standards, and recommended emergency response procedures on a voluntary basis. Each autonomous response to the initial relay issue exacerbated the problems. At the same time, decades of collegial information exchange and collaborative problem solving engendered a sense of unity among most of the utilities sharing power in the northeast. Within days of the blackout, industry leaders offered expertise, information, time, and staff support to the government agencies that investigated the blackout. The Federal Power Commission (FPC) likewise worked hand in glove with state utility commissions and Canadian power authorities to determine precisely how the blackout had happened, where weaknesses lay, and how to best approach strengthening systems in the future.

The FPC, through its National Power Survey, had recently promoted interconnected transmission lines as a technology of conservation, but the blackout immediately pushed conservation concerns into the background as reliability became a top priority. In the aftermath of the massive power failure, the public demanded explanations of how and why the breakdown occurred as well as promises of greater security in the future. Americans expected ubiquitous and reliable electricity by this time, a sign of the power industry's past success. For utilities, the grid had offered the means of ensuring that electricity flowed even if a generator failed or a power line went down. Now consumers questioned the reliability of the grid itself. In fact, many asked if a grid was even necessary.

The Power Industry in 1965

The 1964 National Power Survey provides a benchmark for the status of the power industry at the time of the Northeast blackout.[4] As the survey reports, the United States used 40 percent of the electricity produced in the world, and electric power businesses constituted the largest industry in the US economy by several measures. As had been the case during the prior eighty years, a gaggle of entities electrified North America.[5] There were 3,600 power systems in the United States at this time, a number of which crossed into Canada. As illustrated in figure 7.1, a relatively small number of investor-owned utilities controlled the majority of power generation and sales. More than three thousand small municipal utilities and cooperatives operated a mere 11 percent of the system, providing power to 21 percent of the customers. The federal government owned 13 percent of the system but provided no direct service to retail customers.

Long-distance transmission and interconnection were on the rise in the early 1960s. The industry operated ninety thousand miles of transmission lines, 97 percent of which were interconnected across North America in five large networks. These networks, plus three smaller ones, are depicted in figure 7.2. In 1964, the loosely organized CANUSE pool served customers in New York, New England, Ontario, and Michigan. CANUSE shared power with the Interconnected Systems Group (ISG),

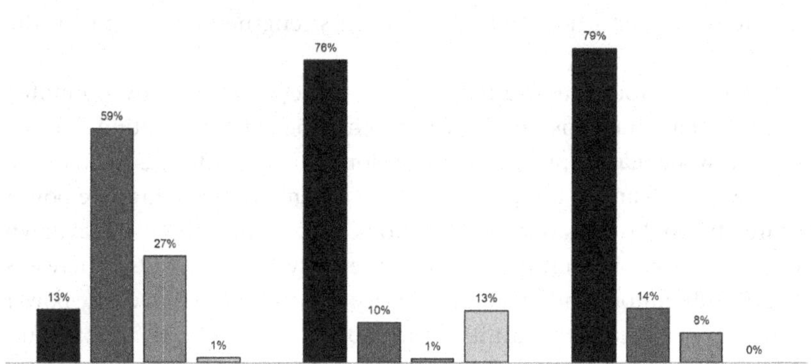

Figure 7.1

Comparison of ownership, generating capacity, and retail service of electric power systems in 1964.

Source: National Power Survey: A Report by the Federal Power Commission, 1964 (Washington, DC: US Government Printing Office, 1964), table 3, 17.

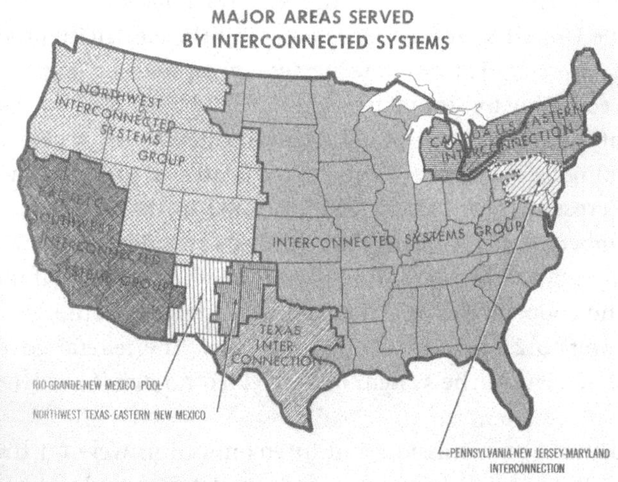

Figure 7.2

Five large networks and three smaller networks identified by the Federal Power Commission in 1964.

Source: National Power Survey: A Report by the Federal Power Commission, 1964 (Washington, DC: US Government Printing Office, 1964), figure 8, 15.

which by this time stretched west to the Rocky Mountains and south to the Gulf of Mexico. CANUSE had also recently closed ties with the Pennsylvania–New Jersey–Maryland Interconnection (PJM). All three of these major pools were composed of numerous smaller pools and operating groups. Within CANUSE, the interties were, in many cases, of limited capacity. CANUSE and PJM covered the states stretching from Maryland and Delaware in the south to Maine in the north and to Michigan in the west. The portion of CANUSE that extended into Ontario, Canada, is also shown in this image.

The blackout area, though geographically smaller than other major power pool areas, represented a significant portion of the continent's electrification. Figure 7.3 indicates the portions of Canada and the northeastern United States affected by the blackout. The utilities within CANUSE generated nearly one-third of the continent's power.[6] Investor-owned utilities owned and operated most of this northeast system. Some of the pools, like the Connecticut Valley Electric Exchange (CONVEX), were highly coordinated. Other utilities maintained loose interconnections, as in New Hampshire and Vermont. Utilities in Maine operated in virtual isolation. In general, electricity rates varied greatly across the country, reflecting regional energy costs, the costs of infrastructure, the presence of federal power sources, and the interplay between utilities and regulators in the rate-setting process. At the time of the blackout, fossil fuels provided 81 percent of the energy for power production, hydroelectric plants provided 18 percent, and nuclear plants provided less than 1 percent, and in that year, the United States shipped a little bit of electricity to Canada.[7] The cost of electricity was highest in the Northeast—nearly 30 percent higher than the rest of the country.[8]

Throughout the history of electrification, there had been many blackouts, and they had merited moderate concern. New York City alone experienced significant blackouts in 1935, 1938, 1959, and as recently as 1961. Overall, however, the public took reliable electricity for granted. The Edison Electric Institute, the New York Public Service Commission, and the FPC all acknowledged the reliability benefits of the grid, but they mainly touted economic and resource conservation functions when promoting national interconnections.[9] Looking back on the 1965 blackout in 1991, a prominent utility executive tried to recapture the "1960s mindset": "As we approached the mid-1960s, the reliability of electric bulk power supply was not the major issue, either within the electric utility industry or within its various publics."[10] No one expected a crippling blackout.

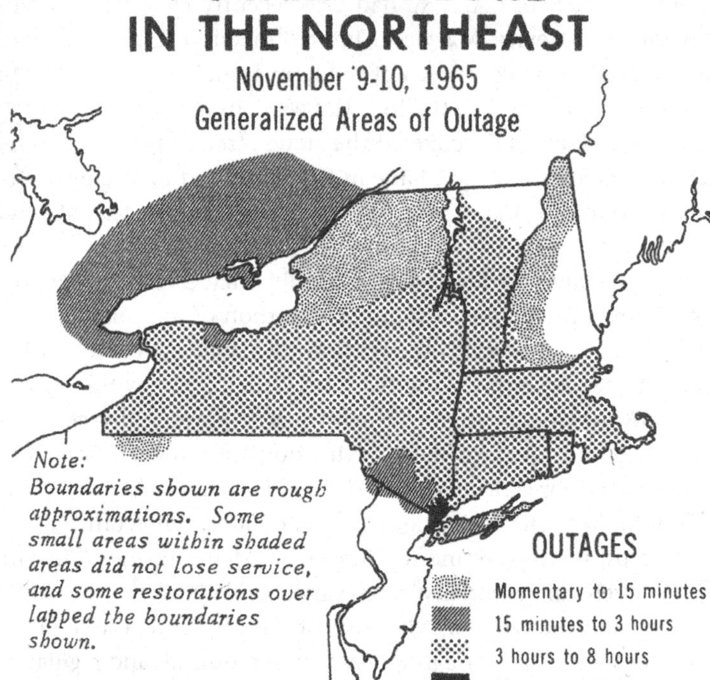

Figure 7.3
Map of areas affected by the 1965 Northeast blackout.

Source: Federal Power Commission, *Prevention of Power Failures: An Analysis and Recommendations Pertaining to the Northeast Failure and the Reliability of U.S. Power Systems, A Report to the President by the Federal Power Commission* (Washington, DC: US Government Printing Office, 1967), figure 1, 7.

November 9, 1965, 5:16 p.m.–5:28 p.m. (EST)

It took only twelve minutes for one unanticipated relay response to bring down the nation's most intensely electrified region. On November 9, 1965, at 5:16:11 p.m. (EST), a relay on the transmission line carrying power from Niagara Falls to Toronto "tripped" and stopped the flow of electricity.[11] At an earlier date, this relay had been set to protect the line from a sudden influx of too much power. In addition, the relay engineer and the system operator had not communicated with each other about the setting. The setting was lower than the amount of power the line could actually carry, and the load had been steadily increasing in prior months. As a result, the relay tripped when the load exceeded the relay setting but was still well below the amount of electricity the line could carry. In other words, the amount of electricity on the line was not enough to threaten the stability of the line, but the relay behaved as if it was. With this line down, the load shifted to four additional power lines connecting Niagara to Toronto, and the relays on those lines tripped as well. In less than three seconds, excess power started flowing into New York State. In one more second, in accordance with predetermined operating procedures and automated responses, the CANUSE system broke into four isolated sections.[12] Maine and part of New Hampshire separated as well and did not lose service. Each of these five areas is approximated by the shading in figure 7.3. There followed a series of events, described as a cascading blackout, which ultimately plunged the Northeast into darkness by 5:28 p.m.

On this cold November evening, in the midst of rush hour, demand for electricity was reaching its peak. In the moments before the relay tripped, the collection of twenty large utility companies and pools that served the majority of the northeastern United States and Ontario generated nearly forty-four thousand megawatts of electricity to meet the area's demand. As excess electricity flowed south, the interconnected system witnessed a variety of responses from individual utilities and pools. As previously noted, automatic relays caused some areas to separate; in other areas, operators made decisions to remain connected and aid neighbors or disconnect and protect their own generators and customers. In a system developed piecemeal and characterized by shared responsibility but divided authority, the lack of a uniform response marked the sequence of events that followed the initial opening of a relay in Ontario.

The automatic relays that caused some areas to separate from CANUSE at the first indication of trouble were of a different kind than the automatic tie-line and load control systems discussed earlier. Automatic tie-line and load control apparatus addressed normal operating conditions and caused adjustments that allowed interconnected power systems to continue operating in parallel. The relay trip that initiated the Northeast blackout caused such a large mismatch of demand and supply that automatic tie-line and load control mechanisms were overwhelmed. On the other hand, relays and governors designed to protect systems from sudden exceptional changes in power flow did act automatically to separate transmission lines and generators from the network.[13] No single approach guaranteed protection from the cascading power failure.

Automatic response systems led to protection of certain areas, like New Jersey and most of Pennsylvania. The Pennsylvania–New Jersey–Maryland

Interconnection (PJM), serving the majority of customers in those states, had a long history of stable operations within its original system and displayed conservative tendencies when it intertied with CANUSE utilities. In the early 1960s, PJM and Consolidated Edison began negotiations to build an intertie between the two systems. PJM engineers, not fully trusting the effects of major system trouble on the Consolidated Edison side of the intertie, installed special relays at the New Jersey–New York boundary and at the Pennsylvania–New York boundary. The day after the blackout, Pennsylvania Power & Light Company, a PJM member utility, reported to its customers, "These ties are designed to disconnect when overloaded in order to prevent extension of any trouble that might arise. Such overloading did occur last night and these ties did disconnect as designed."[14] PJM customers did not experience any power failure.

By contrast, when Ontario automatically separated from the rest of CANUSE, different regions experienced the cascading failures in different ways. Within one-half second of the loss of the five transmission lines from Niagara to Toronto, the province of Ontario was operating on its own and experienced a major deficiency of power. Actions within the province no longer affected New York and the other regions of CANUSE. The Hydro-Electric Power Commission of Ontario (HEPCO) networks automatically separated into three major sectors. The western sector of Ontario received assistance from Michigan and did not lose power. Throughout the other two sectors, major cities went dark within seconds of the start of the cascade. HEPCO restored service to the entire area by 8:30 p.m.[15]

Near the Canadian border in upstate New York, automatic separation protected a discrete group of economically significant power users. The Power Authority of the State of New York (PASNY) planned for automatic separation of its major hydroelectric plant in case of trouble. Following the tripping of the five HEPCO power lines that initiated the blackout, lines connecting the PASNY plant to downstate New York and to New England also tripped out, as did five generators within the plant. The hydroelectric generating station now operated in isolation from the rest of the system and supplied electricity to nearby industrial loads. In this island of power, facilities owned by Aluminum Company of America, General Motors, Reynolds Metals Company, the city of Plattsburg, and Plattsburg Air Base all continued activities unaffected.

Over the next several minutes, system operators on manual control throughout New England and New York faced difficult decisions. In the case of Consolidated Edison, for example, the system operator observed rapidly switching inflow and outflow of power as other utilities separated from the pool. In keeping with the recommended emergency procedures of the North American Power Systems Interconnection Committee (NAPSIC), the operator attempted to provide aid to his neighbors by increasing generation. After a few minutes, however, this effort failed and the system electrically disintegrated, leaving New York City in the dark. During this same period, operators at CONVEX and Long Island Lighting Company, among others, manually separated from neighboring systems, exacerbating the situation for Consolidated Edison. In each case, the system operator had the authority to determine how to balance obligations to his own network and to his interconnected neighbors in order to address the growing crisis.[16]

CANUSE failed in twelve minutes in much the same way interconnections had been assembled over several decades. Each participating entity acted according to its own internal objectives as the system fell apart in pieces. While utilities shared responsibility for managing a stable network, they operated their own subnetworks with autonomy. Some were able to protect their customers from power loss by automatically separating. Others attempted to do the same through manual separation. Still others acted to meet their pooling obligations. With insufficient coordination, lagging communication between systems, and inadequate preparation for failure on this scale, they lost power nonetheless. The immediate effect of the blackout, in addition to causing major problems for millions of power customers, was to force the power industry itself to reassess the nature of the interconnected systems.

The Aftermath: November 9, 1965, to December 6, 1965

The 1965 blackout affected millions of people and institutions across eighty thousand square miles in nine states and in Canada. The power was out in some areas, including most of Manhattan, for more than half a day. It took hours to restore power across the blackout area, days to determine the cause, and weeks to inspect all the affected generation and transmission equipment. The scale of this event prompted President Lyndon Johnson to immediately call on the chairman of the FPC to investigate and ultimately led to a multitude of reports by government and private sector entities documenting every nuance of the outage and its causes.[17]

Reporters looked to the industry and to governments to explain what had taken place. The earliest news briefs expressed palpable relief that no enemy attack had caused the trouble. The media quickly dismissed several other preposterous explanations, including the less likely possibilities of nuclear attack, UFOs, and a little boy in Massachusetts hitting a telephone pole with a stick. On November 10, headlines across the country reported that the cause of the blackout was a mystery. One day later, headlines posited there had been a "quarrel" between generators, thus explaining that generators operating out of synchrony could bring down a power system. In fact, utility representatives understood quite quickly that there had been some sort of problem on the line near Niagara, but investigations lasted days before the public had a clear explanation. One week after the blackout, Ontario officials finally accepted responsibility for the relay action that initiated the cascading failures.[18]

As the popular press pondered the cause of the giant power failure, reporters also discussed what the grid was, why it existed, and whether it was a good thing. The public began to understand that behind the light switch lay "an immensely complex and interlocking network of men, machines, and wires that is not infallible."[19] One critic compared the grid to the skin of a ripe cantaloupe, vulnerable to failure.[20] Another observer noted, "Interest in power system controls has been literally hurled into the public consciousness."[21] The press delineated the inherent conflict of the grid: "The Northeast power system was considered the last word in sophisticated engineering and the product of computer science. Ironically, the interlocking grid system designed to assure a supply of electricity in an emergency helped spread the blackout over the huge area."[22]

The lack of immediate information about the sequence of events and the frightening firsthand experiences of sudden darkness and prolonged service outages shook public confidence in electric utilities and the notion of interconnection. On November 22, *Electrical World* conducted a spot survey of two hundred customers of utilities in the Northeast to determine whether the blackout had tarnished the industry's image.[23] While respondents felt that their utilities did a reasonably good job of restoring service, nearly a third thought their utility was to blame for the power failure. The *Electrical World* journalist reported, "At fault, of course, in the public's view, was the grid—and everyone connected into it. As a Brooklyn housewife put it, 'They (electric companies) shouldn't put all their eggs in one basket that way. Why should we be in the dark because of something that happened in Canada? I've never even been to Canada.'"[24] *Electrical World* noted in an earlier survey that the majority of North Americans had never heard of "the grid" or thought it was a football field. With the blackout, consumers began to understand that the functioning of the living room light switch might be contingent on decisions made several states, or an entire nation, away. Surprisingly, in the November 22 survey, two-thirds of respondents still believed interconnections were a good idea. Consolidated Edison's customers, however, were critical of that utility's long delay in restoring service after the blackout.[25]

Journalists, politicians, and government officials expressed their own concerns. *Electrical World* posited that the massive shutdown "had been considered extremely improbable" before it happened.[26] As the journal reported, from New York governor Nelson Rockefeller to Texas congressman Walter Rogers, many others shared this view. In a hearing called by Governor Rockefeller, utility executives began blaming others as they proclaimed their own systems were not at fault. The spokesman for the American Public Power Association noted that independent municipal electric companies, unlike the linked private utilities, did not lose power. Secretary of the Interior Stewart Udall urged stronger interties. The *New York Times* stated, "The utilities are on trial. They must give a complete account of what went wrong. And they must see to it that the public will never again be faced with the helplessness that comes from a total power failure."[27] International reporting on the blackout also raised the question of whether the US approach to power networks aided or harmed the system. In many countries outside North America, governments owned and operated national grids. To these observers, the hybrid collection of public and private ownership characterized by CANUSE looked particularly unreliable following the blackout.[28]

Closer to home, the blackout represented a breakdown of the industry's organic approach to growth and development. Throughout November, newspaper and magazine editors expressed concern about future blackouts, while power experts offered differing views of interconnection.[29] Joseph Swidler, chairman of the FPC, repeatedly spoke in favor of the grid, and Robert Person, president of the Edison Electric Institute, stated, "The principle of pooling and interconnection, as it has evolved over the years, is basically sound, as indicated by the fact that the kind of massive failure just experienced has rarely occurred."[30] Philip Sporn, by now immediate past president of American Gas & Electric Company, tempered enthusiasm for a national grid in favor of more tightly organized regional pools.[31] Other utility executives argued that interconnections weakened the power system, offering that the "entire nation could have been plunged into darkness in less than a second if a Federally proposed plan had been in effect."[32] Boston Edison's executive called for a thorough study before proceeding with a nationwide grid.[33]

The Report: December 6, 1965

While individuals and organizations debated North America's approach to electrification, the FPC conducted a detailed study of the power failure. With the aid of dozens of private sector utility representatives and several government agencies, the FPC sought to understand precisely what had taken place, how and when each link in the network failed, and what the implications were for future planning. The preliminary results of the investigation, released less than a month after the blackout, reinforced the industry commitment to grid development.[34] The report concluded that the failure was not inevitable and that interconnections added strength and reliability to electric power service. Following a detailed description of how the blackout occurred, the report identified measures to strengthen the grid and confine future outages. The FPC committed to carrying out further studies and offered a set of specific recommendations to President Johnson, Congress, and the industry. With high praise for the industry representatives who assisted with the investigation, the FPC affirmed the strength of the power sector and the benefits of the path to interconnections previously chosen for increased electrification.[35]

The section of the report detailing the cascading failure illustrated the autonomy of each utility and pool in responding to a crisis. While indicating and explaining the instances in which relays tripped, transmission lines fell out of service, and generators slowed or stopped, the report dwelt on the decision making at Consolidated Edison. The FPC noted that individual systems generally followed the recommendation of NAPSIC to maintain parallel operations if at all possible in order to render "maximum assistance to the system in trouble and . . . prevent cascading of trouble to other parts of the system."[36] At the same time, however, the NAPSIC guidelines also suggest that a system should disconnect if an "intolerable overload" threatens the equipment.[37] Beyond these potentially conflicting guidelines, the on-duty Consolidated Edison operator had no specific instructions concerning what particular circumstances should trigger "load shedding" in order to save the remainder of the system.[38] This individual had full authority for making a decision but insufficient information to act quickly and in the best interest of Consolidated Edison's customers. The FPC further explained that each company faced a similar problem, particularly those that were not automatically disconnected by an emergency relay trip.

To remedy this situation, the FPC called on the utilities to reexamine the extent of planning and coordination in place for their interconnected systems. As the report explained, equipment failures must be expected, but system failures can be prevented. Noting the great variety among the "power grids of this nation," the FPC called for several specific measures to minimize the likelihood of a repeat blackout.[39] Among the list of nineteen "partial and tentative" recommendations, the agency highlighted closer coordination between the United States and Canada, both at the government level and at the operating level.[40] Further, the agency called for independent power companies to join power pools for the creation of planning and operating entities with sufficient responsibility to require close coordination within pools, more studies of how to ensure stable operations, more frequent checks of relay settings, and increased reserve capacity in both transmission lines and generators. The FPC also encouraged widespread use of more advanced automated controls and reconsideration of load shedding under emergency conditions. For the most part, the FPC looked to the industry to proceed as it had in the past, only more so. The nineteenth recommendation, however, called for greater regulatory authority at the federal level.

One week later, speaking to a US House subcommittee investigating the blackout, Joseph Swidler expanded on the nineteenth recommendation. Swidler declared that interconnections, at the heart of continuity and reliability in the bulk power supply, were a matter of national interest. He acknowledged that new legislation should "leave upon the shoulders of management" primary responsibility for reliability.[41] But he sought authority for the FPC to set minimum standards for system design and operation and for intersystem coordination. He requested that legislation encourage additional and more fully coordinated interconnection. He urged Congress to establish legislation that covered all entities in the power industry. The press focused on the call for new and stronger regulatory authority at the federal level.[42]

Proposals for Federal Regulation

The press provided widespread coverage of the FPC blackout report, opening further questions about federal authority and the benefits of interconnection. Wire service stories appeared across the country outlining the findings and generally focusing on the FPC's quest for more regulatory power.[43] Taking a different tack, Eileen Shanahan, writing for the *New York Times*, highlighted the FPC's claim that "more, rather than fewer interconnections . . . were needed to provide electrical service."[44] She noted, "Since the blackout, there have been some assertions in Congress and elsewhere that the interconnection system itself is a bad idea, inasmuch as it permits the wide spreading of power failures."[45] She returned to this theme nine days later: "What was more startling to many people was the vigor with which the Government and industry experts who worked on the report reaffirmed their belief in the whole concept of interconnecting power systems."[46] Reflecting the uncertainty of many across the country, Shanahan reported that, on the one hand, a failure of interconnecting systems led to the blackout and, on the other, strengthened interconnections offered the answer.

In the weeks following the FPC report, the US Senate and House of Representatives conducted their own investigations of the blackout. The Senate Committee on Commerce requested information from federal agencies, emergency relief groups, the utility industry, and state and municipal officials. The Senate released the report in March 1966, including correspondence from the constituencies surveyed. All seemed to concur that "this country has the technical talent and facilities available . . . to upgrade the power systems of this country so that power failures of this severity will be extremely improbable."[47] All shared faith in interconnections for both reliability and economy. Witnesses before the Senate Committee on Commerce upheld a belief in technology and the capability of power systems experts to maintain a growing and stable power supply.

The executives of several private utilities extolled what they saw as the strength of their own pooling arrangements to the Senate Committee. Commonwealth Edison boasted that the regional power system in the middle West could "achieve a degree of reliability that will practically rule out a widespread electric shutdown."[48] Florida Power Corporation offered that excellent coordination among that state's utilities minimized the chance of a cascading failure. Pennsylvania Power & Light Company could not "conceive of the occurrence on PJM of a power failure similar in cause and scope to the Northeast power failure."[49] (Ironically, the PJM system experienced a cascading failure on June 5, 1967, little more than a year later.) Pacific Gas & Electric Company assured the Senate that California systems were inherently less vulnerable to major outages, although the executive from Southern California Edison offered some humility: "We want to be as certain as we can be that we are not overconfident in our self-appraisals."[50] The long-standing overconfidence of the industry, however, was in evidence throughout most of the responses to the Senate Committee.

Industry executives responded with caution to Swidler's request for increased FPC oversight of the interconnected power systems. The president of Northern States Power Company, operating in the Upper Mississippi River basin, expressed his belief that responsibility for coordination should remain with local utility operators. The president of Virginia Electric & Power Company likewise discouraged legislation that would increase controls on utility companies. He offered that the highly specialized competence of utility engineers "is the only reason there has never been a shortage of electric energy in this great Nation."[51] The chief of the American Electric Power Company strongly objected to a national grid on the grounds that it would add unmanageable complexity to planning and operation of power systems.[52] The utilities clung tightly to their operating autonomy while praising their ability to coordinate reliable service among themselves.

Government officials and representatives from government-owned and cooperative power companies were less sanguine. For example, the Missouri Basin Systems Group, composed of preference customers of the Bureau of Reclamation, reported to the Senate Committee that joint planning with private utilities left much to be desired. Despite effective relations with the Bureau of Reclamation, which built the transmission network in this area, the rural cooperatives and municipal power companies found the private utilities and large generation and transmission cooperatives to be less forthcoming. They encouraged more federal intervention in the planning process. In the area more directly affected by the blackout, the mayor of New York reported efforts to gain local jurisdiction with the state over Consolidated Edison's operations, while the governor of Vermont claimed the blackout indicated a regulatory vacuum. Amid the consensus favoring continued interconnection, there was a great diversity of opinion as to whether the grid should be "national," how governments should be involved, and how much leeway should be enjoyed by the private sector utilities.[53]

The House Committee on Interstate and Foreign Commerce Special Subcommittee to Investigate Power Failures echoed the findings of the Senate. The subcommittee held hearings on December 15, 1965, and February 24–26, 1966, and solicited responses from each of the fifty states.[54] The report included the hearing testimony, the full text and exhibits of a Stone & Webster study commissioned by utilities in northeastern states, and correspondence from thirty-two states. The responses reflected regional and political differences from across the country. In many states, the regulatory commissions took exception to the pronouncements of the utilities or the FPC, although in several states, commissioners touted regional preparedness for emergencies.

Commissions in areas unaffected by the blackouts generally praised their own exceptional systems and doubted that they would suffer similar outages. They appeared to self-consciously protect systems of shared management and divided authority. They detailed long regional histories of operating interconnected without major power interruptions, strong interties and agreements, effective use of automated controls, access to more stable energy supplies, and better plans for responding to emergencies. Many noted that they were already implementing enhanced digital computing systems to evaluate their networks and plan for and protect against future contingencies.

From Georgia to Idaho, numerous state utility regulators opposed increased federal oversight and resisted the completion of coast-to-coast interties. The utility commission in Florida expressed a preference for remaining autonomous from FPC jurisdiction.[55] The Georgia public service commission expressed concern that more interties would lead to greater system complexity and future cascading blackouts.[56] Idaho lauded its own regional coordination and opposed "unnecessary rigid restrictions" imposed by a federal regulator.[57] Nevada argued that the FPC "invaded and deteriorated intrastate utility regulation" with the preference clause.[58] The Virginia State Corporation Commission expressed faith in state-level regulation to "see that service is reliable."[59] In contrast, several states sought greater oversight and improved coordination through interstate ties. For example, Nebraska laws restricted power service to government-owned utilities and cooperatives, yet the Nebraska Power Review Board found that "on the administrative and planning levels, coordination was found to be sadly lacking."[60] Regional variation proved to be the rule rather than the exception in defining the status of power systems across the nation.

The More Things Change: The Final FPC Report, July 1967

The power industry responded to the 1965 blackout by renewing a commitment to the path it had been following for decades. Through a combination of public expressions of confidence in the system, investment in technology, increased interconnection, and formation of entities that fostered voluntary adherence to reliability standards, electric utilities managed to sidestep the challenges to the status quo brought about by the blackout crisis. The earliest statements from utility executives reflected the hubris of the engineers who had developed the complex, intertied power system. As *New York Times* reporter Gene Smith remarked, before the blackout "the top executives of the utilities . . . would certainly have denied that any blackout such as the one that did occur could ever occur in so vast an expanse of the United States. . . . They would also have argued that it was inconceivable that an area extending from New York City to Quebec to North Bay to the outskirts of Detroit and back to New York City could ever be blacked out short of an enemy attack in wartime."[61] Over the longer haul, the utilities played to their strengths, focusing on technical and organizational solutions to the question of grid instability.[62]

Within months of the blackout, utilities—particularly those in the Northeast—had made a number of technical improvements to the interconnected power system. Advances included improved communications systems; updated displays in control centers; increased use of automated control technologies, including automatic load shedding devices; new system monitoring equipment; backup generators for control centers; and even new turbines. Utilities debated the merits of automatic load shedding versus on-the-spot decision making, yet the industry touted the greatly increased use of technology to separate segments of the power system that were experiencing failures. According to the FPC, "The best insurance against a major power failure is sound planning plus a well-designed and -operated bulk power supply system. . . . Automatic controls are essential."[63] Even more important than more sophisticated instruments and devices, however, was the move toward tightened pooling agreements and shared central control facilities.[64]

Eighteen months after the big blackout, the FPC issued a thorough statement on how to prevent future blackouts.[65] With thirty-four recommendations, the report outlined a path to greater reliability of the power grid. Most of these recommendations addressed expanding the size and strength of the transmission network, improving coordination among participating entities, and upgrading the technologies used for studying and operating the grid. As in the case of prior FPC reports, the commission relied heavily on participation and input from industry representatives. More than seventy-five individuals from across the country, representing private utilities as well as cooperatives, municipal companies, and federal agencies, aided in preparation and review of the three-volume report, thus ensuring that the FPC findings reflected a very broad range of perspectives.

The report called for the creation of coordinating entities to oversee implementation of report recommendations, very much in line with the types of organizations the industry had already created over the prior decades. Just before releasing the report, the FPC had asked Congress to consider a proposed "Electric Reliability Act of 1967" that would establish this approach in federal law. The power industry was already moving in the direction of the report's recommendations but resisted the FPC's push for greater regulatory authority. Shortly after the blackout, the affected utilities in the Northeast formed the Northeast Power Coordinating Council to strengthen planning and operations of the interconnected pools. By the end of 1966, an additional eight regional coordinating councils appeared across the country.[66]

The FPC and supporting legislators introduced eighteen bills during the Ninetieth Congress to implement the proposals for electric reliability, including the controversial Electric Reliability Act of 1967. The Senate Committee on Commerce attempted to garner widespread input on the proposed legislation. The committee held a series of hearings beginning in Washington, DC, in August 1967 and continuing in the Pacific Northwest in December 1967 and in Salt Lake City in April 1968. The testimony was heavily weighted toward representatives from western states, although representatives from national entities also participated. A definite trend opposing federal legislation emerged, with some also offering remarks in favor of greater oversight from the FPC.[67]

At the initial hearing, several federal agencies and legislators presented testimony in support of the act, arguing that the severity of recent blackouts called for greater central control and oversight at both planning and operating levels. In addition to the FPC, the committee heard from the Office of Emergency Planning, the Department of Transportation, the Bonneville Power Administration, and the assistant to the president for Consumer Affairs—all favoring the bill. Senators from Montana, California, Maryland, and Maine also expressed support. The hearing record included editorials from a wide range of publications, from the *Pittsburgh Post-Gazette* and the *Washington Post* to *Life Magazine*, offering further endorsement of legislative action. As the editor of *Public Utilities Fortnightly* remarked, "The FPC produced a bill it believes will help insure the nation against future blackouts and yet one which it thinks the electric industry will be able to live with."[68]

An interesting collection of public interest organizations joined the side supporting new laws. The National Rural Electric Cooperative Association, numerous individual municipal utility districts, environmental and conservationist groups, sportsmen's associations, and several Indian Tribal Councils argued in favor of increased federal authority. These entities sought to defend their own interests in the process of strengthening transmission grid planning, location, construction, and operations. A stronger FPC offered this solution. In addition, the California Public Utilities Commission acted as the lone state regulator in favor of greater federal oversight.

These favorable testimonials contrasted sharply with the opinions of investor-owned utilities, the majority of state regulatory commissions, and active power pools. For the most part, presenters from the private sector, including Commonwealth Edison, Pacific Gas & Electric Company, and even the very small Nevada Power Company, documented the extent to which cooperation marked the industry's practices. These individuals noted that their systems suffered very few outages, stayed on top of current technical innovations, and offered superior service to customers. In like fashion, representatives of interconnected systems, including the Northwest Power Pool, the California Power Pool, and Western Energy Supply and Transmission Associates, offered strong arguments in favor of continuing with voluntary coordination and planning. These entities detailed the history of their interconnections and outlined the nature of their cooperative relations. Many introduced actual contracts and written agreements into the hearing record.

The delicate balance of power between state and national governments figured in the unfolding dispute. Several state utility commissioners defended the autonomy of their regulatory authority and expressed dismay at the possibility that federal regulators would intrude on local sovereignty for protecting reliability and consumer interests. Other voices joined the opposition, including *Electrical World* and a handful of daily papers; several federal agencies, including the Department of the Interior, the Tennessee Valley Authority, and the Rural Electrification Administration; the Los Angeles Department of Water and Power; the United Mine Workers of America; and the International Brotherhood of Electrical Workers. While every one of these hearing witnesses agreed that the industry should move toward greater reliability, all concurred that the federal role should continue to be advisory, along with voluntary industry cooperation on a regional basis. One even went so far as to say, "The legislative proposal, in my opinion, would adversely affect reliability and the future vitality of the electric utility industry."[69]

While the hearings unfolded, the industry moved in 1968 to establish the National Electric Reliability Council (NERC), a totally voluntary organization.[70] Floyd L. Goss, the newly elected chairman of NERC, explained to the press, "The primary purpose of the council will be to continue improvements in reliability of bulk-power supply through exchanging and disseminating information on regional coordination practices; to review, discuss, and resolve matters affecting inter-regional coordination; and to provide an informed and responsible means of communication with the public, as well as with regulatory and governmental authorities in regard to the reliability of electric power."[71] NERC distinguished itself from NAPSIC by focusing on planning rather than technical standards. Unlike other national power organizations, NERC offered a wide umbrella to all classes of electric utilities and explicitly included at least two representatives from each sector, including federal agencies, investor-owned utilities, rural cooperatives, and state and municipal companies. Further, NERC invited the chairman of the FPC to send an observer to every meeting. The utilities formally announced the creation of NERC at a press event in June 1968, two months after the third of the Senate Committee hearings. According to Goss, NERC did not represent an attempt to circumvent pending legislation, yet in the same news conference, he confirmed that legislated reliability controls would now be unnecessary. Goss claimed that the industry had begun developing this national planning group as early as 1965, long before the FPC formulated the proposed Electric Reliability Act.

As NERC gained prominence, the proposed reliability legislation lost traction. The Senate Commerce Committee held no additional hearings on this topic in 1968 and held none in 1969. In January 1970, the Committee on Commerce's new Subcommittee on Energy, Natural Resources, and the Environment convened a hearing on Federal Power Commission oversight, but by this date the focus had shifted away from reliability to the health of the environment and growing demand for electricity.[72] In fact, at this hearing, the FPC dwelt at some length on the significance of the recently signed National Environmental Policy Act and the need for state and federal regulators to expand consideration of environmental issues when addressing the power industry. Reliability held limited interest during the hearing, despite the fact that blackouts continued to plague utilities and pools around the country.

The new focus on environmental concerns marked a shift away from traditional conservationism for the power industry. The 1964 National Power Survey couched plans for industry expansion in terms of resource conservation and gave fleeting attention to emerging environmental concerns such as air and water pollution and plant siting. The survey described pollution issues in terms of technical challenges soon to be solved by engineers. After the 1965 blackout, the industry focused on expanded interconnections as a path to greater reliability to the exclusion of resource conservation and, to some extent, operating economy. Congressional hearings specifically addressing interconnections from 1965 to 1968 ignored the question of resource conservation altogether. When the Senate returned to consideration of FPC oversight and interconnections in 1970, attention had shifted again, away from reliability and toward environmental protection.

During the 1960s, new environmental movements emerged in North America. Beyond protection of scenic beauty, local pollution problems, and resource conservation, advocates pressed on several fronts for greater environmental controls at federal and state levels and conservation at the consumer level. Groups sought preservation of ecosystems; limits on pollution of the air, water, and ground; constrained nuclear power development; and a slower pace of natural resource development. Once a tool for achieving resource conservation and energy efficiency, the grid now represented the path by which very large electric generating plants delivered more and more power to consumers. Regardless of the energy source—falling water, hydrocarbons, or nuclear energy—giant power plants and the transmission lines that linked them embodied the concerns of modern environmentalists. The fraternity of technical experts who had previously taken on the role of designing economical and efficient power systems now found themselves at the bewildering center of controversy.

By 1970, new federal laws and new public discourse shaped the context of electric power system expansion. Following the embarrassment of the 1965 blackout, the power industry faced opposition to nuclear power, opposition to dams, opposition to water storage projects, opposition to the siting of extra-high-voltage power lines, opposition to the use of coal in generating plants, and opposition to rising utility rates in sharp contrast to decades of praise for bringing modern technology to citizens across the land.[73] Further, to protect the environment, Congress enacted laws that placed more aggressive controls on the development of new power plants.[74] The FPC produced a second National Power Survey in 1970, developed in the shadow of the blackout and in response to the blackout reports. This time around, the survey praised interconnections for contributing to the reliability of power supply: "Thus, today it is reliability more than economy that provides the thrust for . . . complete and better interties."[75] The survey opened with a bleak description of the future for electric power: strained power supply in some areas; recurrent and spreading shortages; conditions slowing orderly development; and rising prices due to environmental protection efforts, market pressure on fossil fuels, and inflation. As the survey authors saw it, the core problem for the nation was to "ensure an adequate and reliable power supply without undue adverse impact upon the environment."[76] Environmental protection figured prominently throughout the survey.

During this period, the investor-owned utility sector moved quickly to strengthen its position as a self-regulated, well-coordinated industry. Utilities regularly announced the formation of new power pools and the expansion of existing interconnected networks, frequently touting the increased reliability sure to result. NERC took on the task of monitoring, assessing, and reporting on the adequacy and reliability of future generation and transmission plans across the continent. NAPSIC focused on operating a stable system. Both NERC and NAPSIC encouraged voluntary reliability compliance by all types of power producers, transmitters, and distributors. Through regional pooling agreements and participation in national associations, the industry retrenched behind the idea of shared management and divided authority. The reluctance of state utility regulators to cede authority to the federal government aided the investor-owned utilities in blockading the FPC's legislative moves. Greater public interest in the environmental impact of specific projects rather than the stability of the entire system, even in areas affected by the big blackout, further dulled the FPC's efforts. Beyond that, the mere fact that the lights stayed on nearly all the time reduced the perceived importance of proposed federal reliability oversight.[77]

In the aftermath of the 1965 blackout crisis, the power companies regrouped around tried-and-true techniques for operating interconnected. They shared information about expansion plans, devised agreements for interties, vetted technology through professional associations and well-publicized trials, and coordinated operations and maintenance through power pools and regional councils. At the same time, individual utilities, cooperatives, and government agencies maintained economic autonomy and avoided federal regulation of transmission grid reliability. And every few years, but not terribly often, the public experienced cascading and/or widespread power failures. Table 7.1 offers a partial list of failures significant enough to affect at least one thousand people for more than one hour and to cause more than one million person-hours of disruption. This list includes outages that affected the transmission networks as well as outages due to storms, such as Hurricane Sandy in 2012, which affected primarily the distribution networks.[78] The latter type of outage is different in kind from the cascading failures that affect major segments of the transmission network. From the perspective of a retail customer sitting in a dark house, however, a major blackout is a major blackout.

A National Grid at Last

As the Northeast blackout captured local and international attention, engineers, system operators, and government officials continued the quest for a national grid. From the early decades of the twentieth century, power systems experts and eager politicians had envisioned a coast-to-coast transmission network. Some advocated for a centrally planned and constructed system. Others endorsed private sector development of interconnections. Still others puzzled over the technical ramifications of building such a large and complex electrical network. For more than fifty years, discussions about linking the East and West Coasts with power lines ebbed and flowed. Presidents and administrators argued over authority, members of Congress considered

Table 7.1
Partial list of major North American power outages

Year	Locale
1965	Northeast
1967	Pennsylvania–New Jersey–Maryland
1971	New York City
1976	Utah and Wyoming
1977	New York City
1981	Utah
1982	California
1989	Quebec
1991	Iowa to Ontario
1996	Western North America
1998	San Francisco
1998	Ontario and North Central United States
1999	Northeast
2003	Northeast
2011	Southern California, Arizona, Mexico
2012	New York and New Jersey
2013	New England

Sources: US Department of Energy, *Final Report on the August 14, 2003 Blackout in the United States and Canada: Causes and Recommendations* (https://energy.gov/oe/information-center/library) and "List of Major Power Outages," *Wikipedia* (https://en.wikipedia.org/wiki/List_of_major_power_outages).

proposals for federal ownership, and individual utilities stood on both sides of the question—some hoping to dominate large geographic sectors of the country through interconnections and others demanding local autonomy and control. While the 1965 blackout raised the question of whether interconnected systems offered greater reliability or risk, the technicians pursued tying together the giant eastern and western power networks.

In the early 1960s, President John F. Kennedy gave new life to the idea of a national grid as a tool for achieving greater efficiency and resource conservation, and the industry took the proposition to heart. On March 1, 1962, Kennedy sent a special message to Congress on conservation. Regarding electric power, he stated:

One of the major challenges in resource conservation lies in the orderly development and efficient utilization of energy resources to meet the Nation's electric power needs—needs which double every decade. The goal of this Administration is to ensure an abundance of low cost power for all consumers—urban and rural, industrial and domestic. To achieve this, we must use more effectively all sources of fuel, find cheaper ways to harness nuclear energy, develop our hydroelectric potential, utilize presently unused heat produced by nature or as a by-product of industrial processes, and even capture the energy of the tides where feasible.

The ability to make long-range plans for the expansion of our Nation's electric power supply required by constantly growing power needs will be enhanced by a comprehensive nationwide survey to be undertaken by the Federal Power Commission. Under existing authority contained in the Federal Power Act, the Commission will project our national power needs for the 1970's and 1980's and suggest the broad outline of a fully interconnected system of power supply for the entire country. This information will encourage the electric power industry—both private and public—to develop individual expansion programs and intertie systems permitting all elements of the industry—and more importantly the consumers—to benefit from efficient, orderly planned growth.[79]

The 1964 National Power Survey illustrated the opportunities for saving energy by moving electricity across the Rocky Mountains to take advantage of seasonal demands and resource availability. Not long after the survey appeared, power pools in the northern and southern states west of the Rocky Mountains shared power, reducing the number of grids serving North America from five to four. By early 1965, representatives of public and private utilities began working on plans to achieve coast-to-coast power transmission. In addition to the potential conservation benefits of this truly giant grid, engineers relished the challenge of bringing numerous very large systems into parallel and operating them without serious mishaps. The Soviet Union had built the only other power system comparable in scale to North America's, but this had been achieved under central command and control that greatly contrasted with the gaggle of interconnected companies operating in the capitalist West. A successful closing of the North American ties represented a technical, organizational, and political accomplishment that was international in scope.[80]

The Bureau of Reclamation, working with the East-West Intertie Closure Task Force, moved steadily ahead on closing the ties. The task force included representatives from the Bureau of Reclamation and several utilities based in states directly affected by the project. During 1965 and 1966, in spite of blackouts, discussions of the value of interconnections, FPC and state investigations and reports, and congressional hearings, the planning and technical installation continued apace. In November 1966, Secretary of the Interior Stewart Udall distributed a press release announcing the plan to test the closure the following February. The announcement was met with a small flurry of news reports, mostly neutral, although the *Chicago Tribune* accused Udall and the Bureau of Reclamation of "empire-building."[81] The *Tribune*'s editor suggested that the linkup was of interest only to the Interior Department, which sought to co-opt private power markets in the central part of the country. By contrast, *Electrical World* offered praise for the "history-making interconnection of systems east and west of the Rockies."[82] *Electrical World* cautioned that this should not be interpreted as the completion of a "fully integrated, monolithic power grid" but rather as another phase in a process that had been under way for years. The project quickly faded from public view as the year drew to a close.[83]

Udall piqued public interest again in late January 1967 with a lengthy press release detailing the date of the first major test of closure. He highlighted the cooperation between industry and government and suggested that east-west links would both improve the operating economy of power systems and grant greater reliability from coast to coast. The Bureau of Reclamation planned to close four tie lines on February 7, 1967 (in northeastern Montana; south-central Montana; Gering, Nebraska; and North Platte, Nebraska). With this announcement, a handful of newspapers offered brief reports, primarily listing details of the coming event.[84] The *New York Times* did counter the *Chicago Tribune*'s earlier claim that the closure represented a power grab on the part of the Bureau of Reclamation. The *Times* noted that this event "has long been sought by engineers" and by former FPC chair Joseph Swidler.[85] The report further offered that most in the utility field expected a full national grid to "do much towards avoiding" future major blackouts.

In line with the hope that the larger interconnections would minimize blackouts, the task force took care to prepare for closure-related problems. Task force chair and Bureau of Reclamation Power Systems Operation Officer Frank Lachicotte distributed a four-page document delineating steps to take in case of automatic separation. Of note, the eastern system was nearly four times larger than the western system, each carrying 170,000 and 45,000 megawatts of capacity, respectively. By comparison, the interties could carry only a small fraction of that capacity. The potential for severe power swings between the two areas was significant. Engineers in the broader industry shared heightened concern about the results of linking these two systems and further wondered what effects trouble on one coast would have on the other. In anticipation of problems, the participating utilities and power pools agreed that any prolonged or serious difficulties would lead to opening of the ties until the problems were resolved. The task force arranged for all participating utilities affected by the closure to receive information during the February test and to receive operating data and analyses in the ensuing weeks. Leeds & Northrup Company (L&N) engineers, with deep interest in the process of managing interties and the physical control apparatus involved, participated as observers of the test.[86]

The big event, soon to be hailed by power engineers as "Driving the Golden Spike," approached quietly. On February 7, 1967, at 9:49 a.m. (MST), the eastern and western interconnections shared electricity. "Actually, the East-West closure itself was almost without unusual incident," announced Lachicotte. "After a 19-minute delay during which the two massive power interconnections lazily drifted into synchronism, the connecting circuit breakers were closed . . . establishing the tie."[87] Lachicotte directed the closure from the Watertown, South Dakota, office of the Bureau of Reclamation. One trade magazine editor in attendance described the scene: "All we got on film was intent expressions on the faces of twelve men gazing at a group of electrical meters which recorded nothing at all unusual."[88] For the occasion, L&N set up a special frequency recorder in Philadelphia to observe and report the oscillations between the systems as they came into parallel. As the image in figure 7.4 illustrates, power systems experts followed the event closely, using both traditional means of communication (the telephone) and the most modern data collection and recording techniques (graphing recorders). The event seemed to bring a sense of somber anticipation to those observing and enormous relief when the ties held.

Walt Stadlin, a power engineer present at the L&N observation locale, shared some recollections of the significance of the closure. The observers were especially concerned with frequency fluctuations as the two giant grids came into synchrony. The lower recorder in figure 7.4 measured the frequency on the eastern connection. The upper recorder may have been tracking the difference between eastern and western frequencies. Nathan Cohn was likely talking on the phone to a colleague on the West Coast, who was likewise measuring the frequency of the western connection, and the two shared information. The engineers were vitally interested in the viability of the largest AC network in the world. Stadlin recalls, "Since this was a 'proof-of-concept,' everyone was hoping for the best but was prepared for the unknown and potential accompanying problems that need to be resolved."[89] Stadlin noted that the results of the test led to quick recognition that weak AC interties were limited and could not support the exchange of large trades of power from coast to coast without affecting system stability.

News reports captured the sentiments of power engineers, utility operators, and government officials. Lachicotte shared a sampling of headlines and highlights with employees of the Bureau of Reclamation in an internal newsletter: "An unprecedented accomplishment of public and private

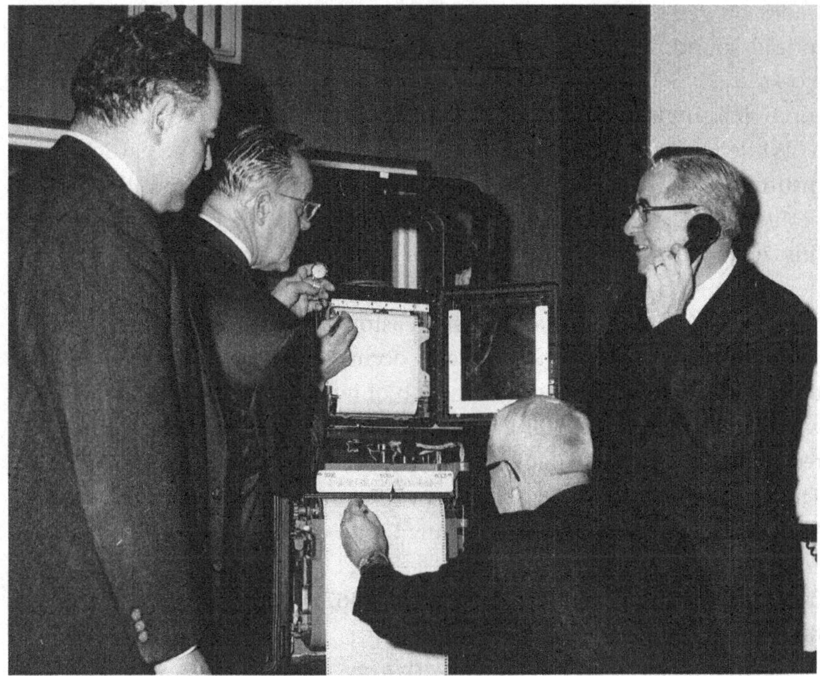

Figure 7.4
Leeds & Northrup engineers witness the closing of the ties, February 7, 1967: "On the occasion of the closing of the East-West Ties, Leeds & Northrup set up a special ultra narror [*sic*] range frequency recorder in their R & D center to observe and report to utility operators what they saw. In photos are Nathan Cohn (on phone), S. B. Morehouse (with watch), and others."

Source: W. Spencer Bloor Collection, now in Electric Control Systems Records, Hagley Library, Wilmington, Delaware.

power groups working cooperatively together," "two massive power interconnections consisting collectively of over 209 public and private electric systems in Canada and the United States joined into one big system for the first time," "94% of the nation's electrical [*sic*] might now be joined into one vast interconnected power network," "a culmination of longtime dreams of engineers," "the Golden Spike operation, connecting East and West."[90] The popular press hailed this event as the test of a huge grid intended to prevent blackouts. *Electrical World* offered the utility perspective that this was a successful process of coordination built on past attempts. Two days of connecting and disconnecting different regions, sending power east to west and then west to east, and changing up the schedule of electricity trades proved that the concept of integration across the continent was indeed sound. In the past, generators delivered electricity through networks and pools and very large regional systems; now electricity flowed through a single grid.[91]

While the moment of closure brought great delight to the participating institutions and individuals, operations over the ensuing months proved problematic. Lachicotte reported to colleagues that he opened the ties on July 20 at the request of three western companies. Both instability on the network and large inadvertent interchanges of electricity proved onerous to local and regional operations, and customers experienced several power outages. All but one of the problems occurred on the western portion of the grid. The task force took on the job of testing the system, determining what to change, and deciding where to make changes. Lachicotte hoped to reclose the ties by mid-August, in part to protect the prestige of the industry. The task force developed a set of nine recommendations and enlisted the Western Systems Coordinating Council, one of NERC's signatory regional councils, to persuade the utilities to cooperate. The Bureau of Reclamation finally reclosed the ties on December 3, 1967, with new operating guidelines and monitoring apparatus in place.[92]

North America's grid operated with very weak links between east and west for eight years, and then the power companies permanently opened the AC interties in 1975. When high-voltage direct current (HVDC) interties replaced the original AC ties in the 1980s, it was no longer necessary to keep both the eastern and western systems in synchrony. With DC interties, each interconnected AC network operated in synchrony internally, the energy was converted to direct current using AC-to-DC transformers before entering the transmission line, and the energy was reconverted using transformers to alternating current before entering the other network. This would be much like two objects pivoting on a swivel at each end of a still pole—the AC-to-DC-to-AC transformers served as the swivels. As one observer ruefully remembered in 2000, "We . . . had some fancy, brilliant schemes to close the East-West ties. But there were problems with them. If something would happen on one side or the other, you wound up tripping the lines."[93] Stadlin explained, "The solution to the interconnection problem (in all large countries) has been to interconnect the AC networks by means of high-capacity HVDC interties that have very fast controllability, in order to maintain the maximum power exchange and stability of the overall grid in each country or region."[94] In practical terms, the HVDC interties allowed the four major interconnected systems in North America—East, West, Texas, and Quebec—to operate with links to each other but not in synchrony. Nonetheless, the symbol of a single network carried far more significance than the challenges of maintaining stable links between the two major eastern and western systems. From the time of the 1967 East-West Intertie forward, the interconnected power systems of North America have been referred to as "the grid."[95]

8 Reaching Maturity

Integration, Security, and Advanced Technologies, 1965–1990

The 1965 Northeast blackout and the 1967 closure of ties between the east and west proved to be seminal events in the development of North America's power grid. The disastrous blackout marked a major setback for the power industry, which had enjoyed more praise than criticism through most of its history. Just as significantly, the blackout intensified a focus on reliability across all sectors of the industry. Reliability meant that the system had enough generation and transmission to meet customer demand and remained stable in real time. Power system experts redoubled efforts to secure the grid against failures, and in fact, the term "security" gained usage to describe a stable and reliable system.[1] The closing of ties between the east and west served as a countervailing event. It was a celebration rather than a disaster, but it took place with almost no public fanfare or notice. Nonetheless, by linking the two giant systems, the industry succeeded in creating the world's largest interconnected machine. From that date forward, despite the permanent opening of the alternating current interties in 1975, Americans conceptualized the collection of interconnected power companies as a single system.

The newly unified grid operated within changing constraints in the United States. The industry confronted several firsts during the 1970s: the first major federal environmental legislation, the first worldwide energy crises, and the first slowdown in the rate of increase of demand for power.[2] In addition, inflation rates increased dramatically during the 1970s through 1980 to the highest levels since 1947. And the overall regulatory environment was changing. By the end of the twentieth century, the power industry had substantially reorganized to meet new regulatory, market, and consumer regimes. Yet the grid remained in place as the backbone of electrification in North America.

The Changing Context for Electric Power

For decades prior to the 1970s, power companies forecasted demand for electricity to increase at the rate of about 7 percent every year. This provided the basis for planning in the 1960s as companies designed new generation and transmission facilities, anticipated construction costs, committed to construction plans, and prepared future rate schedules to match the expected needs of American consumers. The trend changed in the 1970s. In 1974, power sales were flat, and they flattened again following increases in 1976 and 1977.[3] They remained nearly flat through most of the 1980s, with an actual drop in 1982. These dramatic shifts are visible in figure 8.1. Between 1974 and 1990, the per-customer rate of electricity use slowed as well, with actual drops in 1974, 1980, 1981, 1982, and 1985.[4] The changes in rates of usage are illustrated in figure 8.2, with a small dip in the chart appearing in the early 1980s. Several factors influenced the slowing of consumer demand, including fears prompted by repeated international energy crises, the rising cost of electricity, an overall increased cost of living, and calls for conservation of resources through reduced use.[5]

Rising consumer prices and international energy crises had an overarching influence on the US economy. In 1973, the Organization of Petroleum Exporting Countries (OPEC) imposed an embargo on oil shipped to the United States, driving the price of oil up steeply. The Arab State members of OPEC took this action in response to US support of Israel during the 1973 Arab-Israeli war. In the United States, this led to regional petroleum shortages, long lines at gas stations, and problems for power companies with oil-burning generators. During the Iranian revolution, in 1979, oil production in that country fell and prices again rose, causing another energy panic in the United States.[6] Overall, the price of oil increased by nearly 700 percent between 1970 and 1980.[7] Coal and natural gas prices rose as well. The broader economy reflected these rising fuel costs. While the rate of inflation barely reached 2 percent in the early 1960s, it averaged 4.9 percent in the early 1970s and then jumped to 11 percent in 1974.[8] Electricity prices, which remained relatively unchanged through the 1960s, began to increase in the 1970s, nearly doubling between 1970 and 1975.[9] As can be seen in figure 8.3, consumer prices were much more volatile than electricity prices, which were subject to regulation at both the state level for retail customers and the federal level for wholesale interstate transmission. As electricity and other goods became more expensive, consumers cut back, and this flummoxed the power system planners, as did broader trends in the social discourse.

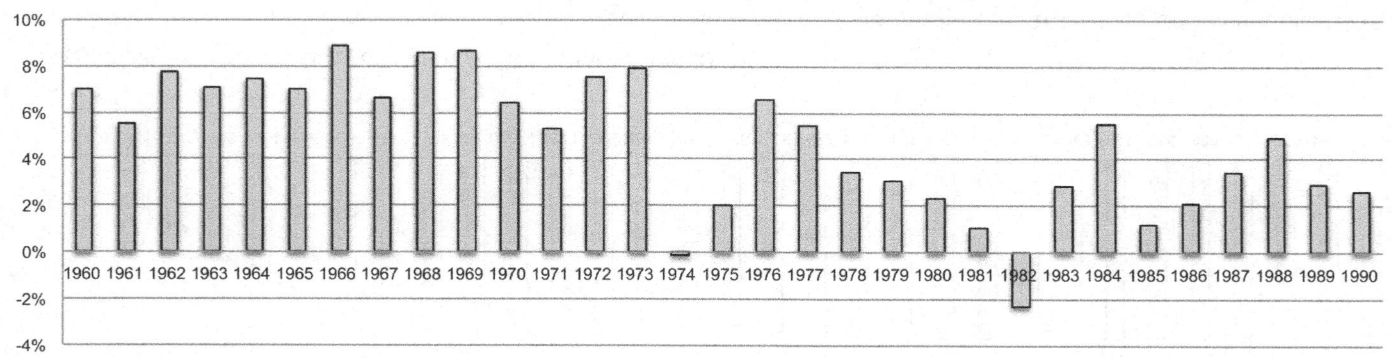

Figure 8.1

Year-over-year percent change in electricity sales, 1960–1990.

Source: Lizette Cintrón, ed., *Historical Statistics of the Electric Utility Industry through 1992* (Washington, DC: Edison Electric Institute, 1995), table 38.

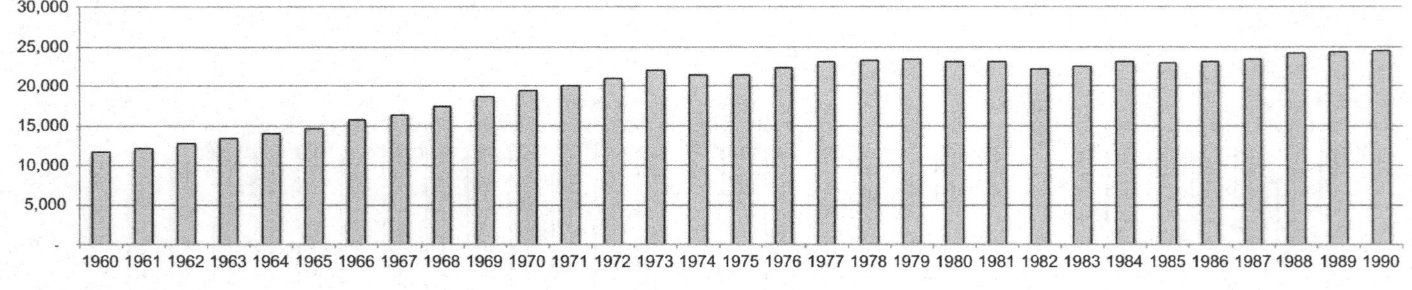

Figure 8.2

Average annual use of electricity in kilowatt-hours per customer, 1960–1990.

Source: Lizette Cintrón, ed., *Historical Statistics of the Electric Utility Industry through 1992* (Washington, DC: Edison Electric Institute, 1995), table 45.

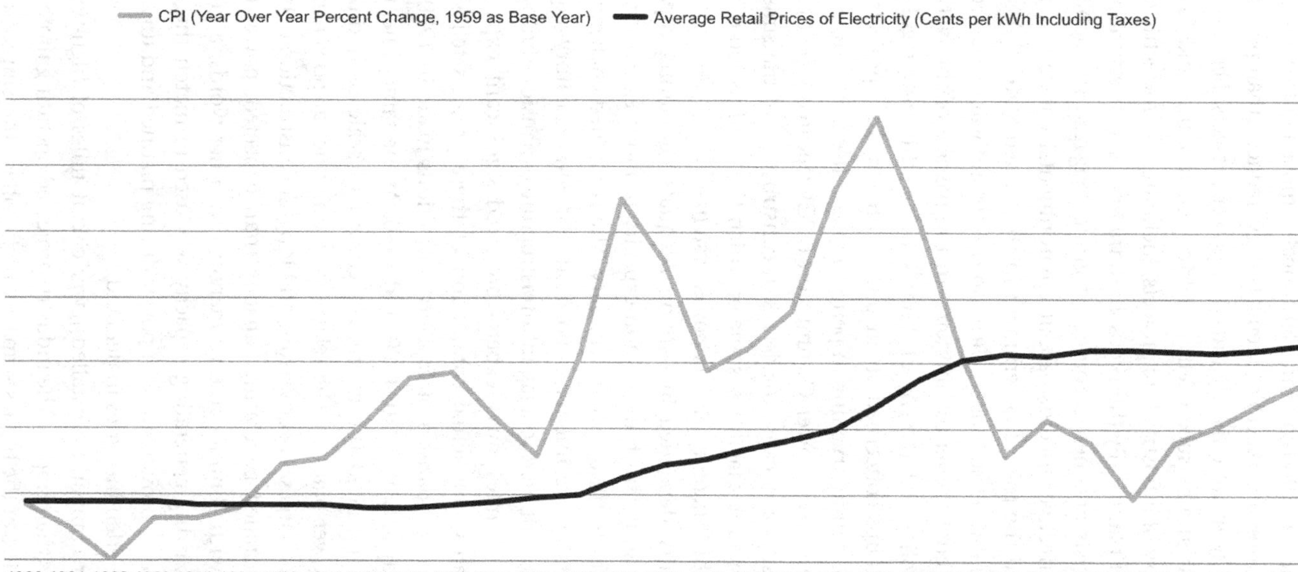

Figure 8.3

Chart showing year-over-year percent change in the consumer price index and nominal average electricity prices across the United States.

Sources: Bureau of Labor Statistics, "Table 24. Historical Consumer Price Index for All Urban Consumers (CPI-U): U.S. City Average, All Items," *CPI Detailed Report: Data for December 2015*, 72, 74 (http://www.bls.gov/cpi/cpid1512.pdf); US Energy Information Administration, "Table 9.8 Average Retail Prices of Electricity," *January 2016 Monthly Energy Review*, 141 (http://www.eia.gov/totalenergy/data/monthly/previous.php).

Increasing environmental activism significantly affected the process of electrification from the 1960s onward. Historically, movements to conserve natural resources, preserve places of scenic beauty, reduce urban pollution, and protect ecosystems were tightly linked to power projects. Beginning in the 1950s, activist groups succeeded in halting, postponing, and forcing revision of a variety of infrastructure projects, including hydroelectric dams and nuclear power plants.[10] By the 1960s, both traditional conservation and preservation activists and newly composed groups engaged the public in an increasingly effective discourse about environmental concerns, industrial activities, and the role of government in the United States.[11] A series of state- and federal-level actions to clean up the air and water and protect and preserve the environment culminated in the passage of the National Environmental Policy Act (NEPA) of 1969.[12] Under this new regulatory regime, citizen groups concerned about power infrastructure projects had even greater success in intervening in permitting and licensing processes.[13] In the 1970s, the Federal Power Commission (FPC) repeatedly pointed to environmental issues as a cause for delays in construction of transmission lines and new generating facilities, some of which the commission considered essential to the reliability of bulk power systems.[14]

System reliability depended in part on adequate generating capacity to meet demand, sufficient transmission capacity to move electricity, and substantial reserves to fill in when necessary. The combined challenges of high material costs, environmental opposition, and new regulatory requirements certainly contributed to a lag in infrastructure expansion during the 1970s and 1980s. While power experts projected significant expansion of generating plants, the actual completion of these projects fell short. Figure 8.4 offers a comparison of capacity projections made in 1970 and actual installed capacity in 1970, 1980, and 1990. At the same time, technical limitations inhibited the continued growth in scale of traditional fossil-fuel-fired power plants.[15] The relatively young nuclear power industry experienced the largest cost overruns and longest construction delays.[16] With less new installed generating capacity than expected, power companies operated with slimmer reserve margins. In other words, a greater amount of the existing generating capacity was used to match the load, with less available to provide backup power during planned and unscheduled outages or sudden increases in demand.

Further, power companies installed fewer circuit miles of high-voltage transmission lines than hoped. The industry projected operating more than 118,000 circuit miles of the highest-capacity high-voltage transmission lines (243 kV or above) by 1980 and nearly 160,000 circuit miles by 1990.[17] In

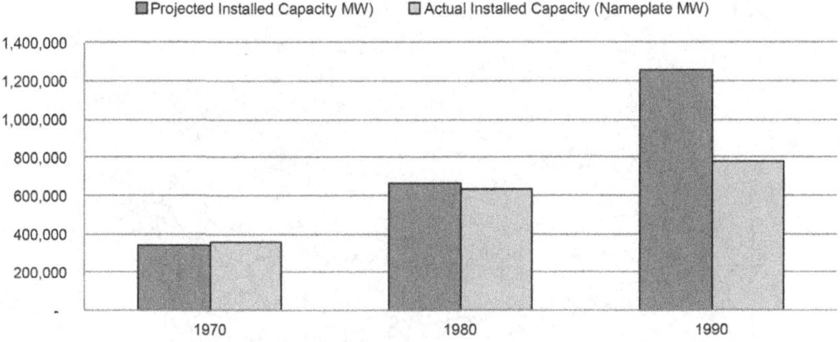

Figure 8.4

The difference between anticipated and actual growth of generating capacity, 1970–1990.

Sources: The 1970 National Power Survey of the Federal Power Commission (Washington, DC: US Government Printing Office, 1970), table 1.2, I-I-17; *Historical Statistics of the Electric Utility Industry through 1992* (Washington, DC: Edison Electric Institute, 1995), table 1.

fact, the addition of new transmission lines fell somewhat short of those goals—with 116,000 circuit miles in 1980 and 140,000 in 1990—but the industry did install more circuit miles at higher voltages than anticipated.[18] For interconnected systems, progress was mixed. Companies continued to expand the scale of power pools, and the level of interconnection achieved by the mid-1970s is depicted in figure 8.5. With less generating capacity than planned, smaller reserves than needed, and a smaller margin for failure, the power system experts addressed new approaches to system reliability.

The Focus on Reliability

Before the 1965 Northeast blackout, power system experts applied their considerable know-how to quotidian matters of grid operations. While they contended with disturbances and local outages, they generally dismissed the threat of a major cascading power failure. Utilities installed a wide array of controls and relays to facilitate the work of system operators as they balanced generation and demand and maintained steady frequency across networks. The engineers certainly distinguished between apparatus intended to measure and control small-scale instabilities and devices that isolated equipment

Figure 8.5
Major interconnections of the United States as of 1978.

Source: Nathan Cohn, "Power-System Interconnections: Control of Generation and Power Flow," reprinted from *Standard Handbook for Electrical Engineers, 11th Edition,* ed. H. Wayne Beaty and Donald G. Fink (New York: McGraw-Hill, 1978), section 16-6. Reproduced with permission of McGraw-Hill Education.

experiencing a major problem. The Northeast blackout introduced a new level of concern, causing the experts to refine notions of stability and security, redouble efforts to deploy new technologies—especially computers—to assess and operate power networks, and participate in new organizations to share knowledge and standardize practice across the country.

The striking change in energy priorities—from economy and resource conservation to reliability—is evidenced by a quick comparison of the 1964 and 1970 National Power Surveys, both produced by the FPC with extensive cooperation from the industry. The 1964 survey invokes the 1935 Federal Water Power Act, "which directs the Commission to 'promote and encourage . . . interconnection and coordination' for . . . 'the purpose of assuring an abundant supply of electric energy . . . with the greatest possible economy and with regard to the proper utilization and conservation of natural resources.'"[19] In contrast, the 1970 survey notes a rededication to interconnection, but with a different focus: "The searching examination that followed the Northeast blackout led to the conclusion that an even greater benefit of system interconnection is that, when properly conceived and executed, it can contribute importantly to the reliability of power supply."[20] Between the publication dates of these two surveys, the FPC issued "new regulations (Order No. 331, 36 FPC 1084) requiring all electric utilities to report promptly any major power failures (18 CFR 141.58)."[21] The regulations required both public and private entities to report, by telephone, losses of service that lasted for fifteen minutes or longer or affected loads of more than two hundred thousand kilowatts (kW) and to report smaller losses by telegraph. In this way, the FPC elevated the importance of reliability and established a more formal role as the repository for reliability data. Between 1970 and 1977, utilities reported an average of thirty-seven interruptions per year, with a trend toward increasing numbers of interruptions over the years.[22] Following the reorganization of federal energy agencies at the end of 1977, utilities reported to the Federal Energy Regulatory Commission (FERC).[23] Industry experts and journalists looking back agreed that "the Northeast Blackout probably did more than any other single event to advance the concept of reliability in the interconnected system."[24]

Immediately after the big blackout, power experts faced the task of sorting out what happened and explaining the process to the public. The experiences at Leeds & Northrup Company (L&N) serve as a case in point. Companies like L&N manufactured automatic control devices that were used by companies affected by the blackout. Immediately after the failure, L&N fielded inquiries concerning whether and how their devices were implicated. L&N had provided a wide array of telemeters, recorders, transducers, and controllers to the key utilities involved in the blackout, including Consolidated Edison and Niagara-Mohawk. The instruments assisted with daily operation of the interconnected networks. In addition, L&N supplied automatic relays to the Pennsylvania–New Jersey–Maryland Interconnection (PJM) to address system stability at that power pool's boundary with New York. These relays acted to isolate PJM just as the cascading power failure began, and customers of PJM member power companies enjoyed continuous electrical service throughout the event.

In the short term, L&N carefully navigated inquiries from both within and outside the company. The day following the blackout, the company placed test facilities at the disposal of Consolidated Edison and Niagara-Mohawk.[25] Consolidated Edison later declined the offer.[26] In response to an in-house advertising manager, the company determined not to make any public statements regarding the performance of its equipment. The company's executives agreed that "under no conditions should a supplier to an industry be a spokesman for that industry."[27] By November 11, 1965, L&N engineers obtained, and shared internally, frequency charts from recorders at key locations.[28] One day later, numerous press reporters had contacted the company with inquiries about the blackout, and the president distributed a management bulletin explaining the "no comment" policy.[29] In part, attention was drawn to L&N by a recently published article in *Spectrum Magazine*, compellingly titled "The Automatic Control of Electric Power in the United States," by Nathan Cohn.[30] Cohn had to go to some lengths to clarify to interested parties that "the paper would not be helpful in explaining the blackout problems since it reviews normal and not emergency control conditions."[31] Cohn established an in-house task force to examine the actions of all L&N equipment in service during the blackout and to respond to further inquiries.[32] In general, this type of response reflected the challenge to the industry to explain why the technologies of interconnection were sound if they also failed to protect customers against a massive power failure.

The process of clarification continued even after public agencies and private utilities issued their first investigative reports and Congress held hearings on the blackout. L&N highlighted the control of electric power systems at its September 14, 1966, shareholder's meeting. The company displayed maps, charts, and control instruments in an exhibit area adjacent to the meeting room, and Mr. Cohn made a special presentation about the company's involvement in system control. He paused partway through the talk to focus on the blackout, noting that L&N was asked frequently, "'Did equipment of our kind cause the blackout?' and 'Could equipment of our type have prevented it?'"[33] Cohn offered that the answer to both questions was no. He then went on to describe the fundamentals of power system operation:

> An important characteristic of electrical energy on power systems is that for all practical purposes it must be made as it is needed. It cannot be stored or warehoused. A second important matter is that all of the generators are tied together by what may be regarded as "elastic bands". Normally, all generators run at the same effective electrical speed. They are then said to be "in step", or to be running "synchronously" and the system is said to be "stable". . . .
>
> When customer demand changes up or down, supply must promptly be changed correspondingly, up or down. All during the course of any typical day, as customers flip their switches on and off, demand continually changes, and there are corresponding automatic adjustments to the total power being supplied. All of this is normally done in a manner that does not disturb the "stability" or the "synchronous" operation of the system. The function of our automatic control equipment is to see that such changes in demand are met by corresponding changes in supply in those locations and on those generators which at that particular time should be taking these changes.[34]

Cohn explained to the shareholders that sudden large losses of generation or load "shocked" a power network, straining the "elastic bands" that tie it together. Most of the time, the system can absorb the shock without losing the synchronous operation of its component parts, with which automatic controls assist. But if the shock exceeds the capacity of the system to absorb the change in generation or load, the "elastic bands" separate, emergency conditions develop, and customers lose power. This is what occurred in the case of the Northeast blackout. Cohn referenced the FPC report, stating that the mismatch between supply and demand was of such size and suddenness that the system could not absorb the impact of the initial shock. He explained that it was a situation "beyond the scope or capability of present-day automatic control equipment, which, as I have said, can do its job only when system stability prevails in its area."[35] Cohn wanted the shareholders to understand that there was a line to draw between normal operations and a sudden, unanticipated, and major change to network activity.

System Security and Operations

Cohn's analysis served as a precursor to a broader trend toward distinguishing among different types of threats to power systems and classifying types of power control. In the aftermath of the blackout, the power system experts began to discuss control of the system as a whole rather than addressing frequency and load control separately from economy loading and separately from automatic relaying to isolate trouble areas. They defined states of the whole system and established a hierarchy of control activities. The experts built on the growing capabilities of computers to gather and analyze data and model system behavior. And they conferred both in traditional ways and through new and existing organizations established specifically to improve power system reliability. The evolution of control techniques advanced the art of automation, facilitating the increasingly complex job of operating a power network. At the same time, computer manufacturers offered digital machines capable of managing larger quantities of data, more quickly, in a variety of parallel operations, and at lower prices than analog machines. Finally, theorization of systems advanced in the mid-twentieth century in several related and mutually influential fields, including control engineering, cybernetics, general system theory, systems engineering, and system dynamics.[36]

By 1960, the power industry trended both toward physical integration of systems and operating practices and also toward conceptual integration of the various control challenges of interconnection. Traditionally, power system experts identified several objectives for system control, including (1) matching total generation to total demand, (2) maintaining economic allocation of generation, and (3) maintaining the scheduled power flows over ties between systems.[37] During the 1950s, utilities had begun to integrate automatic frequency and load control (designed to meet objectives 1 and 3) with economy loading (designed to meet objective 2).[38] For a system operator controlling power on an interconnected system, it was advantageous not only to make generating adjustments that addressed frequency regulation and planned power exchanges but also to do it as economically as possible.[39] The increasing sophistication of computers and control apparatus facilitated this process. In the early 1960s, engineers began to call the integration of objectives 1 and 3 "automatic generation control."[40]

In 1970, the Institute of Electrical and Electronics Engineers (IEEE) published standard definitions for various terms related to power control.[41] The IEEE responded to requests from operating groups for standardization and also determined that engineers used multiple definitions in their technical publications. By this date, the notion of automatic generation control was indeed a standard part of the power system engineer's lexicon. It comprised the title of the publication and was defined therein: "The regulation of the power output of electric generators within a prescribed area in response to changes in system frequency, tie-line loading, or the relation of these to each other, so as to maintain the scheduled system frequency and/or the established interchange with other areas within predetermined limits."[42] Automatic generation control was a somewhat more integrated approach to system control and an essential tool for maintaining a stable and reliable network.

Following the blackout, the FPC appointed a special Advisory Group on Reliability of Electric Bulk Power Supply to investigate reliability problems and to recommend guidelines for planning and operation of the nation's transmission networks.[43] The advisory group's report, released in 1967, defined *reliability* as "the ability of a utility system or group of systems to maintain the supply of power. Reliability is gauged by the infrequency of interruption, the size of the area affected, and the quickness with which the bulk power supply is restored if interrupted."[44] In an interview preceding release of the report, the chair of the advisory group explained that customers expected bulk power systems to withstand "tornadoes, airplanes, or hunters taking pot-shots at insulators" without cascading outages.[45] In other words, system reliability now encompassed a variety of matters—frequency control, vandalism, equipment failures, natural disasters, and even human error.

The advisory group articulated a focus on the concerns of electricity customers. In the final report to the FPC, the group outlined the industry's primary objective: "the complete avoidance of widespread outages and cascading interruptions" across bulk power systems.[46] Importantly, the advisory group noted that any piece of equipment in a power system is likely subject to failure at some point. The industry, in addition to minimizing the likelihood of equipment damage or failure, had to address preventing an isolated disturbance—say, a storm that took down a transmission line—from affecting a widespread area. Interconnection was seen as key to improved reliability, and the benefits of interconnection were to be realized through planning, pooling, and coordination on a regional basis.

The advisory group's report included the notion of "system security," a new description of the combined concerns of equipment reliability and service reliability. Engineers had earlier referred to the security of systems and in the early 1960s began to conduct "security assessments" of power systems. For example, in 1965, engineers working for the Central Electricity Generating Board in England described a computer that conducted a security assessment of that nation's power grid.[47] A central computer used telemetered data to model system response to changes in power flow and then simulated failures to determine what might happen next and whether additional failures might occur. The computer also controlled the economy loading of generators. The objective was to "obtain the maximum of security at as economic a price as possible."[48] By the 1970s, some utilities routinely conducted computer-based modeling of power systems to assess system security.[49] But at the time of the 1965 blackout, security assessment was a new activity and addressed a narrow set of concerns on a grid.

To secure a reliable bulk power supply, the advisory group urged utilities to undertake comprehensive planning. Design decisions should incorporate high-voltage and high-capacity interconnections; sufficient generating reserves; relay systems; thoughtful location and concentration of generating, transmission, and switching facilities; continuing study of system behavior; and examination of contingencies—that is, the myriad events threatening system stability. Plans for emergencies should include criteria for load shedding, for dropping generators, for opening interconnections, and—only in a very small number of cases—for separating a network into islands of matched demand and generation. The newly defined problem of system security integrated the physical, technical, and operating aspects of system design, with a heavy emphasis on operations.

Contrary voices weighed in as well. In 1970, for example, consulting engineer C. F. Paulus suggested that the consuming public was not satisfied with the explanations for the 1965 blackout proffered by system experts.[50] He claimed that the customer cared far more about how long a blackout lasted than the fact that it had occurred. He urged the industry to focus on immediately available reserve capacity and to restore service locally as swiftly as possible rather than focusing on complex operating procedures. While addressing a different aspect of power failures—namely, customer satisfaction—Paulus pointed at disjointed technical investigations as a problem. He advocated greater coordination and communication between the engineers who manufacture power equipment and those who operate it.

Efforts to make operations more reliable, nonetheless, received a great deal of attention. In 1967, Tomas Dy Liacco, then working for Cleveland Electric Illuminating Company (CEI) and also teaching at Case Western Reserve University, published a paper calling for the design of "a total control system for the improvement of the reliability of the generation-transmission system."[51] Dy Liacco advocated for a control system that combined the most advanced information technologies, automatic and manual control techniques, and better-informed decision making. As Dy Liacco explained, system operation involved both day-to-day decisions and emergency management, and the best new controls should be able to optimize the former while recognizing and responding very quickly to the latter. He then described the design and testing of such a control system in Cleveland.

Dy Liacco opened by explaining, "The electrical operation of the generation-transmission system may be viewed as a series of control actions taken to maintain continuity of service at standard frequency and voltage."[52] CEI sought to improve the reliability of its system by automating all transmission substations and by developing a better control system to improve reliability and minimize catastrophes. Dy Liacco approached the design hierarchically by delineating multiple levels of the control problem and breaking down the elements under control into subsystems. He proposed that a comprehensive information system was essential and provided measurements, communication, information processing, and information display and reporting: "The aim is to be as comprehensive as possible so that all of the requirements within the framework of the approach are covered even though not all of the solutions have as yet been formulated."[53] Dy Liacco promised mathematical explanations in a separate publication.

The comments on this paper indicated how influential Dy Liacco's ideas were to become. Cleveland-based consulting engineer John R. Linders found the paper inspiring yet wondered who would be responsible for integrating the different control functions. He noted that almost all the tools described were already in use by power companies, but each fell into the domain of a different actor, ranging from the system planner, to the substation engineer, to the relay engineer, to the system operator, to the equipment vendor. Referring to the 1965 Northeast blackout, Linders suggested, "The needed leadership for moving such an across-the-board program forward lies in the operating arm of the utilities."[54] He projected that Dy Liacco's proposed system control approach would have a great impact on future system design and operation. A Westinghouse engineer noted that many had been working toward Dy Liacco's methodology "on a step-by-step basis without defining it."[55] As a commenter from General Electric explained, Dy Liacco's approach was evolutionary rather than revolutionary but provided a "unifying concept" that would be one of the "most valuable contributions of the paper."[56] He went on to say that the work had "the potential of showing to the industry where the next great strides are to be taken in terms of system operation and security."[57]

Two years later, Dy Liacco pushed the notions of system security and security control further. In "The Emerging Concept of Security Control," he described the need to integrate traditional approaches to operations—matching generation to demand—with the requirements of a secure system.[58] As he explained, the industry had made "dramatic technological and analytical improvements" in the areas of generation, transmission, and interconnection, but the security of operations was "left to the intuition of human operators guided (or misguided) by grossly inadequate system information and by predetermined operating rules and management policies."[59] The FPC report of how the Consolidated Edison system operator responded to the cascading failure in 1965 serves as a telling example of this.[60] Dy Liacco advocated for security control that integrated automatic and manual controls and maintained continuous power service under all operating conditions. In 1974, he delineated the turning point in system control:

> Over the years the improvements in generation control and in supervisory control had been in the hardware used and in the efficiency of the control techniques. The basic objectives had remained the same, i.e., to control generation and to control remotely located devices or equipment. Near the end of the 1960's, however, power-system engineers began analyzing the entire system operation problem from a systems-viewpoint, motivated by the evident need for a more comprehensive and more effective operating control than had been conventionally available to the power-system dispatcher. We are now witnessing in the decade of the 1970's the beginnings of a new wave of power-system control systems, much broader in scope of system monitoring and control due to the integration of operating functions, and the addition of a new dimension—"system security."[61]

Dy Liacco's papers, along with the incremental work of numerous other engineers, like those at the Central Electricity Generating Board in England, formed the basis for the next generation of automated control of power systems. To further the reliability of the nation's power system, engineers focused on comprehensive and coordinated planning, assessment, and operations with increasingly heavy reliance on new computer technologies to get the job done. The direction of this work was clearly and specifically influenced by the 1965 Northeast blackout, which had focused industry attention on both operations and reliability of power systems.[62]

As the experts moved toward a more comprehensive approach to reliability, they also invoked greater stratification when describing power networks and control systems. In the 1967 report *Prevention of Power Failures*, the advisory group named two stability states for power networks: "steady state" and "transient state," with "dynamic state" as a subcategory of the latter.[63] The report explained that system stability reflected the ability of a system to return to normal operations following a disturbance, such as a sudden change in demand or generation or a short circuit on a transmission line. For any given system at any given time, a set of analyzed disturbances reflected the contingencies under which the system operated. Thus the contingencies determined whether or not a power system was in a steady, transient, or dynamic state. Yet the report writers were careful to note, "There is no precise delineation between the steady state, transient state, and dynamic state of operation."[64]

Dy Liacco likewise described the state of power systems. In his 1967 paper, he named three operating states that were similar to those identified in the advisory group report: "preventive" (or "normal"), "emergency," and "restorative."[65] These terms were more reflective of the operating response demanded from the person controlling the network than were the terms used by the advisory group. In the preventive (normal) state, a power system operated securely. The system operator successfully ensured that power reached customers on demand and at the standard frequency with desired voltages. The objective of the operator was to continue indefinitely in the normal state without interruption and at minimum cost. In the emergency state, any number of conditions could threaten the normal operations of the system. For example, the emergency ratings of equipment might be exceeded, the voltage on the system might be unsafe, the frequency might be dropping and motors stalling, or two interconnected networks might be falling out of synchrony with each other. When a power system entered the emergency state, the operator's objective was to relieve the cause of distress and prevent further degradation of the system while meeting customer demand. Economic considerations were secondary. In the restorative state, service to customers had been lost, and the operator's objective was to safely return the system to full service as quickly as possible. Dy Liacco then explained that operating the system was a matter of time frame—"'before, during, and after' a system emergency."[66]

Dy Liacco also redefined the terminology of control in order to more successfully integrate operating strategies. Power system operators deployed multiple control techniques in response to any given system state. The experts long understood that control took place on multiple levels within a power network. For example, governors controlled generator speed at the most local level—physically connected to the generators themselves—while human system operators controlled which generators carried how much load and when, often at a location central to the larger system. Dy Liacco outlined three levels of control—direct, optimizing, and adaptive—and explained, "The controls during each of the three operating states should all be designed to accomplish the overall objective" of keeping the system in the secure normal state.[67] Direct control took place locally and resulted in high-speed decisions based on logic. Following Dy Liacco's rubric, speed governors exerted direct automatic control; they were distributed locally, responded to changes in generator speed (frequency), and immediately acted to adjust turbine valves to speed up or slow down. Optimizing control took place centrally and employed computation and modeling of a system in order to produce an optimal solution. Computers running economy loading calculations, for example, were optimizing machines; they computed data about generator efficiency and transmission line losses and provided a scheme for distributing the load among the generators to achieve the greatest possible energy efficiency. Computers calculating economy loading generally appeared in control centers, not at generating stations. Adaptive control took place centrally as well and involved monitoring and adjustment of the settings of the devices exerting optimizing and direct control. At this level of control, human operators assessed computer-generated information about system performance and, preferably with the aid of computers, adjusted the other devices as needed. Adaptive control took place at the human-machine interface. A system operator might be employing any combination of the three types of control to manage a power system in any of the three system states.

The Pennsylvania Power & Light Company (PP&L) was an early adopter of Dy Liacco's scheme. In 1974, Pennsylvania Power & Light integrated online and offline computer analyses of the company's power network. A new Power Control Center provided centralized control of a system serving ten thousand square miles in northeastern Pennsylvania as well as two-way data exchange with the Pennsylvania–New Jersey–Maryland Interconnection.[68] The utility installed a computer scheme that predicted and displayed the effects of an unexpected change on the system—for example, a dangerously high megawatt overload on any 69 kV or greater transmission line or a possible single failure of any other transmission line, transformer, or generator. This was known as single contingency overload analysis. An offline computer ran calculations of how each possible contingency might affect the current system, initially assumed to be in a normal state. The offline computer produced output information (a table of numbers) that indicated how each generator or transmission circuit outage might result in redistributed power flow on the other transmission circuits. The table of numbers was saved in an output file and then used as input data for the online computer. The online computer, using the table of numbers from the offline computer and online telemetered data, then calculated the predicted power transmission overloads, if any, and displayed the results every thirty minutes for the system operators. Thus if the calculation showed that the transmission system was heading for a problem after any given single contingency (in other words, if the system was moving from a normal state to an emergency state), the operators had time to exercise adaptive control by making appropriate adjustments and hopefully averting a problem. This was an example of Dy Liacco's hierarchies of states and controls at work. According to James Robinson, who at that time was a PP&L operations coordination engineer assigned to this project, "For PP&L, this was the first 'real time' pre-contingency blackout prevention technique."[69]

Before the installation of the new computer scheme, the control room real-time data at PP&L "was limited to the current state of the power system with no ability to examine possible contingencies."[70] Robinson recalled, "Prior to that time during real time daily operation there was no way to anticipate the effect of losing a 230 kV line in the immediate future. Offline studies in the 1950s and early 1960s were done on Network Analyzers. In the early 1960s and 1970s offline studies were conducted on digital mainframe business computers, mostly by the planning department getting ready to add new facilities."[71] If a utility planned to add a new generating facility to the network, for example, this might entail shutting down a connecting transmission line for a period of time during construction of the new plant. The individual operating the system in real time would have no idea what might happen if a second transmission line suddenly failed during construction or maintenance outages. This changed with new computing software and hardware made available in the 1970s.

According to Robinson, the integration of online and offline computing tools was extremely valuable in the 1970s: "The new tool was initially given to operations planning technical engineers (Operations Coordinators) . . . in the back offices of the control room. The 'front desk' real-time operators needed this tool immediately," but everyone knew both the real-time data and the offline data "needed to be 'scrubbed' for errors." In other words, the back room operations planning engineers spent several weeks ensuring that both telemetered data and human-entered data were accurate and that "the new tool was ready for 'front desk' real-time operations." Robinson remembered that the technical support staff "had a new surprise every day of a predicted emergency overload." He elaborated, "Often enough it was a revelation to us how many contingency analysis overloads were out there that we were blind to before while facilities were out of service for construction or maintenance outages." Just as importantly, the integration of online and offline computing tools led to greater interaction among planning department engineers, operations planning engineers, and real-time operators: "Until that time, operations planning engineers were rarely present in the control room. Most of their work was done in a back office," and "coffee break time was the majority of the occasions when they chatted with 'front desk' real-time operators."[72]

The PP&L case exemplified the focus of power system experts on improved operating reliability and security. In addition to using new tools that brought data analysis closer to real-time decision making, the PP&L operations planning engineers and operators bridged the divide between the control room and the "back room." The conceptualization of the system security problem in terms of hierarchies of system state and of system control worked in practice. Echoing past experiences, however, new tools that brought about tighter control of the flow of electricity also revealed previously undiscovered potential overloads on the power lines. As Robinson noted, the experts had been blind to the multitude of minor trips and potential emergencies that could take place on any given day.

Research, Another Blackout, and Digital Computing

Into the 1970s, the 1965 Northeast blackout continued to influence research programs in both the private and public sectors. Earlier, in April 1965, the Edison Electric Institute had responded to pressure from the Federal Power Commission by establishing the Electric Research Council to investigate and research technical problems.[73] Indicating that the work of the Electric Research Council fell short of the country's needs, in 1971, the

Senate considered establishing a federal research program financed by the utilities. The industry, acting through the Electric Research Council, formed the Electric Power Research Institute (EPRI) with funding from voluntary member utilities. Organized as a not-for-profit corporation, EPRI became an important research arm for the industry. While historian Richard Hirsh notes that EPRI's contributions to technical innovation were minimal through the 1970s, the institution eventually engaged in research across the spectrum of power systems concerns from generation to environmental protection.[74] For power system experts concerned with interconnected systems, EPRI projects offered another avenue for sharing information and vetting power control techniques.

In 1974, the US Department of the Interior initiated a research program to address power system security, control, and development. The Energy Research and Development Administration, created later that year, assumed responsibility for the program. The lead researchers, Lester Fink and Kjell Carlsen, expressed concern that "the rapidly growing complexity and magnitude of systems problems in electric power were rapidly outstripping the capabilities of existing tools of analysis and synthesis."[75] With the aid of more than two dozen engineers working across the industry, the researchers published a report on the state of the art of system control in 1975. Among numerous topics, the collection of papers that comprised the report updated notions of automatic generation control, security assessment, system security, stability, system state, and central control of power systems. Several referenced 1965 as a turning point in addressing system security.[76] It is noteworthy that a few discussions of dispatch, load control, economy loading, and computer-aided assessment and control also addressed environmental concerns.[77]

Among other findings in the 1975 report, researchers noted that the practical application of new approaches to power system reliability lagged behind the theoretical. In "Security Assessment of Power Systems," A. S. Debs of the Georgia Institute of Technology and A. R. Benson of the Bonneville Power Administration traced the history of security concerns.[78] They acknowledged that system planners had long incorporated security considerations into their work, in the sense that they provided margins of error when designing generators, transmission lines, and interconnection ties. Debs and Benson identified the work of Charles P. Steinmetz, published in 1920, as the earliest to investigate the effect of a disturbance on the overall security of a power system.[79] Debs and Benson then specifically mentioned the 1965 blackout as the moment when the industry understood that security assessment had to extend from planning to operations. It was Dy Liacco's work, they explained, that established the "new philosophy of secure operation."[80] Debs and Benson went on to discuss different approaches to security assessment, describing the available technologies and operating practices. They surveyed sixty utilities in the United States and abroad to understand security assessment activities then under way. The results were mixed—most conducted offline studies, but those who attempted to implement online security assessments met with a variety of challenges, from inadequate data to inaccurate results. An appendix to this paper listed 343 references, all but twelve of which were published after 1965. Thus the theoretical art of security assessment had indeed advanced dramatically over the prior ten years, but the practical had yet to catch up.

System operators hoping to install new system control instruments faced significant financial hurdles. Comprehensive control apparatus cost as much as $25 million in the 1970s.[81] During that time, utilities were notoriously strapped financially.[82] Company managers looked for cost-benefit analyses to justify new capital expenditures, but costs and benefits were difficult to compare for system security equipment. If the equipment worked well, there were indeed economy gains to be made. But the more significant benefits came from improved reliability—in other words, preventing unwanted events from happening—and this was not a quantifiable result. Operators were hard-pressed to persuade managers to invest in these technologies.

Power system experts marked the tenth anniversary of the Northeast blackout with a mixed analysis of progress made. At the January 1976 meeting of the IEEE Working Group on Dynamic System Performance, engineers disputed the state of the control art.[83] The engineers at this meeting lauded the creation of the National Electric Reliability Council (NERC) in 1968, through which twelve regional reliability councils, later consolidated into nine, formed and adopted uniform criteria for system planning and design and operation. Several noted, however, that the definition of reliability was inconsistent from one region to the next, and as one explained, the three terms used to define a major failure—*widespread, cascading,* and *uncontrolled*—were not only vague but also impossible to quantify. The use of differing reliability criteria led some areas to strive for "unrealistic and economically unjustified levels of reliability."[84] The working group pushed for greater cohesion across the industry. The group organized follow-on symposia at which engineers from across the industry could discuss reliability criteria.

The shortcomings in practical application of system security philosophy became evident on July 13, 1977, when the New York area suffered a second major blackout. This outage took place entirely within the Consolidated Edison system. The causes of the blackout brought attention to additional complexities of interconnected power systems.[85] Two lightning strikes triggered the initial disturbances. Protective equipment then failed to operate correctly, thus forcing a major generator and several transmission lines out of service. From that point forward, additional equipment malfunctions and human operator decisions brought about another massive cascading outage. As a result of the investigations that followed, FERC underscored operating inadequacies and suggested that the blackout could have been prevented. FERC's principal recommendations for improving system reliability focused on system control, operating procedures, transmission protection systems, and system restoration plans.

The 1977 New York blackout was the type of event that gave system operators ammunition to persuade utility managers to install more advanced computing and control apparatus. One IEEE engineer noted that circumstances, including economics, had "mitigated against the rapid and widespread application of the computer to systems control," especially after more than a decade of relatively reliable operations (from 1965 to 1977).[86] In the mid-1970s, a steadily increasing number of power companies had adopted advanced system controls. At the time of the 1977 blackout, thirty-seven control centers in North America included automatic generation control integrated with some combination of economy dispatch control, contingency evaluation, security monitoring, and other advanced control technologies.[87] Notably, Consolidated Edison was not on this list. By 1985, power companies, including Consolidated Edison, had installed 137 digital system control centers—nearly a fourfold increase in just eight years. Yet even in 1988, consulting engineer John Undrill noted that not all computer-based research and modeling activities had successfully transitioned to useful operating applications.[88] In Undrill's view at that time, dramatic improvement in computing capacity and new synergy between utility engineering and processing departments promised to bring even more widespread application of digital computer models to power system control.

The importance of new computing technologies to integration of power system control activities after 1965 cannot be overstated. As one engineer reflected in 1994, "Then, in November 1965, the great Northeast Blackout set the stage for perhaps the most interesting development period to date in the application of computers to power system problems."[89] From economy loading calculations to economy loading control, for example, computers provided fast and accurate solutions to increasingly complex problems, and computers swiftly activated automatic responses. Whereas the 1950s "Early

Bird" analog computer integrated the calculation of generator costs and transmission line losses to optimize economy loading, by the 1970s, a dispatch computer might calculate and control automatic loading on the basis of not only economy but also security and environmental considerations.[90] More broadly, advanced system control computers estimated the effects of multiple influences on power system stability to aid in both planning decisions and, with a few hours of lead time, operating decisions as well.[91] Larger and more powerful machines brought offline analysis and real-time control closer together.[92] In addition, new computers included enhanced displays that improved the human-machine interface and facilitated rapid response by operators to changing system conditions.[93] Internationally, computers enabled utilities to address system security as well as more traditional operating concerns.[94]

While capacity, speed, price, and display represented significant advances in the physical components of computers used by power companies, innovations in programming mattered deeply as well. In the early 1960s, power companies had used the newest technologies to "solve" power networks—in other words, to model and predict power flow and voltage under various scenarios. While network analyzers had solved these load and flow problems in the past, digital computers promised to get the job done more quickly and accurately, with a greater complexity of parameters included. Yet the calculations were too large and took too long for computers in the 1960s to complete in any useful manner. Engineers employed several options for programming a digital computer to analyze the load and flow conditions on a power network.[95] In general, the computer ran through a series of equations using historical data to predict load and flow outcomes. One group of methods, termed *direct* (not to be confused with "direct" power system control), provided useful results with very few repeated iterations on the computer but used up inordinate computer memory and processing time if the power network—and hence the amount of data—was large.[96] The other group of methods, termed *iterative*, did not use a lot of memory, but the number of iterations of the equations increased rapidly for larger and more complex power systems. In other words, both sets of methods became too unwieldy for digital computer processing if the power system under study was large and complex.

In the early 1960s, engineers at the Bonneville Power Administration provided the first step forward in making digital machines useful for solving large and complex power networks. William Tinney and Clifford Hart experimented with a variation on the direct method developed by engineers at Northwestern University and General Motors.[97] Tinney and Hart were able to reduce both the processing time and the related memory problem that had overburdened digital computers. In 1967, they shared both their mathematical method and their programming approach with the wider engineering community, and this work was considered a major breakthrough.[98] Yet the processing still took a very long time. Tinney reflected in later years, "It was a terrible method, but I didn't know that I'd later discover the right way to do it; I was missing it too."[99] He eventually realized that the program solved the equations out of order and would work significantly more rapidly by correcting this problem. In 1968, Tinney and another colleague, Hermann W. Dommel, published another landmark article detailing an optimal power flow solution that allowed digital computers to solve the load and voltage problems even more rapidly.[100] Eventually power companies across the industry deployed digital computers using the Tinney method.

The technical literature and activities of professional organizations reflected the rising importance of digital computers. While ten to twenty technical papers about computers and power systems appeared each year during and shortly after World War II, they numbered in the hundreds in the 1950s and exceeded one thousand each year by the mid-1970s.[101] In 1958, the IEEE launched a biannual conference on the Power Industry and Computer Applications, held in North America.[102] At the first meeting, engineers focused on the role of the relatively new digital computing devices in the power industry, not the more familiar analog computing, controlling, and modeling apparatus. In 1962, a group of academics, engineers, and system operators, based mostly in Europe, followed suit by holding triennial international conferences on power systems computation.[103] The first Power Systems Computation Conference took place in London in 1963. In 1974, the IEEE published a special issue of its proceedings dedicated narrowly to the use of computers for the analysis, design, planning, operation, and control of power systems.[104] Just as engineers conceptually integrated control problems, new computers actually integrated offline and online functionality. This marked a new era of power system control.

The Integrated Unified Grid

The move toward integration of power system control matched the attempt to physically interlink nearly all the power companies in North America into a single network. Between 1960 and 1967, the number of major power networks dropped from six to two, as listed in table 8.1. Following the

Table 8.1
Number of major interconnected systems in North America, 1960 to present*

Year	Number	Systems
1960	6	CANUSE ISG Northwest Power Pool Pacific Southwest Power Pool Texas Interconnection PJM
1964	5	ISG Northwest Power Pool Pacific Southwest Power Pool Texas Interconnection Rio Grande–New Mexico Pool
1966	4	ISG Northwest Power Pool Pacific Southwest Power Pool Texas Interconnection
1967 (preclosure)	3	ISG Western Interconnected System Texas Interconnection
1967 (postclosure)	2	East-West System Texas Interconnection
1975	3	Eastern Interconnection Western Interconnection Texas Interconnection
2006	4	Eastern Interconnection Western Interconnection Texas Interconnection Quebec

* NERC recognized the Quebec transmission system as a full interconnection in 2006.

closure of interties between the eastern and western interconnections in 1967, power system experts depicted just two grids serving the country: one network that integrated 94 percent of the systems and another that encompassed most of Texas.[105] But this representation was misleading. The alternating current links between the east and west failed regularly, and by 1975, the utilities abandoned them. Over the next two decades, companies replaced the four AC ties with six DC ties, allowing the two giant eastern and western systems to operate out of synchrony with each other. On the one hand, the notion of a single system serving the entire continent endured in descriptions of electrification; on the other, experts carefully returned to a three-grid representation of the US power system—four grids when they included Canada.

Government reports documented these consolidations. The 1964 National Power Survey explained, "Today, 97 percent of the industry's generating capacity is to a greater or lesser degree interconnected in five large networks."[106] The 1970 National Power Survey offered, "With the exception of the Texas Interconnected Systems Group, virtually all of the large utilities in the United States operate synchronously. . . . In recent years the eastern two-thirds of the contiguous United States was interconnected with the western interconnected system through what have become known as the 'east-west ties' in the Rocky Mountain Area."[107] In other words, the country's power companies operated in two interconnected systems. Notably, the survey went on to state, "These ties have very limited capacity and are vulnerable to frequent tripping on power swings."[108] Most of the nation's power system appeared as a single grid with physical but tenuous links between east and west.

The engineers who worked on controlling the movement of power on interconnected systems highlighted the aggregation of east and west, although the achievement was short-lived. In Nathan Cohn's presentations for both technical and nontechnical audiences, the maps of interconnections are telling. In publications before the east-west closure, he used maps similar to the one in figure 8.6. This illustration reflects the four major linkages operating through 1966, with the small Southwestern Public Service System shown in isolation. In his remarks, Cohn described this as five separate systems.[109]

After the east-west closure, Cohn shifted his focus to the very large system that reached from coast to coast. In his first publication following the closure, he proudly noted, "At the time of printing this Handbook, the ISG-PJM-CANUSE interconnected system, the Northwest-Rocky Mountain interconnected system, and the Pacific Southwest interconnected system have all been tied together for test periods, operating successfully as a single nationwide interconnected system, encompassing within it about 94% of the country's generating capacity."[110] For the next few years, Cohn used a map like the one in figure 8.7 to illustrate the reach of this network.

By 1978, the system had broken apart into three networks, and Cohn's publications reflected this change. In the revised edition of the *Standard Handbook for Electrical Engineers*, for example, Cohn presented a map showing three large systems, reproduced in figure 8.8, and this is how the systems are represented today. Without referring to the closure experiment, nor the instability of the east-west ties, Cohn stated, "There are . . . three major synchronized interconnected areas in the United States and Canada."[111]

Despite this careful acknowledgement of the functional relationships among power systems across the continent, engineers continued to refer to a single power system. For example, at a major 1977 conference assessing

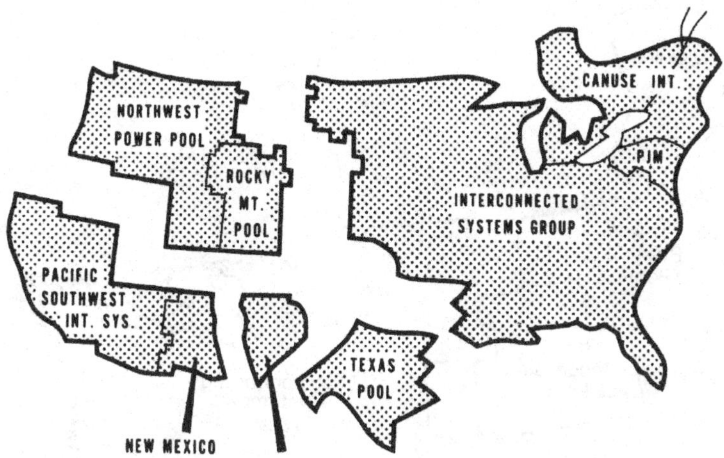

Figure 8.6
Map showing the five major interconnected systems in the United States, 1966.

Source: Nathan Cohn, "Automatic Control of Power Systems" (paper presented to IEEE International Convention, Part 12—Large Interconnected Power Systems, New York, New York, 1966), 1.

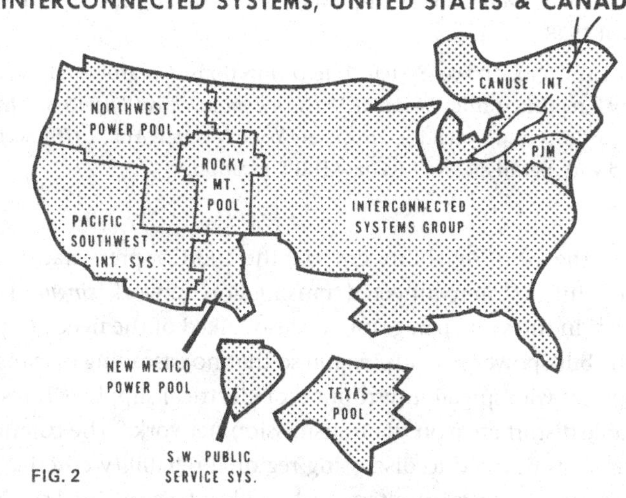

Figure 8.7
Map showing interconnected systems in the United States as of 1971.

Source: Nathan Cohn, "Area Generation Control—a Basic Factor in System Security" (essential text of talk presented at the Minnesota Power Systems Conference, October 12, 1971), i.

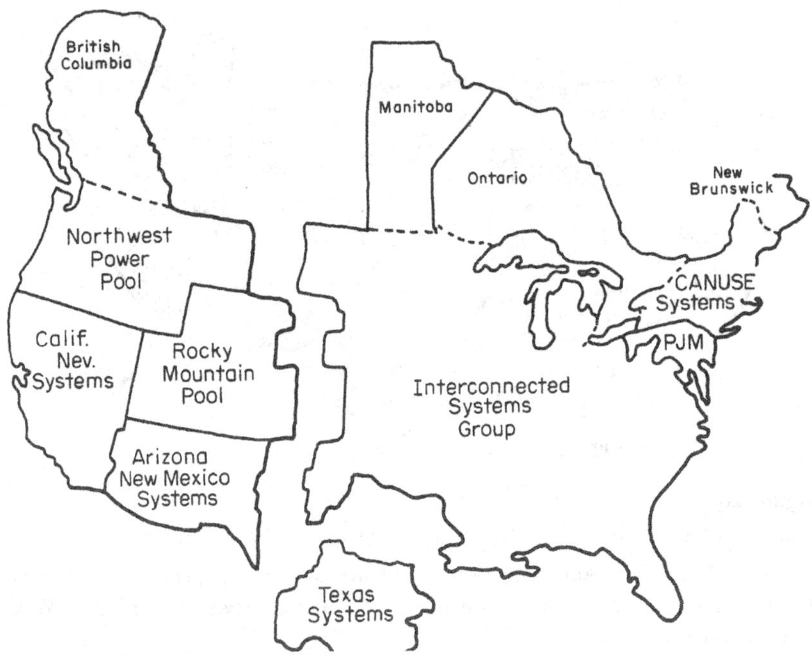

Figure 8.8
Map of three major interconnected power systems serving the United States and Canada as of 1978.

Source: Nathan Cohn, "Power-System Interconnections: Control of Generation and Power Flow," reprinted from *Standard Handbook for Electrical Engineers, 11th Edition*, ed. H. Wayne Beaty and Donald G. Fink (New York: McGraw-Hill, 1978), section 16-6. Reproduced with permission of McGraw-Hill Education.

the state of the art of reliability criteria, the NERC representative spoke of "the present highly interconnected transmission network [*singular*] in North America."[112] In the same paragraph, he also talked of the need "to plan and operate the bulk power systems [*plural*] so that no cascading of transmission line outages or widespread interruption of electrical supply will result from a predictable disturbance on the transmission network." The balance of the symposium was devoted to discussing regional reliability criteria. The participating system experts simultaneously spoke of a combined system and a collection of separate systems with geographical, technical, and economic differences in the considerations affecting stability.

In popular usage, a single grid still serves the power needs of Americans. The term "America's power grid" appears frequently and describes the collection of generating plants, transmission lines, and often distribution networks that carry electricity to consumers.[113] Significantly, this reflects the notion that all the nation's (and sometimes continent's) many public and private power-producing and power-delivering entities are connected and perhaps operate in unison. This is a conceptual artifact from the 1967 east-west closure. As the power experts had discovered, the four relatively small AC ties were inadequate, despite technical innovations, to synchronize the very large eastern and western interconnections. In reality, there are four interconnections, or grids, linked with HVDC ties stretching across most of North America.

The Return of Direct Current

While alternating current served as the gold standard for building long-distance and high-voltage transmission networks through most of the twentieth century, innovations in direct current transmission technologies drew the attention of utilities for particular applications. For especially long distances and for connections between extremely large networks, DC transmission lines provided stable links that allowed the world's largest interconnected machine to maintain steady frequency within each network, without requiring synchronization from coast to coast. This proved essential to reestablishing functional links between the eastern and western interconnected systems after power companies opened the AC ties. The old was new again.

The AC interties that linked east and west in 1967 produced frequent outages, particularly on the western side of the grid. In the first months after the February closure, utilities on the western system experienced numerous disturbances and outages. Slightly different underlying frequency oscillations of the two systems contributed significantly to the problem. As a result, each network maintained a steady frequency of 60 Hz, but they tended to move out of phase with each other. In effect, this was similar to two large balls spinning at the same speed side by side but slightly out of step with each other. If the balls were connected by a slender rubber band, the link between them would experience stress due to the different behavior of each ball. In the case of the east-west interconnections, the ties, like the imagined rubber band, were much too small to support the oscillations between the two systems.

The East-West Intertie Closure Task Force anticipated this problem while planning the closure. In advance of February 7, 1967, the task force installed an experimental relay at the Yellowtail-Custer tie line in

Montana to cause the two systems to separate when the oscillations grew too large.[114] But this relay did not work as hoped. The test apparatus, as well as commercially produced apparatus at the other three tie locations, tended to try to maintain the interconnection and regulate the phase and frequency between the two networks when each system would have operated more smoothly had they been allowed to separate. The Bureau of Reclamation's Denver-based power system operator, Frank Lachicotte, opened the ties in July to alleviate these problems and reclosed them in December 1967. A redesigned relay, installed for the December reclosing of the ties, proved more effective. The refined device had a very high sensitivity to changing oscillations and small or gradually increasing amplitude, which had been troublesome on the western interconnection. It was designed to sense these changes and cause the tie to open when the interchange was approaching difficulty. In addition, when the relay at Yellowtail-Custer tripped, it transferred signals to the other three ties to trip as well. In the first eight months after the ties were reclosed, the relays produced 381 separations between the eastern and western interconnections.

An innovation in relays improved the operability of the "national" grid, but challenges continued. At the time, system engineers concluded that this new type of relay made interconnection between east and west possible during "otherwise unsatisfactory conditions."[115] This allowed for "interchange of valuable energy that would have been otherwise wasted or of decreased value."[116] Yet the inherent instability between the two large networks caused continued disconnections over the next several years. Of the four AC ties, a NERC official reflected in 1985, "They weren't strong enough to hold the two masses of energy together. As loads changed, the power would swing, and the lines would trip out, separating the networks. It got to the point where it was happening as many as a dozen times a day."[117] The power companies opened the ties in 1975 and did not anticipate linking the two networks with AC ties again in the near future. As of 1977, the western and Texas systems operated apart from the eastern system "with no major interregional interconnections planned."[118]

Around the world, engineers worked to address the difficulty of maintaining steady frequency and synchrony between very large interconnected systems. At the time of the east-west closure, all the North American transmission networks relied on AC lines. Cost-effective DC lines, however, offered benefits for high-voltage long-distance transmission, especially the ability to execute straightforward control of the power flow. On a DC line, the electricity moves in only one direction, in the quantity and at the frequency determined by the operator. Just as importantly, at high voltages, less energy is lost in DC transmission than in AC transmission.

In the early decades of electrification, using the transformer design first introduced by Nicola Tesla, power companies changed the voltage between generators, AC transmission lines, and customers. But without a similar technology for direct current, DC transmission lines fell out of favor in the United States. In Europe, a few companies continued to employ direct current transmission well into the early twentieth century. Beginning in the 1920s, the Swedish company Allmanna Svenska Elektriska AB (ASEA) focused on designing converters and valves that allowed for the use of high voltages on transmission lines and lower voltages at the consumer end.[119] With steady advances through the 1930s and 1940s and experiments on AC and DC systems, the company introduced the first high-voltage long-distance DC transmission line in 1954. The DC connection between the Swedish mainland and Gotland demonstrated the efficacy of high-voltage direct current (HVDC) transmission and opened the door to future applications around the world.

For American companies building larger interconnected systems, this DC technology looked promising. Across very long distances, HVDC offered substantial cost savings. With DC-to-AC transformers at each end of a DC transmission line, operators had the ability to maintain a common frequency within each network without attempting to keep very large interconnected systems spinning in parallel. By 1970, ASEA, the Soviet Union, and English Electric had installed nine HVDC lines in Russia, Sweden, New Zealand, Italy, Canada, and under both the North Sea and the English Channel.[120] California was the site of the first installation in the United States, when one of four links forming the Northwest/Southwest Intertie went into service in 1970.

By 1976, one year after power companies opened the east-west ties for the foreseeable future, *IEEE Spectrum* published an article assessing the potential for additional use of HVDC in North America. The authors gave a "thumbnail review of 22 years of progress" and reported on all planned new installations in the United States.[121] At this point, the Northwest/Southwest Intertie was still the only HVDC line in the country. Four additional projects were under construction: one in western Nebraska to provide a stable link between the eastern and western interconnected systems; two specifically designed to carry electricity from mine-mouth generating plants to load centers; and one to serve as a research facility for the Electric Power Research Institute, the General Electric Company, and Consolidated Edison of New York. Three HVDC lines operated in Canada. The authors suggested that this application of HVDC technology would be beneficial for linking very large interconnected systems: "There is also the coming possibility of strong regional interconnections (east-west, north-south), some of which may be HVDC."[122]

As in many aspects of grid development, neither the investor-owned utilities nor the federal government necessarily led technical innovation. Tri-State Generation and Transmission Association, a power supplier for rural electric cooperatives in Colorado, Nebraska, New Mexico, and Wyoming, installed the first HVDC intertie between the eastern and western systems. In 1974, the association board approved a turnkey contract with General Electric Corporation to construct the line.[123] The David A. Hamil tie started up in March 1977, for the first time allowing operators to link the two networks and maintain stable frequency but ignore synchrony. Tri-State was the first utility to supply electricity to customers in both grids across a single tie.[124] This HVDC tie in Stegall, Nebraska, served as a bridge between the two large systems. By 1987, utilities effectively replaced the earlier AC connections by installing additional HVDC ties in Eddy County, New Mexico; Miles City, Montana; and Sidney, Nebraska.[125] With these ties, the east and west were connected but did not operate in parallel.

During the late nineteenth century "battle of the currents," a key innovation—the rotary converter—allowed the integration of AC and DC elements into a single network. In the 1890s, this enabled utilities to operate AC and DC system elements together and proved instrumental in the flourishing of AC systems around the world.[126] Over the next eighty years, while piecing together the grid, power companies employed AC technology to link across the country. Direct current technology proved critical to allowing power companies to build even larger networks in the latter decades of the twentieth century. The HVDC transmission technology developed within existing technical systems, allowed networks to grow beyond previously delimited boundaries, and facilitated cooperation among multiple types of power companies, including investor-owned utilities, rural cooperatives, and federal agencies.

The relatively mature grid of the 1980s reflected a century of aggregation of technologies, practices, physical infrastructure, and popular conceptions. It had become the backbone of a complex and opaque network of public and private companies, governments, nonprofit organizations, and consumers. Despite the dramatic changes ahead for the industry, and even despite evidence of significant vulnerabilities in interconnected systems, the grid persisted as the platform for ensuring instant electricity at the flick of a switch.

9 Deregulation and Disaggregation

A Brief Overview, 1980–2015

Between 1980 and the first decades of the twenty-first century, the power industry experienced significant change. Environmental concerns surrounding electrification expanded to include heightened worries about accidents at nuclear power plants, evidence of more subtle effects of generating station pollution, and large-scale apprehensions about climate change and global warming. The trend toward deregulation at both the state and federal levels undermined the dominance of regulated monopolies in some regions of the country. Companies outside the traditional power industry introduced to the market new generating technologies, including cost-effective solar and wind generation. Some states dramatically revised rate-making approaches. As digital technologies grew both more powerful and more affordable, grids became "smarter." Entities including industrial manufacturers, defense agencies, and small communities touted "microgrids"—discrete, localized groups of generators and loads that can operate apart from the main transmission networks—as a safer and more efficient approach to electrification. Amid this fairly dramatic reorganization at all levels, the industry continued to count on the grid as the backbone of the system and to rely on system operators to keep the power moving.

While the grid infrastructure was a constant throughout this period of transition, the relationships among power generators, transmission line owners, and electricity distributors changed dramatically.[1] As this book documents, until the 1980s, a combination of regulated investor-owned utilities, municipal power companies, rural cooperatives, and federal agencies generated, transmitted, and delivered power to customers. Together these entities ensured that the right amount of power, at the right frequency, arrived at the instant the consumer turned on the switch, and this was accomplished through a variety of informal and formal agreements, but without any central oversight. Organizational and regulatory restructuring began in 1978 with the passage of the Public Utility Regulatory Policies Act (PURPA) and continued into the twenty-first century. In response, the private sector reorganized in a number of ways that matched the goals of individual owners and investors. What had once been an industry dominated by large vertically integrated utilities that participated in a utility consensus with customers and government regulators now is different and perhaps even more diverse than before.

The regulatory changes unfolded over nearly three decades as industry leaders, nonutility power producers, politicians, and legislatures at every level of government renegotiated the utility consensus. Beginning with PURPA, Congress created an opportunity for nonutility power generators to sell electricity into the grid.[2] An industrial plant, for example, that used excess steam to generate electricity qualified under PURPA to sell electricity into the transmission network. During the 1980s, states increasingly permitted nonutility generators to contribute to the grid, most notably in California.[3] The 1992 Energy Policy Act gave the Federal Energy Regulatory Commission (FERC), the successor agency to the Federal Power Commission, the right to impose "wheeling authority."[4] With this authority, FERC was able to require a transmission line owner to provide transmission services to a nonutility power generator. Through the 1990s, FERC issued orders that made transmission networks increasingly accessible to any power generator and enabled states to establish competitive wholesale power markets. In Order No. 888, FERC required vertically integrated utilities to functionally unbundle generation and transmission activities in order to ensure fair access to transmission service.[5] Between 1995 and 1999, twenty-two states and the District of Columbia passed laws to deregulate the power industry and introduce retail choice for electricity consumers. By 2016, seventeen states and the District of Columbia maintained deregulated electricity markets in some degree. This overview of regulatory change suggests the dramatic difference between a century of electrification dominated by regulated investor-owned monopolies and a new era of multiple power system players.

The private sector responded to the changing regulatory climate in numerous ways.[6] Some vertically integrated investor-owned utilities divested generating plants, while others merged with each other or with gas utilities. New power marketers, independent generating companies, and autonomous transmission companies entered the market. Power companies formed Independent System Operator (ISO) entities, different in kind from power pools or informal interconnection groups, to oversee transmission network activities.[7] In general, the structure of ISOs intentionally fostered competitive wholesale markets in power generation and opened access to the physical transmission networks to all generators. Organizationally, the ISOs involved multiple types of entities, including regulated utilities, independent generators, and municipal power companies, depending on the region in which they operated. The ISOs also ensured that decision making took place without undue influence from any one stakeholder. Some ISOs, like Electric Reliability Council of Texas (ERCOT), operated wholesale markets as well as physical transmission networks. In Order 2000, released in 1999, FERC introduced Regional Transmission Organizations (RTOs) to manage segments of the nation's grid.[8] This order provided that an ISO that met a number of delineated principles—addressing financial conflict of interest, among other matters—could act as the RTO in a given area.[9] The current geography of RTOs and ISOs is depicted in figure 9.1. Importantly, Order 2000 resulted in the complete separation of grid operations from grid ownership for about 60 percent of the US power supply.[10]

Not until 2005 did the federal government have the formal ability and duty to ensure reliable grid operations. With the passage of the Energy Policy Act of 2005, Congress gave FERC jurisdiction to certify and oversee an electric reliability organization (ERO) that would establish and enforce reliability standards.[11] As early as 1997, both the North American Electric

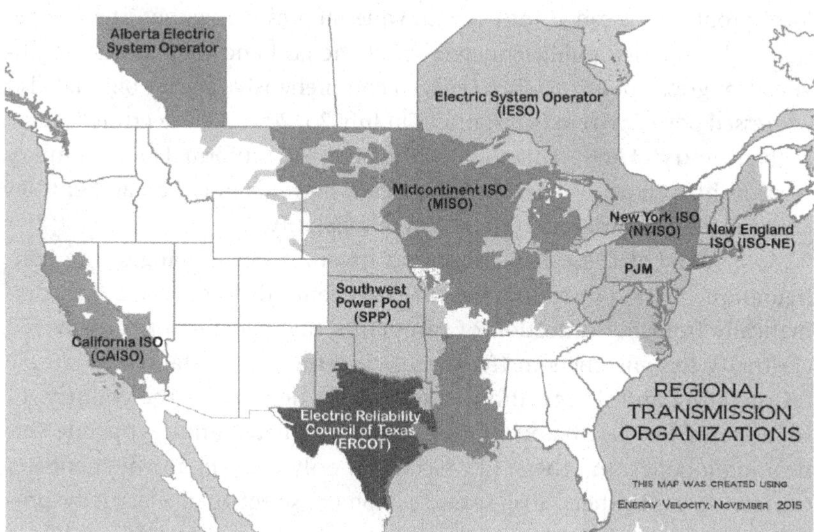

Figure 9.1
Regional Transmission Organizations and Independent System Operators as of November 2015.

Source: Federal Energy Regulatory Commission website (https://www.ferc.gov/).

Reliability Corporation (NERC) and the US Department of Energy recognized the need for mandatory grid reliability rules.[12] Former NERC executive David Nevius recalled, "As the electric industry began its transition from a regulated, vertically integrated structure to a competitive structure with functional unbundling and disaggregation, it became apparent that it was no longer clear who had the responsibilities for reliable planning and operation of the grid, who should pay for reliability, who should enforce reliability protocols, and what obligations were required of market participants to ensure that system reliability was not compromised."[13] NERC and the Department of Energy offered Congress legislative proposals beginning in 1999. Two catastrophic events affecting the grid, the 2000–2001 California electricity crisis and the 2003 Northeast power failure, further influenced the decision of lawmakers to address reliability. The former occurred as a result of a perfect storm of weather conditions, natural gas prices, deregulated electricity pricing in California, and market manipulation by Enron and others.[14] In short, the effect on Californians was dramatic—high electricity costs, rolling blackouts, and fear that the state's power system might be fiscally and morally bankrupt. On August 14, 2003, the northeastern states and part of Canada experienced a third major cascading power failure. This event echoed the region's prior blackouts, in which local problems cascaded into system-wide failures.[15] Together, these experiences heightened public concern about the grid and no doubt strengthened Congress's political will to enact a comprehensive energy bill that also addressed power system oversight.[16] On July 20, 2006, FERC certified NERC as the country's ERO.[17] This codified the heretofore informal and voluntary relationships among government agencies, power companies, and entities created by the utilities to oversee system reliability.

As a result of these regulatory and organizational changes, the configuration of the power industry in the twenty-first century differs dramatically from the collection of companies that generated and delivered electricity to consumers in the twentieth century. The status of industry restructuring and deregulation varies a great deal across the country. In some states, like North and South Carolina, regulated utilities operate vertical monopolies. In others, like Nebraska, only consumer-owned entities operate. In still others, like Texas, customers select their electricity provider in a competitive market, generators sell wholesale electricity into a network owned by transmission companies, and distribution companies deliver the electricity to the customers through a monopoly distribution system. Notably, apart from power generating entities in the ERCOT region of Texas, companies trade power across state lines, and with companies in Canada and Mexico, regardless of the different regulatory regimes.[18] Power producers now include regulated investor-owned utilities, independent power generators, industrial cogenerators, federal agencies, rural cooperatives, municipal companies, and commercial and residential electricity customers themselves.

To implement reliability standards, NERC works through eight Regional Entities, delineated in figure 9.2, which monitor and enforce compliance. The Regional Entities provide reliability oversight for both the United States and Canada. Within, and in some cases crossing between, these segments of the continent, seventy-four "Balancing Authorities" operate segments of the North American transmission network. (Both the Regional Entities and the Balancing Authorities are depicted in figure 9.2.) Balancing Authorities match generation and load, and control frequency, within a designated geographic area and also manage power interchanges and frequency with intertied areas. In other words, the Balancing Authorities replaced the past century's power pool operators in order to manage a control area. The RTOs and ISOs monitor loads, operate transmission facilities, determine load distribution among generators, establish operating limits, and prepare and oversee emergency and contingency plans.

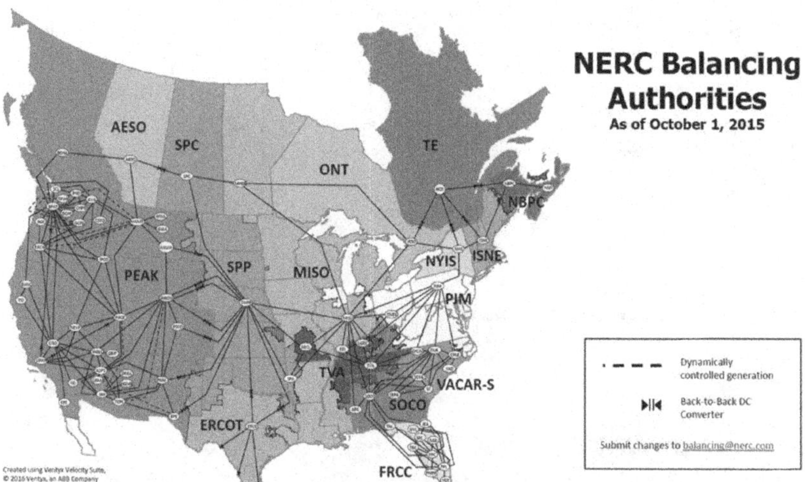

Figure 9.2
NERC Balancing Authorities as of October 1, 2015.

While perhaps confusing to the lay reader, the array of organizations moving electricity across the continent operates fairly seamlessly. By identifying reliability functions and defining the entity responsible for each function, NERC established a governance and operating structure that ensures that each physical asset of the grid has only one entity coordinating reliability, one Balancing Authority, and one transmission operator—although all three might be the same entity.[19] This functional model for ensuring grid reliability is depicted in figure 9.3. Rather than defining the various entities by their organizational names, NERC assigned functional names, thus creating both clarity and flexibility in determining who is responsible for what activity. A single company—like Duke Energy of Florida, for example—might fulfill ten different functional capacities of the functional model. In contrast, the city of Bartow, coincidentally also located in Florida, fulfills only one. (Bartow is the Distribution Provider for its service area.)[20] Beginning in 2006, NERC required all entities involved in ownership, operation, and use of the

Functional Model Diagram

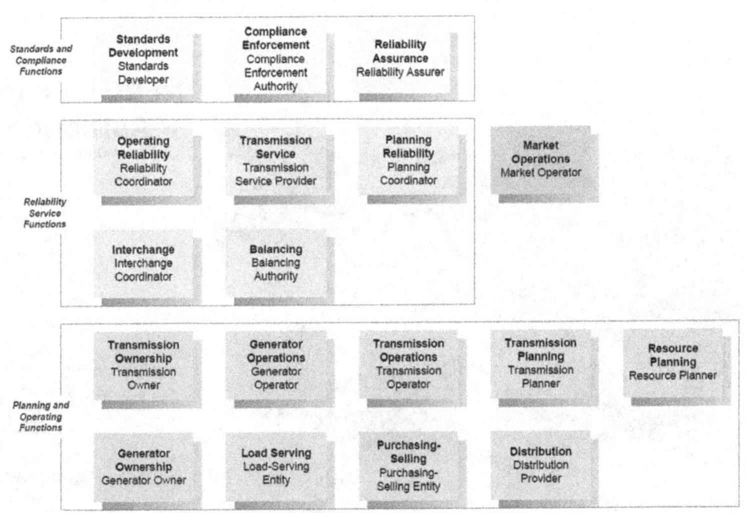

Figure 9.3
Functional Model Diagram.

country's bulk power transmission system to register, and additionally, NERC certified the Reliability Coordinators, Transmission Operators, and Balancing Authorities (functional entities identified in figure 9.3).[21] More than 1,400 entities, from large federal government agencies to tiny towns, play a role in ensuring grid reliability.

As this short overview illustrates, the power industry has reorganized substantially over the past four decades, yet a wide diversity of entities operates the interconnected power system. Congress formalized federal oversight of reliability on the transmission networks yet placed responsibility for policing operations squarely on the shoulders of industry stakeholders. Within a new hierarchy of quasi-governmental organizations, the system operators and the fraternity of experts to which they belong continue the project of balancing generation and demand on a moment-to-moment basis.

10 Conclusion

On a cold January day in 2016, a system operator enters the control room of the Electric Reliability Council of Texas (ERCOT). In so many ways, this moment is unlike the start of the day in control rooms across North America during most of the twentieth century. In the first place, this individual is a woman, and before 1980, very few women operated power networks. Second, this system operator has tools at her fingertips that her predecessors could only dream about. Her workstation, seen in figure 10.1, sports eight monitors that provide instant data about demand, generation, and a multitude of related metrics on the Texas network. Above her, dynamic data images are projected across a wall-sized screen, including a giant map of the state that illustrates the current status of every transmission line. By selecting a single line, she can zoom in on her monitor and see how much electricity is flowing, in which direction, at that moment, on that line. Third, within Texas, as in many other regions of the country, there is a wholesale market for power, and the ERCOT data screens indicate the price of electricity at every generating location. Through telemetry and very advanced digital computing, it takes only seconds for the computers to collect and display data for the operator as she monitors the network and ensures that the system is working as intended. Finally, this individual works for an Independent System Operator (ISO), not a power company, and her obligation is solely to the successful operation of the grid.[1]

While still a system of shared management and divided authority, the reliability structure within which ERCOT operates is significantly different from the world of the system operator before the 1990s. At one time, most system operators worked for the investor-owned companies that controlled power generation, transmission, and distribution within monopoly regions across most of North America. Change began in the wake of the 1965 blackout and continued through the years of deregulation and

Figure 10.1
ERCOT control room operator.
Courtesy of the Electric Reliability Council of Texas.

industry restructuring. Some aspects of operating the grid remain the same in the twenty-first century. Like many before her, the system operator at ERCOT says the first thing she does when she walks into work is to check the frequency on the grid.[2] If it is very, very close to 60 Hz, which is almost always the case, the network is stable. Keeping the network at 60 Hz continues to be one of several key interconnection parameters in the control room. The system operator has the 1950s Area Control Error equation (depicted in its original form in figure 6.6) in mind throughout the day. If she is maintaining an error close to zero, she is doing her job well. Many operators jokingly say their primary thought at work is "what's for lunch," but in fact, this reflects the nature of modern system control.[3] So much is automated that operators *monitor* generation and load distribution rather than *direct* it. On this morning, there is a minimum of tension in the control room; it is a sunny, clear, and quiet day in Texas. If there is an emergency, however, system operators will still rely on phone calls to discuss adjustments with individuals at the generating stations, just as they did in the early 1900s.

In other ways, the grid has advanced. While engineers like Philip Sporn, Robert Brandt, Nathan Cohn, and Charles Concordia could only dream of a control room like the one at ERCOT, their solutions of power system control challenges laid solid groundwork for high-tech grid management. Visualization tools, maps, trending displays, real-time analytics, and so forth inform an operator of problems that previously were too subtle to detect. For example, where new measuring instruments, called synchrophasors, have been installed and displayed, an operator sees quickly when oscillation is occurring in a particular region of the grid, and he or she can also drill down to discover which facility is causing the problem.[4] With a network of synchrophasors, an operator has the opportunity to detect and address small changes in system behavior very quickly and well before the small changes escalate into serious problems, promising even closer control of the grid.[5]

At ERCOT, and elsewhere, the system operates close to the edge. With advanced measuring and controlling technologies, operators can maximize efficiency to a degree not possible twenty or thirty years ago. The competitive wholesale markets for power, in part, have created the pressure to achieve these efficiencies. The modern grid also incorporates a wide variety of new generating technologies. Very large wind farms and solar installations, for example, have added a degree of unpredictability to the grid not present in the past. System operators have to be prepared to manage sudden unexpected losses or gains of power as the wind changes, and they have to manage the more predictable end of solar generation as the sun sets each day. In addition to these new large-scale contributions to power networks, grid operators juggle a new relationship with customers. When a homeowner installs solar panels, for example, he or she becomes both a power producer and a power customer. On this more diverse grid that is also a marketplace, there is a smaller margin for error but more opportunity to understand and control the system in real time.

A fragmented industry built this network infrastructure, the interconnected power system, over the course of a century and negotiated the technical and social terms of operation through the informal alliances of a fraternity of experts. The issue of control was central to the process. In 1899, the arrangement was simplicity itself. The San Gabriel Electric Power Company owned the generating plant, the transmission lines, and the distribution system that together converted falling water into electricity and delivered it to customers. To manage energy resources, hold down costs, build a bigger market, and maintain a steady supply of power, San Gabriel simply bought power from the Los Angeles Railroad when needed and sold power to the railroad when it was advantageous. The railroad controlled its own system and benefited from the deal in the same way. On direct current transmission lines, the investor-owned companies sent and received power without difficulty. No state or federal agencies intervened in any way.

Twenty years later, power companies sharing electricity on alternating current networks found control slightly more complex. In a typical central office, a load dispatcher (later called a system operator) attempted to keep up with varying customer demand, conditions on several subnetworks, a growing body of data, and the need for constant frequency across the system. The load dispatcher answered to his employer, usually an investor-owned utility, and did his work primarily with manual tools, pieces of paper, and simple communication devices (levers on the generators, charts and tables detailing past demand and system behavior, and a telephone). In more than half the US states and some Canadian provinces, regulators determined regions of monopoly operations and rates charged for service but did not concern themselves with reliability of interconnections. An international treaty governed trades across the Niagara River, but the US federal government had little else to do with linked power networks.

By the 1930s, both the physical and social demands of operating interconnected had increased. Investor-owned utilities shared power with federal agencies and rural cooperatives as well as each other, the newest technologies to maintain steady frequency upset the economic arrangements between companies, the networks spread across very large regions, and the system operators calculated and analyzed unprecedented quantities of data. In addition, new laws required some restructuring of the private sector industry, and at the same time, federal policies favored increased interconnection between companies. Engineers and operators experimented with control technologies that variously centralized and dispersed authority while addressing shared responsibility among autonomous network participants. During the Second World War, the historically independent power companies relinquished a degree of control for several years and cooperated with each other and federal agencies to meet the energy demands of the defense industries. The fraternity of experts tested and improved many cutting-edge control techniques, while the power company owners realized the advantages of extended interconnected transmission networks and operations at maximum capacity. By 1950, Nathan Cohn called simultaneous control of load and flow, within the parameters of technical capability, power pool arrangements, and economic requirements, "the nub of the problem."[6] The solution developed by the fraternity of experts during the 1950s—net interchange tie-line bias frequency control—remains an industry standard in the twenty-first century.

In the 1960s, 1970s, and 1980s, power companies, government agencies, and the public continued to negotiate technical and social control of the industry. Responding to blackouts, the industry redoubled efforts to technically strengthen the grid while Congress, the media, and the public questioned the role of the private sector in controlling the essential service of power delivery and questioned the grid itself as the supporting infrastructure. As hundreds of power companies of every size and description joined ever-larger power pools, engineers deployed increasingly sophisticated computing tools and organizational strategies to integrate control activities from the generating plant to the distribution network. During the decades of deregulation, federal and state laws recognized an ongoing natural monopoly status for transmission networks while establishing markets for wholesale power and for customer sales in various regions. The fraternity of experts continued to control the movement of power much as they had for the prior hundred years, but through increasingly formalized arrangements of shared management and divided authority. Interoperability among the newly segmented pieces of the industry proved crucial for both deregulation and restructuring.

Across much of North America, the newest control arrangements separate technical control of the grid from the financial deals among the now disaggregated power producers, transmission line owners, distribution companies, and customers. In many instances, the independent entities charged with operating interconnected power networks also maintain marketplace information for wholesale power trades and integrate cost data into decisions about how much power is generated by whom and where and when it is generated. As reflected in the NERC Functional Model Diagram (figure 9.3), the system operators maintain unity within a seemingly chaotic collection of moving parts and diverse monopolies and markets.

As this history illustrates, experts designed control technologies to reflect and support the social arrangements among power companies. But at the same time, the technical choices influenced the next generation of social arrangements. For example, in the early 1930s, system operators learned that new frequency control apparatus upset scheduled load distribution. In order to address this problem, operators within the continent's largest network, the Interconnected Systems Group, organized a Test Committee. This committee developed standards, albeit voluntary, for operating interconnected, and companies across North America tended to adopt the standards. In the late 1950s, the fraternity of experts acknowledged that coast-to-coast interconnections appeared inevitable and identified new technical control problems likely to result. The Test Committee took the lead in forming the North American Power Systems Interconnection Committee (NAPSIC), which included representatives from all potential national grid participants, to address control challenges on this largest of networks.

San Gabriel Power Company, the Test Committee, NAPSIC, and individual engineers like Nathan Cohn all engaged with the many tensions of electrification. The ERCOT system operator likewise balances similar tensions. In the twentieth century, interconnected power systems aided both conservation of natural resources and consumption of more energy. In the twenty-first, the grid facilitates replacement of fossil-fuel generating plants with wind and solar farms and also ensures that Americans have as much electricity as they want or need at the flip of a switch. Beginning in the 1890s, Americans wrestled with the nature of electricity and whether it should be sold as a commodity or a service. If a commodity, investor-owned companies rightly controlled the industry. If a service, government rightly ensured that citizens had equal access at a fair price. Each of the major technical milestones covered in this history—frequency and load control, the bias setting debate, economy loading, and system security analysis—hinged on the commodity/service balance. The ERCOT system operator today integrates commodity considerations, such as the cost of electricity at each generating site in the network, with service considerations while striving to achieve an Area Control Error of zero.

Whether the transmission grid strengthens power systems or represents a point of weakness is a continued debate as well. The grid offers redundancy, and the operators use improved technologies and techniques to keep the system whole. At the same time, those very technologies represent new points of possible breakdown, both because the technologies allow systems to operate with a smaller margin of error and because the networked information systems through which operators work present opportunities for outsiders to break in and wreak havoc. Tensions between public and private sectors, commodity and service, consumption and conservation, unity and autonomy, robustness and fragility have inhered in American power systems since the very beginning of electrification and will continue to play out going forward.

At the core, the grid represents the ability of many actors to develop and manipulate technologies for one common goal—that is, providing a steady supply of electricity. At the same time, the stakeholders serve a variety of other objectives, including profitability, environmental protection, service equity, expanded government authority, public accolades, and so on. It is the very diversity of participants, consistent over more than a century of technology development, that characterizes the grid. Thus while it may be described as a (single) large-scale technological system, or a (single) critical infrastructure, or the (single) backbone of the power industry, the variety of the grid's parts is the most salient feature of this network. Further, while individual companies and agencies intentionally built and expanded networks, most of the actors never envisioned creating the world's largest interconnected machine. In a different way, the consistent physical properties of electricity delimit what is possible in future power systems. New paradigms for generating, accessing, and using power are both plausible and likely. But to make a shift, Americans will have to contend with the many social, physical, technical, economic, environmental, and control elements that historically have played a role in building the grid.

In the United States, the power grid is unique when compared to transmission networks in other countries and when compared to other types of networks. The origins of the power industry in the for-profit sector, the evolution of electricity from luxury to necessity, the absence of a central controlling authority, the multiple roles of governments, and the importance of the fraternity of experts all frame America's interconnected power system. With other power networks, the American grid shares the physical characteristics of electricity itself. Despite the invisibility of electric power, it is central to both the quotidian and the extraordinary in American life. In the midst of a changing energy landscape, a glance back to the development of the grid is useful. Understanding what the grid is, why it exists, and how it evolved sheds light on technological development in American history. The reach of the grid illustrates the tight link between energy history and environmental history. Just as importantly, the grid will be central in future energy decisions in the United States.

Appendix

Abbreviated principles from FERC Order 888, Section IV.F.4. Bi-lateral Coordination Arrangements: ISO Principles, pp. 279–286.

1. The ISO's [Independent System Operator's] governance should be structured in a fair and non-discriminatory manner.
2. An ISO and its employees should have no financial interest in the economic performance of any power market participant. An ISO should adopt and enforce strict conflict of interest standards.
3. An ISO should provide open access to the transmission system and all services under its control at non-pancaked rates pursuant to a single, unbundled, grid-wide tariff that applies to all eligible users in a non-discriminatory manner.
4. An ISO should have the primary responsibility in ensuring short-term reliability of grid operations. Its role in this responsibility should be well-defined and comply with applicable standards set by NERC and the regional reliability council.
5. An ISO should have control over the operation of interconnected transmission facilities within its region.
6. An ISO should identify constraints on the system and be able to take operational actions to relieve those constraints within the trading rules established by the governing body. These rules should promote efficient trading.
7. The ISO should have appropriate incentives for efficient management and administration and should procure the services needed for such management and administration in an open competitive market.
8. An ISO's transmission and ancillary services pricing policies should promote the efficient use of and investment in generation, transmission, and consumption. An ISO or an RTG [regional transmission group] of which the ISO is a member should conduct such studies as may be necessary to identify operational problems or appropriate expansions.

9. An ISO should make transmission system information publicly available on a timely basis via an electronic information network consistent with the Commission's requirements.
10. An ISO should develop mechanisms to coordinate with neighboring control areas.
11. An ISO should establish an ADR [administrative dispute resolution] process to resolve disputes in the first instance.

Notes

1 Introduction

1. Hertz (Hz) is the term used to indicate the measurement of a unit of frequency equal to one cycle per second. In the case of electricity, the term is used with reference to alternating current to indicate the speed with which the electrons change direction as the electricity flows.

2. For earlier studies of the history of electrification, please see the selected bibliography.

3. "The Transmission Systems of the Great West," *Electrical World* 59, no. 22 (1912): 1142.

4. S. Ruttenberg, "Letter to Nathan Cohen [*Sic*]," April 22, 1927, Nathan Cohn Papers, MC 317, MIT Libraries Institute Archives and Special Collections (hereafter cited as Nathan Cohn Papers, MIT). Nathan Cohn (1907–1989) was an MIT-trained electrical engineer who spent his career at the Leeds & Northrup Company, starting as a salesperson in 1927 and retiring in 1972 as executive vice president. He consulted for the industry until 1988. Cohn received numerous awards and recognitions for his contributions to power systems control, including the Institute of Electrical and Electronics Engineers (IEEE) Lamme and Edison Medals, the Franklin Institute Wetherill Medal, and the Instrument Society of America Sperry Medal. He was a life fellow of the IEEE and the Franklin Institute and a member of the National Academy of Engineering. He was also the author's father.

5. In the early years of electrification, entities that generated, transmitted, and delivered electric power were typically called "companies," regardless of the nature of ownership. Thus a municipality, a rural cooperative, or a group of investors might own a power company. Throughout this manuscript, the term "power company" will likewise carry this general meaning. The term "utility" is also used to refer to multiple types of power companies, including investor-owned entities operating regulated monopolies, federal agencies generating and transmitting power, and municipal power companies. For purposes of clarity, the terms "investor-owned

utility" or "for-profit company" will indicate that the entity is privately owned and operated for profit. The adjectives "municipal," "cooperative," and "federal" will be used to indicate other types of ownership, which are also not for profit.

6. Throughout this book, "the grid" refers to the collection of interconnected systems in North America, except when specifically delineated as a particular network.

7. Thomas Parke Hughes, *Networks of Power: Electrification in Western Society, 1880–1930* (Baltimore: Johns Hopkins University Press, 1983), 353.

8. "Electricity Explained: How Electricity Is Delivered to Consumers," US Energy Information Administration website, accessed March 5, 2013, http://www.eia.gov/energyexplained/index.cfm?page=electricity_delivery. Depending on the voltage used, some entities state that there are more than two hundred thousand miles of high-voltage power lines comprising the grid.

9. "Understanding the Grid: *Reliability Terminology*," North American Electric Reliability Corporation website, accessed March 5, 2013, http://www.nerc.com/page.php?cid=1|15|122.

10. David Nevius, personal communication with the author, January 9, 2013. A graduate of Drexel University, Mr. Nevius began his professional career with Public Service Electric and Gas Company (later Public Service Enterprise Group, or PSE&G) in 1969 as part of the transmission expansion planning team. From 1977 to 2013, Mr. Nevius worked in various capacities for NERC and retired from the position of senior vice president in 2013. He continues to serve as an industry consultant. As NERC's representative to the US Department of Energy's Electric Reliability Panel, Mr. Nevius helped draft the legislative language that established the electric reliability organization (ERO) provisions of the Energy Policy Act of 2005. He also led NERC's investigations of the 2003 Northeast blackout and the 2011 Arizona/Southern California blackout.

11. "Energy in Brief," US Energy Information Agency website, accessed January 20, 2013, http://www.eia.gov/energy_in_brief/article/major_energy_sources_and_users.cfm.

12. Congress established the electric reliability organization as an entity to be certified by the Federal Energy Regulatory Commission to "establish and enforce reliability standards for the bulk-power system, subject to Commission review." Energy Policy Act of 2005, Pub. L. No. 109-58, 115 Stat. 594, Title XII—Electricity, Subtitle A. Reliability Standards, Section 1211 (a) (2). The regional reliability entities monitor and enforce reliability compliance, assess future reliability, develop region-specific standards, and analyze disturbances and other events for lessons learned. Of the eight regional entities, only two are contiguous with two of the four major interconnected systems. The Western Electricity Coordinating Council (WECC) oversees the same region served by the Western Interconnection, and the Texas Reliability Entity

(Texas RE) oversees the Texas Interconnection, also known as the Electric Reliability Council of Texas (ERCOT).

13. Hughes, *Networks of Power.*

14. Ibid., x, 1; Olivier Coutard, *The Governance of Large Technical Systems* (New York: Routledge, 1999), 10.

15. Geoffrey C. Bowker et al., "Toward Information Infrastructure Studies: Ways of Knowing in a Networked Environment," in *International Handbook of Internet Research*, ed. Jeremy Hunsinger, Lisbeth Klastrup, and Matthew Allen (London: Springer, 2010), 97.

16. Wiebe E. Bijker, Thomas Parke Hughes, and T. J. Pinch, *The Social Construction of Technological Systems: New Directions in the Sociology and History of Technology* (Cambridge, MA: MIT Press, 1987).

17. Paul L. Joskow and Roger G. Noll, "The Bell Doctrine: Applications in Telecommunications, Electricity, and Other Network Industries," *Stanford Law Review* 51, no. 5 (1999): 1249–1315. For a recent study of the origins of the ideas of "natural monopoly" and "public utility," see Adam Plaiss, "From Natural Monopoly to Public Utility: Technological Determinism and the Political Economy of Infrastructure in Progressive-Era America," *Technology and Culture* 57, no. 4 (2016): 806–830.

18. Joskow and Noll, "Bell Doctrine," 1303.

19. Douglas D. Anderson, *Regulatory Politics and Electric Utilities: A Case Study in Political Economy* (Boston: Auburn House, 1981); Richard N. L. Andrews, *Managing the Environment, Managing Ourselves: A History of American Environmental Policy*, 2nd ed. (New Haven: Yale University Press, 2006); Stephen G. Breyer, *Regulation and Its Reform* (Cambridge, MA: Harvard University Press, 1982); Thomas K. McCraw, *Prophets of Regulation: Charles Francis Adams, Louis D. Brandeis, James M. Landis, Alfred E. Kahn* (Cambridge, MA: Belknap Press of Harvard University Press, 1984). For examples of regulation of power systems in other countries, see Christopher Armstrong and H. V. Nelles, *Monopoly's Moment: The Organization and Regulation of Canadian Utilities, 1830–1930*, Technology and Urban Growth (Philadelphia: Temple University Press, 1986).

20. Richard R. John, *Network Nation: Inventing American Telecommunications* (Cambridge, MA: Belknap Press of Harvard University Press, 2010), 209.

21. Ibid., 208.

22. Janet Abbate, *Inventing the Internet* (Cambridge, MA: MIT Press, 1999), 96.

23. Ibid., 4.

24. David E. Nye, *Electrifying America: Social Meanings of a New Technology, 1880–1940* (Cambridge, MA: MIT Press, 1990).

25. Paul N. Edwards et al., "Introduction: An Agenda for Infrastructure Studies," *Journal of the Association for Information Systems* 10, special issue (2009): 367.

26. *IEEE Standard Computer Dictionary: A Compilation of IEEE Standard Computer Glossaries* (New York: Institute of Electrical and Electronics Engineers, 1990), 114.

27. John G. Palfrey and Urs Gasser, *Interop: The Promise and Perils of Highly Interconnected Systems* (New York: Basic Books, 2012), 10.

28. "A Long-Distance Test," *Electrical World* 30, no. 17 (1897): 489; "Pacific Coast Notes: An 81-Mile Transmission Line in Successful Operation," *Electrical World* 33, no. 6 (1899): 188; "Power Transmission in Utah," *Electrical World* 37, no. 15 (1901): 587.

29. William Todd, "The James River, Virginia, Water-Power Development," *Electrical World* 33, no. 18 (1899): 573–574; "Greatest Engineering Achievements of the 20th Century," *National Academy of Engineering*, copyright 2016, accessed December 14, 2016, http://www.greatachievements.org/.

2 The Birth of the Grid, 1899–1918

1. Anon., "San Gabriel Electric Company," *Engineering* (1899): 61; Carl Hering, "San Gabriel–Los Angeles Transmission," *Electrical World* 33, no. 1 (1899): 24; "A Look Back: Our History," Edison International, accessed January 9, 2017, http://www.edison.com/home/about-us/our-history.html#27764; William B. Friedricks, *Henry E. Huntington and the Creation of Southern California*, ed. Mansel G. Blackford and K. Austin Kerr, Historical Perspectives on Business Enterprise (Columbus: Ohio State University Press, 1992), 48–67.

2. "Los Angeles 33,000-Volt Transmission Plant and Electric Railway," *Electrical World* 37, no. 26 (1901): 1113–1115; Hering, "San Gabriel–Los Angeles Transmission," 24.

3. Hering, "San Gabriel–Los Angeles Transmission," 24.

4. *Los Angeles Express*, May 2, 1901, as quoted in Friedricks, *Henry E. Huntington and the Creation of Southern California*, 55, 181n15.

5. "A Look Back: Our History," Edison International, accessed January 9, 2017, http://www.edison.com/home/about-us/our-history.html#27764.

6. "Pacific Coast Notes: An 81-Mile Transmission Line in Successful Operation," *Electrical World* 33, no. 6 (1899): 188; Carl Hering, "83 Miles of Power Transmission," *Electrical World* 34, no. 20 (1899): 750.

7. "A Notable Transmission System," *Electrical World* 45, no. 8 (1905): 375.

8. Ibid.

9. Ibid.

10. Utility experts Leonard S., Andrew S., and Robert C. Hyman offer a straightforward explanation of AC and DC current: "Electric currents come in two kinds: *alternating* current (AC) and *direct* current (DC). Of the two, DC is the simpler. The current flows in one direction. Batteries produce DC, which flows from one pole, through the electrical circuit, to the other pole. AC, on the other hand, changes its direction on a regular basis. Each complete trip, before reversing direction, is called a *cycle*. The frequency of the alternation is measured in cycles per second, or in hertz (Hz). That is, 30 Hz means 30 cycles per second." Leonard S. Hyman, Andrew S. Hyman, and Robert C. Hyman, *America's Electric Utilities: Past, Present and Future*, 8th ed. (Vienna, VA: Public Utilities Reports, 2005), 16.

11. The Pearl Street Station relied on coal-fired generators to produce electricity. As Hyman et al. explain, "Most electricity is generated by burning a fossil fuel (coal, oil, or natural gas), or from the burnup of nuclear fuel, or from the force of water (hydroelectricity)." Ibid., 19–24. In very rough terms, both fossil fuel and nuclear generators heat water to produce steam, which then turns the blades of a turbine. A magnet surrounded by coils of wire is attached to the turbine, and the spinning creates a magnetic field, which then induces electric current into the wire. In the case of hydroelectricity, falling water rather than steam causes the turbine to spin.

12. Paul Israel, *Edison: A Life of Invention* (New York: John Wiley, 1998), 303–337. Both private investors and municipalities organized power companies all over the world. Those that franchised the Edison system typically included "Edison" in the company name. Thomas Edison, however, had little to do with the operations of these franchises, particularly after the merger of Edison General Electric and Thompson Houston in 1892. The merger resulted in the creation of the General Electric Company, and Edison held a position on the board, but he directed his attention to areas other than electric power in the ensuing years.

13. For narratives describing the Edison system and related technologies, see Richard F. Hirsh, *Technology and Transformation in the American Electric Utility Industry* (Cambridge: Cambridge University Press, 1989); Thomas Parke Hughes, *Networks of Power: Electrification in Western Society, 1880–1930* (Baltimore: Johns Hopkins University Press, 1983); Hyman, Hyman, and Hyman, *America's Electric Utilities;* Israel, *Edison*; David E. Nye, *Electrifying America: Social Meanings of a New Technology, 1880–1940* (Cambridge, MA: MIT Press, 1990).

14. Both arc lights and incandescent lights, with innumerable technical and material variations, are still in wide use. An arc lamp consists of two electrodes separated by gas. The lamp produces light when the gas in the gap between the electrodes is electrified. Arc lamp technology in the nineteenth century produced very bright lights that were difficult to regulate and was used mostly in outdoor or theatrical lighting. An incandescent bulb produces light when a metal filament is heated by electricity until it glows. In the nineteenth century, designers and producers of incandescent bulbs successfully introduced softer and longer-lasting lighting that could serve a wide variety of indoor and outdoor functions.

15. For more detailed descriptions of experiments in electrification and lighting, see Hughes, *Networks of Power*; Jill Jonnes, *Empires of Light: Edison, Tesla, Westinghouse, and the Race to Electrify the World*, 1st ed. (New York: Random House, 2003); Michael B. Schiffer, *Power Struggles: Scientific Authority and the Creation of Practical Electricity before Edison* (Cambridge, MA: MIT Press, 2008); Marc J. Seifer, *Wizard: The Life and Times of Nikola Tesla: Biography of a Genius* (Secaucus, NJ: Citadel, 1996).

16. "High Voltage Direct Current," Wikipedia, last updated December 20, 2016, https://en.wikipedia.org/wiki/High-voltage_direct_current. Wikipedia offers an explanation of the relationship between voltage and the size of a power line: "For a given quantity of power transmitted, doubling the voltage will deliver the same power at only half the current. Since the power lost as heat in the wires is proportional to the square of the current for a given conductor size, or inversely proportional to the square of the voltage, doubling the voltage reduces the line losses per unit of electrical power delivered by a factor of 4."

17. Paul A. David and Julie Ann Bunn, "The Economics of Gateway Technologies and Network Evolution: Lessons from Electricity Supply History," *Information Economics and Policy* 3, no. 2 (1988): 165–202; Hughes, *Networks of Power*, 14, 81–105; Israel, *Edison*, 318–335; Harold C. Passer, *The Electrical Manufacturers, 1875–1900: A Study in Competition, Entrepreneurship, Technical Change, and Economic Growth*, Technology and Society (New York: Arno Press, 1972), 123.

18. W. Bernard Carlson, *Tesla: Inventor of the Electrical Age* (Princeton, NJ: Princeton University Press, 2013); Watson Davis, "Strange Electrical Genius," *Science News Letter* 70, no. 1 (1956): 10–11; Jonnes, *Empires of Light*; Eliot Marshall, "Seeking Redress for Nikola Tesla," *Science, New Series* 214, no. 4520 (1981): 523–525; Seifer, *Wizard;* Kenneth M. Swezey, "Nikola Tesla," *Science* 127, no. 3307 (1958): 1147–1159.

19. Two years earlier, Swiss engineer and cofounder of the firm Brown-Boveri Company Charles Brown demonstrated the successful transmission of one hundred horsepower of AC electricity over a distance of more than one hundred miles from Lauffen to the International Exposition in Frankfurt. In his talk "High Tension Currents," he forecasted the potential to deliver large amounts of electricity over great distances. Jonnes, *Empires of Light*, 285; Ignacio J. Pérez-Arriaga, Hugh Rudnick, and Michel River Abbad, "Electric Energy Systems—an Overview," in *Electric Energy Systems: Analysis and Operation*, ed. Antonio Gómez-Expósito, Antonio J. Conejo, and Claudio Cañizarez (New York: CRC Press, 2009), 5.

20. Steven W. Usselman, "From Novelty to Utility: George Westinghouse and the Business of Innovation during the Age of Edison," *Business History Review* 66, no. 2 (1992): 251–304.

21. Hughes, *Networks of Power*, 125; Jonnes, *Empires of Light*, 135–139, 283–306; Harold L. Platt, *The Electric City: Energy and the Growth of the Chicago Area, 1880–1930* (Chicago: University of Chicago Press, 1991), 80–81.

22. "Pacific Coast Notes," 188; "Transmission System of the Bay Counties Power Company, California," *Electrical World* 37, no. 7 (1901): 273–274; "Power Transmission in Utah," *Electrical World* 37, no. 15 (1901): 587, 593–594; J. R. Cravath, "Extension of the 40,000-Volt Lines of the Telluride Power Transmission Company in Utah," *Electrical World* 37, no. 8 (1901): 307; Hering, "83 Miles of Power Transmission," 750.

23. "The Concentration of Philadelphia Lighting Stations," *Electrical World* 35, no. 8 (1900): 276–277; "The Growth of Central Station Practice," *Electrical World* 40, no. 22 (1902): 842; "Electrical Transmission in Boston," *Electrical World* 42, no. 10 (1903): 377–379; Alton D. Adams, "Development of a Great Water Power System at Hartford, Conn.," *Electrical World* 39, no. 10 (1902): 427–434; "Montreal, the Greatest Centre of Transmitted Power—I," *Electrical World* 42, no. 23 (1903): 905–909; "Montreal, the Greatest Centre of Transmitted Power—II," *Electrical World* 42, no. 24 (1903): 957–960; "Montreal, the Greatest Centre of Transmitted Power—III," *Electrical World* 42, no. 26 (1903): 1037–1042.

24. "Storage Batteries Discussed at the Western Society of Engineers," *Electrical World* 49, no. 13 (1907): 625; "Recent Developments in Storage Battery Applications," *Electrical World* 55, no. 15 (1910): 927; S. H. Sharpsteen, "The Storage Battery for Web Printing-Press Control," *Electrical World* 52, no. 11 (1908): 580–581.

25. "Operation of the Boston Edison Company's L Street Station," *Electrical World* 53, no. 22 (1909): 1283–1289; "Practical Operation of Fisk Street and Quarry Street Stations in Chicago," *Electrical World* 53, no. 22 (1909): 1289–1294.

26. "Electrical Transmission on the Pacific Coast," *Electrical World* 39, no. 1 (1902): 25.

27. Ibid.

28. *Transactions of the International Electrical Congress, St. Louis, 1904*, Google Books online ed., the International Electrical Congress, St. Louis (St. Louis, MO: J. B. Lyon Company, 1904), 2:213. The phrasing of this sentence is a bit confusing; the author is stating that the companies used transformers at the points where the generated electricity entered the transmission network.

29. "Power Transmission in Utah," 587.

30. "The Value of Water Storage," *Electrical World* 44, no. 20 (1904): 810–811.

31. "Economics of High Voltage Transmission," *Electrical World* 44, no. 19 (1904): 756–757.

32. "The Year in Power Transmission," *Electrical World* 45, no. 1 (1905): 5.

33. *Compendex*, s.v. "electric," and "long distance transmission," Engineering Village, accessed January 14, 2011, https://www.engineeringvillage.com/. The *Compendex* online database is a comprehensive international engineering index dating back to 1884. For the search variable (s.v.) "electric" in each year from 1890 to 1910,

there was a definite spike between 1902 and 1905. For s.v. "long distance transmission," the peak years were 1904 and 1905.

34. "Fifth International Electrical Congress, St. Louis, MO, Sept 12–17, 1904," *Electrical World* 44, no. 12 (1904): 465–476; *Transactions of the International Electrical Congress, St. Louis, 1904*, vol. 2, 316–318.

35. *Transactions of the International Electrical Congress, St. Louis, 1904*, vol. 2, 199–200.

36. "Transcendental Power Transmission," *Electrical World* 44, no. 27 (1904): 1121.

37. *Transactions of the International Electrical Congress, St. Louis, 1904*, vol. 2, 213; "Frederic A. C. Perrine, A. M., D. Sc.," *Journal of Electricity, Power and Gas* 21, no. 18 (1908): 286–287.

38. "Power Transmission before the International Electrical Congress," *Electrical World* 44, no. 13 (1904): 507–509.

39. *Transactions of the International Electrical Congress, St. Louis, 1904*, vol. 2, 443–444.

40. "Personal," *Electrical World* 46 no. 17 (1905): 716.

41. *Transactions of the International Electrical Congress, St. Louis, 1904*, vol. 2, 445.

42. The "load factor" is the average load on a power system divided by the maximum, or "peak" load during a specified time period. The calculation is used to determine whether or not a generator, or system, is used to its greatest efficiency. The closer the load factor is to one, the more even the demand is on the system.

43. *Transactions of the International Electrical Congress, St. Louis, 1904*, vol. 2, 447.

44. Ibid.

45. "Fifth International Electrical Congress, St. Louis, MO, Sept 12–17, 1904," 460.

46. "The Tendency of Central Station Development—III," *Electrical World* 30, no. 26 (1897).

47. Howard S. Knowlton, "The Storage Battery in Transmission Plants," *Electrical World* 41, no. 20 (1903): 831.

48. "The Year in Power Transmission," 5.

49. "Station Efficiencies," *Electrical World* 52, no. 22 (1908): 1158.

50. Thank you to Joseph Pratt for suggesting several years ago the phrasing "fraternity of experts" to describe the engineers, system operators, managers, and academics who worked on power networks.

51. T. E. Young, "Opening Address by the President," *Journal of the Institute of Actuaries* 33, no. 2 (1897): 97.

52. "The Latest: A 'Notched Collar' for Men of Fashion—Figured Silk on Evening Dress This Winter," *New York Times*, September 7, 1902, 33.

53. Roy Lubove, "Economic Security and Social Conflict in America: The Early Twentieth Century, Part II," *Journal of Social History* 1, no. 4 (1968): 335.

54. "Girl Golfer Wins Honor: Virginia Van Wie Acclaimed as Most Outstanding Female Star," *Los Angeles Times*, December 21, 1934, 9; Diane White, "She Preserves Centuries-Old Fabrics," *Boston Globe*, January 19, 1967, 29; Pam Proctor, "Ladies of the Board Room," *Boston Globe*, July 28, 1974, 16c.

55. Carl Hering, "Women in Electricity," *Electrical World* 34, no. 24 (1899): 904. Edith Clarke was an important exception. She obtained an MS in Electrical Engineering from MIT in 1918. During her professional career, she taught engineering at numerous institutions and worked as an electrical engineer for General Electric Company. In 1926, she was the first woman to deliver a technical paper to the American Institute of Electrical Engineers. Clarke's work contributed to the development of long-distance transmission lines and interconnected power systems, and her 1943 coauthored book *Circuit Analysis of A-C Power Systems* became a standard textbook for power systems engineering students through the mid-twentieth century. J. E. Brittain, "Edith Clarke and Power System Stability," *IEEE Industry Applications Magazine* 9, no. 1 (2003): 9–10; J. E. Brittain, "From Computor to Electrical Engineer: The Remarkable Career of Edith Clarke," *IEEE Transactions on Education* 28, no. 4 (1985): 184–189; Edith Clarke, "Steady-State Stability in Transmission Systems Calculation by Means of Equivalent Circuits or Circle Diagrams," *Transactions of the American Institute of Electrical Engineers* 45 (1926): 22–41; Edith Clarke, *Circuit Analysis of A-C Power Systems* (New York: J. Wiley & Sons, 1943).

56. Amy Sue Bix, *Girls Coming to Tech!: A History of American Engineering Education for Women* (Cambridge, MA: MIT Press, 2013), 12.

57. *American Newspaper Directory (Issued Quarterly)* (New York: George P. Rowell, 1898). Circulation for *Electrical World* in the United States exceeded ten thousand per week by the mid-1890s—higher than other similar journals—and continued to grow to more than eighteen thousand by 1922. For example, between 1895 and 1920, *Electrical World* circulation regularly exceeded *Cassier's Magazine*, *Electrical Record*, *Electricity*, and *Proceedings of the American Institute of Electrical Engineers*.

58. "The Visit of the American Institute of Electrical Engineers to Europe," *Electrical World* 36, no. 9 (1900): 9; "Future of Canada," *Electrical World* 50, no. 17 (1907): 800.

59. "Central Station Management, Policies and Commercial Methods," *Electrical World* 52, no. 2 (1908): 88–93.

60. For example, the *Electric Journal* began publication in Chicago in 1895, and the *Journal of Electricity* began publication in San Francisco that same year. (Thank you to Richard Hirsh for bringing these regional publications to my attention.)

61. Chi-nien Chung, "Networks and Governance in Trade Associations: AEIC and NELA in the Development of the America Electricity Industry 1885–1910," *International Journal of Sociology and Social Policy* 17, no. 7/8 (1997): 57–110. Chung argues that a close-knit group of executives dominated both NELA and AEIC and thereby shaped the direction of industry development through the early decades of electrification.

62. Ibid., 66.

63. "The Electric Club Journal," *Electric Club Journal* 1, no. 1 (1904): ii.

64. Edwin T. Layton, *The Revolt of the Engineers: Social Responsibility and the American Engineering Profession* (Baltimore: Johns Hopkins University Press, 1986).

65. David F. Noble, *America by Design: Science, Technology, and the Rise of Corporate Capitalism*, 1st ed. (New York: Knopf, 1977).

66. Andrew L. Russell, *Open Standards and the Digital Age: History, Ideology, and Networks*, ed. Louis Galambos and Geoffrey Jones, Cambridge Studies in the Emergence of Global Enterprise (New York: Cambridge University Press, 2014), 22.

67. Paul N. Edwards et al., "Introduction: An Agenda for Infrastructure Studies," *Journal of the Association for Information Systems* 10, no. 5 (2009): 364–374.

68. For narratives in which technology users and designers are closely linked, see Janet Abbate, *Inventing the Internet* (Cambridge, MA: MIT Press, 1999); Joanne Yates, *Structuring the Information Age: Life Insurance and Technology in the Twentieth Century* (Baltimore: Johns Hopkins University Press, 2005).

69. "A Plea for Standardization," *Electrical World* 36, no. 17 (1900): 632.

70. "The Great Barrington Convention," *Electrical World* 39, no. 26 (1902): 1150.

71. "International Electrical Standardization," *Electrical World* 42, no. 5 (1903): 167.

72. Ibid.

73. "American Practice in Power Transmission," *Electrical World* 44, no. 14 (1904): 550–551.

74. Stephen Salsbury, "The Emergence of an Early Large-Scale Technical System: The American Railroad Network," in *The Development of Large Technical Systems*, ed. Renate Mayntz and Thomas Parke Hughes (Boulder, CO: Westview Press, 1988), 37–68.

75. Ibid., 59. Congress passed the Safety Appliance Act in 1893, which required railroads to adopt braking standards established by the Master Car-Builders Association.

76. Russell, *Open Standards and the Digital Age*, 26–27.

77. Ibid., 52.

78. Craig N. Murphy and Joanne Yates, *The International Organization for Standardization (ISO): Global Governance through Voluntary Consensus*, ed. Thomas G. Weiss

and Rorden Wilkinson, Routledge Global Institutions (New York: Routledge Taylor and Francis Group, 2009); Stefan Timmermans and Steven Epstein, "A World of Standards but Not a Standard World: Toward a Sociology of Standards and Standardization," *Annual Review of Sociology* 36 (2010): 69–89.

79. Russell, *Open Standards and the Digital Age*, 39.

80. Robert F. Durden, *Electrifying the Piedmont Carolinas: The Duke Power Company, 1904–1997* (Durham, NC: Carolina Academic Press, 2001), 11–23.

81. "The Great Southern Transmission Network," *Electrical World* 63, no. 22 (1914): 1201–1202.

82. "Interconnected Systems of the South," *Electrical World* 63, no. 22 (1914): 1235–1243.

83. "The Growth of a Transmission Network," *Electrical World* 51, no. 16 (1908): 799.

84. "World's Largest Transmission System," *Electrical World* 59, no. 22 (1912): 1197–1204.

85. "High Tension Transmission Experience in Central Colorado," *Electrical World* 58, no. 15 (1911): 871–875.

86. "Interconnected Electric Service at Warren, Ohio," *Electrical World* 63, no. 4 (1914): 197–202.

87. "The Growth of a Transmission Network," 799.

88. Much of the next section appears in Julie Cohn, "Utilities as Conservationists? The Paradox of Electrification during the Progressive Era in North America," in *Green Capitalism? Exploring the Crossroads of Environmental and Business History*, ed. Hartmut Berghoff and Adam Rome (Philadelphia: University of Pennsylvania Press, 2017).

89. For a selection of works on resource depletion, river development, expert planning and efficiency, smoke pollution, and preservation of nature during the Progressive Era, see Robert Gottlieb, *Forcing the Spring: The Transformation of the American Environmental Movement*, rev. and updated ed. (Washington, DC: Island Press, 2005); Samuel P. Hays, *Conservation and the Gospel of Efficiency: The Progressive Conservation Movement, 1890–1920* (Pittsburgh: University of Pittsburgh Press, 1999); David A. Hounshell, *From the American System to Mass Production, 1800–1932: The Development of Manufacturing Technology in the United States*, Studies in Industry and Society (Baltimore: Johns Hopkins University Press, 1984); Martin V. Melosi, *Effluent America: Cities, Industry, Energy, and the Environment* (Pittsburgh: University of Pittsburgh Press, 2001); Martin V. Melosi, *Garbage in the Cities: Refuse, Reform, and the Environment*, rev. ed., History of the Urban Environment (Pittsburgh: University of Pittsburgh Press, 2005); Robert W. Righter, *The Battle over Hetch Hetchy: America's Most Controversial Dam and the Birth of Modern*

Environmentalism (New York: Oxford University Press, 2005); Joel A. Tarr, *Devastation and Renewal: An Environmental History of Pittsburgh and Its Region* (Pittsburgh: University of Pittsburgh Press, 2003); Thomas Raymond Wellock, *Preserving the Nation: The Conservation and Environmental Movements, 1870–2000*, American History Series (Wheeling, IL: Harlan Davidson, 2007).

90. For investigations into other business and industry initiatives that aligned with Progressive Era movements, see Christine Meiner Rosen, "Business Men against Pollution in Late Nineteenth Century Chicago," *Business History Review* 69, no. 3 (1995): 351–397; Hugh S. Gorman, "Efficiency, Environmental Quality, and Oil Field Brines: The Success and Failure of Pollution Control by Self-Regulation," *Business History Review* 73, no. 4 (1999): 601–640; Christine Meisner Rosen and Christopher C. Sellers, "The Nature of the Firm: Towards an Ecocultural History of Business: [Introduction]," *Business History Review* 73, no. 4 (1999): 577–600; David Stradling and Joel A. Tarr, "Environmental Activism, Locomotive Smoke, and the Corporate Response: The Case of the Pennsylvania Railroad and Chicago Smoke Control," *Business History Review* 73, no. 4 (1999): 677–704; Frank Uekoetter, "Divergent Responses to Identical Problems: Businessmen and the Smoke Nuisance in Germany and the United States, 1880–1917," *Business History Review* 73, no. 4 (1999): 641–676.

91. "The Value of Water Storage," *Electrical World* 44, no. 20 (1904): 810–811.

92. "A Notable Transmission System," 375.

93. "A Wasteful Century," *Electrical World* 36, no. 8 (1900): 272; "The Age of Reckless Waste," *Electrical World* 39, no. 16 (1902): 674–675; "Typical American Waste," *Electrical World* 45, no. 9 (1905): 426.

94. "The Value of Water Storage," 810–811.

95. "Large Power Trunk Line Transmission under Way in New York State," *Electrical World* 53, no. 1 (1909): 10.

96. Ibid.

97. "The Debate on Central Stations," *Electrical World* 41, no. 19 (1903): 782; "The Central Station of the Future," *Electrical World* 42, no. 11 (1903): 422–423.

98. "The Growth of a Transmission Network," 799.

99. "Getting After the 'Electric Trust,'" *Electrical World* 54, no. 23 (1909): 1327.

100. "The Great Southern Transmission Network," 1201–1202.

101. For a more detailed development of this finding, please see Cohn, "Utilities as Conservationists?"

102. "The Conservation of Natural Resources," *Electrical World* 51, no. 11 (1908): 550.

103. "Electricity and the Conservation of Energy," *Electrical World* 53, no. 19 (1909): 1195.

104. "Twenty Sixth Annual Convention of the A.I.E.E.," *Electrical World* 54, no. 1 (1909): 4.

105. "Economic Limitations to Aggregation of Electrical Systems," *Electrical World* 57, no. 8 (1911): 468.

106. "Modern Transmission Problems," *Electrical World* 57, no. 12 (1911): 709–710.

107. *Electric Power Development in the United States. Letter from the Secretary of Agriculture Transmitting a Report . . . as to the Ownership and Control of the Water-Power Sites in the United States*, 3 vols. (Washington, DC: Department of Agriculture, 1916).

108. "Economic Limitations," 468.

109. "Proposed National Commission to Solve Water-Power Problems," *Electrical World* 60, no. 17 (1912): 859.

110. "Steinmetz on the Future of the Electrical Industry," *Electrical World* 60, no. 18 (1912): 911–912. Called "the wizard of Schenectady," Charles P. Steinmetz was an influential mathematician and engineer at General Electric Company whose insights into electrical phenomena propelled the electric power business forward during the early twentieth century. Steinmetz headed the General Electric Research Laboratory from its founding in 1900 and drew fame for his innovations in alternating current machines and circuitry, hysteresis, and long-distance transmission. Ronald R. Kline, *Steinmetz: Engineer and Socialist* (Baltimore: Johns Hopkins University Press, 1992); Hughes, *Networks of Power*, 164.

111. "Central Stations and the War," *Electrical World* 64, no. 8 (1914): 357–358; "The War's Effect on the Electrical Industry," *Electrical World* 64, no. 9 (1914): 414–415; "Effect of War on the Industry," *Electrical World* 64, no. 8 (1914): 360–366; "Attention to Export Trade," *Electrical World* 64, no. 9 (1914): 405; "Prospects for Domestic and Foreign Business," *Electrical World* 64, no. 11 (1914): 509–511.

112. "The Industrial Power Problem," *Electrical World* 48, no. 19 (1906): 927; "Small Isolated Plants," *Electrical World* 50, no. 18 (1907): 837–838.

113. Bureau of the Census, *Abstract of the Census of Manufactures 1919* (Washington, DC: Government Printing Office, 1923), 460. The number of units of horsepower used in manufacturing grew from 492,936 in 1899 to 16,317,277 in 1919. William H. Stuart, "Isolated Plants," *Electrical World* 50, no. 5 (1907): 241; W. F. Lloyd, "Isolated Power Plant Costs and Their Relation to Central Station Service," *Electrical World* 52, no. 22 (1908): 1182–1183; "Co-operation with the Isolated Plant," *Electrical World* 62, no. 3 (1913): 113; Bureau of the Census Department of Commerce, *Abstract of the Census of Manufactures 1914* (Washington, DC: Government Printing Office, 1917), 491. By 1919, key markets for electricity use included the transportation sector,

textiles production and related industries, mills and food preparations, foundries, iron and steel industries, lumber and timber industries, and paper goods and printing.

114. "Isolated Plant Economy," *Electrical World* 61, no. 8 (1913): 383.

115. "Electricity Directly from the Coal Mine in Pennsylvania," *Electrical World* 59, no. 19 (1912): 1002–1003; "Competition of Coal-Mine Power Plants with Central Stations," *Electrical World* 60, no. 9 (1912): 460; "Station Efficiencies," *Electrical World* 52, no. 22 (1908): 1158. The term "mine-mouth plant" refers to the strategy of placing a generating plant close to the mouth of an active coal mine, then transporting the electricity from the mine to a load center such as an industrial complex or an urban area.

116. "Niagara Falls a War-Load Center," *Electrical World* 73, no. 20 (1919): 996–1000; "Niagara Power in War Industries," *Electrical World* 72, no. 3 (1918): 109–111.

117. Col. Charles Keller, *The Power Situation during the Wa*r (Washington, DC: Government Printing Office, by Authority of the Secretary of War, 1921), 28.

118. Ibid.

119. Platt, *The Electric City*, 202–203; Robert Cuff, "Harry Garfield, the Fuel Administration, and the Search for a Cooperative Order during World War I," *American Quarterly* 30, no. 1 (1978); "Wartime Lighting Economies," *Electrical World* 72, no. 19 (1918): 885–887; Charles E. Stuart, "War Conservation of Power and Light," *Electrical World* 72, no. 10 (1918): 451.

120. For more detailed analyses of energy shortages, the power industry, and the war years, see Hughes, *Networks of Power*; Nye, *Electrifying America*; and Platt, *The Electric* City.

121. C. S. Cook, "The Future Power Station," *Electric Journal* 15, no. 6 (1918): 185.

122. "Present War Production Made Possible by Utilities," *Electrical World* 72, no. 12 (1918): 541–544; "Spirit of War in the Central West," *Electrical World* 71, no. 1 (1918): 31–34; "Electrical Interconnections to Conserve Fuel," *Electrical World* 71, no. 1 (1918): 12–14; "New England–Boston Power Interconnection," *Electrical World* 72, no. 13 (1918): 612–613; "How the South Handled War-Time Loads," *Electrical World* 73, no. 20 (1919): 1022–1030; "Niagara Falls a War-Load Center"; Gustave P. Capart, "Use of Electricity in the European War," *Electrical World* 70, no. 18 (1917): 861–862; "War-Time Service Problems in New England," *Electrical World* 73, no. 20 (1919): 1007–1019.

123. Rudolph F. Schuchardt, "The Significance and the Opportunities of the Central Station Industry," *Electric Journal* 16, no. 5 (1919): 166– 168.

124. B. G. Lamme, "The Technical Story of the Frequencies," *Transactions of the American Institute of Electrical Engineers* 37, no. 1 (1918): 65–89; P. M. Lincoln, "Choice of Frequency for Very Long Lines," *Transactions of the American Institute of Electrical Engineers* 22 (1903): 373–376; Samuel Sheldon, "Discussion on 'Frequency' (Rushmore), Schenectady, N.Y., May 17, 1912," *Transactions of the American Institute of Electrical Engineers* 31, no. 1 (1912): 973–983; P. Mixon, "Technical Origins of

60 Hz as the Standard Ac Frequency in North America," *Power Engineering Review, IEEE* 19, no. 3 (1999): 35–37. Engineers debated the merits of various frequencies for many years and settled on different solutions in different countries and for different uses. No governing body or industry group has ever formally adopted 60 Hz as a standard for North America, although the vast majority of installations operate at this frequency.

125. Guy E. Tripp, "A Central Station Opportunity," *Electric Journal* 16, no. 2 (1919): 45–46; Keller, "The Power Situation during the War."

3 Contests for Control, 1918–1934

1. F. H. Bernhard, "New Electric Power Plant at Windsor," *Electrical Review and Western Electrician* 70, no. 21 (1917): 878.

2. Ibid., 876.

3. "Power Transmission from Windsor," *Electrical Review and Western Electrician* 70, no. 22 (1917): 926–930.

4. Bernhard, "New Electric Power Plant at Windsor," 876.

5. "Power at the Mine's Mouth," *Electrical World* 71, no. 6 (1918): 291.

6. "Bulk-Supply Generating Station at Mouth of Mine," *Electrical World* 71, no. 7 (1918): 344–347; "Canton Substation for Receiving Power from Windsor Plant," *Electrical Review and Western Electrician* 70, no. 23 (1917): 963–969.

7. "Windsor–Canton Transmission Line," *Practical Engineer*, March 15, 1917, 258.

8. E. H. McFarland, "Three Years Operation at Windsor," *General Electric Review* 24, no. 6 (1921): 572–578.

9. Col. Charles Keller, *The Power Situation during the War* (Washington, DC: Government Printing Office, by Authority of the Secretary of War, 1921), 11.

10. Ibid.

11. Ibid., 14.

12. Ibid., 135.

13. McFarland, "Three Years Operation at Windsor," 572.

14. Keller, "The Power Situation during the War."

15. Ibid., 6.

16. "New England Interconnection Plan Takes Shape," *Electrical World* 72, no. 6 (1918): 257–259.

17. Keller, *The Power Situation during the War*, 13–14.

18. "Power Transmission and Industrial Development," *Electrical World* 75, no. 1 (1920): 31.

19. Ross J. McClelland, "Electric Power Supply for War Industries," *Electrical World* 72, no. 3 (1918): 100–105. General Electric spun off EBASCO in 1906 in response to political pressure to separate electrical manufacturing from power plant financing, design, and ownership. EBASCO functioned as a holding company and as a financial and engineering consultant to investor-owned utilities large and small, including the holding company AGE.

20. Ibid., 105.

21. Keller, *The Power Situation during the War*, 21.

22. Report of the Smithsonian Institution, *The Mineral Industries of the United States: The Energy Resources of the United States: A Field for Reconstruction* (Washington, DC: Government Printing Office, 1919).

23. "Northwest Power Pool," Northwest Power Pool, accessed November 6, 2007, http://www.nwpp.org/; Gus Norwood, *Columbia River Power for the People: A History of Policies of the Bonneville Power Administration* (Portland, OR: US Department of Energy, Bonneville Power Administration, 1981); Edward Wilson Kimbark, *Power System Stability*, vol. 2 (New York: Wiley, 1950).

24. William S. Murray, "The Primaries of Today the Secondaries of Tomorrow," *Electric Journal* 16, no. 5 (1919): 168–170; W. S. Murray, "Economical Supply of Electric Power," *Transactions of the American Institute of Electrical Engineers* 39, no. 1 (1920): 101–166. Over time, the term "Superpower," sometimes appearing as "superpower," "super-power," or "super power," came into wide use referring in general to the combination of very large generating plants, located at the energy source, and transmission over long distances on high-voltage lines, usually as part of an interconnected system.

25. William S. Murray and et al., *A Superpower System for the Region between Boston and Washington*, ed. United States Geological Survey Department of the Interior (Washington, DC: Government Printing Office, 1921), 9.

26. "Secretary Lane's Proposal for Power Resource Survey," *Electrical World* 73, no. 6 (1919): 282.

27. "Good Meeting of A.I.E.E. in Boston," *Electrical World* 73, no. 12 (1919): 593; "Third General and Executive Session," *Electrical World* 73, no. 21 (1919): 1079; W. S. Murray, "The Superpower System as an Answer to a National Power Policy," *General Electric Review* 25, no. 2 (1922): 72–76; George Otis Smith, "National Planning for Electric Power," *Electrical World* 73, no. 23 (1919): 1210–1211.

28. "Hoover on Power Survey," *New York Times*, October 5, 1920, 10; "Super-Power Development," *Wall Street Journal*, May 20, 1922, 9; "Advocates Super-Power," *New*

York Times, April 16, 1924, 37; "Northeast Calls for Electric Power," *Wall Street Journal*, July 29, 1924, 9; Herbert Hoover, "Superpower and Interconnection," *Electrical World* 83, no. 21 (1924): 1078–1080; Herbert Hoover, "Superpower and Its Public Relations," *United States War Department—Military Engineer* 16, no. 88 (1924): 278–282; Murray, "Economical Supply of Electric Power," 101–166; Bayla Schlossberg Singer, "Power to the People, the Pennsylvania—New Jersey—Maryland Interconnection, 1925–1970" (PhD diss., University of Pennsylvania, 1983), 28, 47.

29. "Gov. Pinchot Backs Big Power Project," *Christian Science Monitor*, March 7, 1923, 1; "Starts Great Power Plan: Pinchot to Name Board to Inventory Pennsylvania's Resources," *New York Times*, March 8, 1923, 9; "Pinchot Signs Bill for Power Survey," *New York Times*, May 27, 1923, 9.

30. Singer, "Power to the People," 28, 47.

31. Jean Christie, "Giant Power: A Progressive Proposal of the Nineteen-Twenties," *The Pennsylvania Magazine of History and Biography* 96, no. 4 (1972): 496; Thomas Parke Hughes, *Networks of Power: Electrification in Western Society, 1880–1930* (Baltimore: Johns Hopkins University Press, 1983), 310; Thomas Parke Hughes, "Technology and Public Policy: The Failure of Giant Power," *Proceedings of the IEEE* 64, no. 9 (1976): 1361–1371.

32. Leonard DeGraaf, "Corporate Liberalism and Electric Power System Planning in the 1920s," *Business History Review* 64, no. 1 (1990): 8.

33. *Historical Statistics of the United States, Colonial Times to 1970, Bicentennial Edition* (Washington, DC: Government Printing Office, 1975), 821.

34. Federal Trade Commission, *Electric Power Industry, Control of Power Companies*, 69th Cong., 2nd Sess., 1927, S.Doc. 213, 29.

35. Norman S. Buchanan, "The Origin and Development of the Public Utility Holding Company," *Journal of Political Economy* 44, no. 1 (1936): 48.

36. Ibid., 31–53; E. O. Malott, "Technology and the Widening Market for Electric Service," *Journal of Land & Public Utility Economics* 4, no. 2 (1928): 147–156; quote in Hilmar Stephen Raushenbush and Harry Wellington Laidler, *Power Control* (New York: New Republic, 1928), 20.

37. "Sees Conspiracy of 'Power Trust,'" *New York Times*, October 31, 1923, 18; "Advocates Super-Power," 37; "Norris to Press Inquiry," *New York Times*, December 31, 1924, 4; "Asks 'Power Trust' Inquiry," *New York Times*, December 30, 1924, 3; "Coolidge Attacked on Muscle Shoals," *New York Times*, December 18, 1924, 1, 8; "La Follette Sees Power Combination," *New York Times*, October 30, 1924, 7; "General Electric to Be Investigated," *New York Times*, February 10, 1925, 1, 3; "Asks Direct Inquiry of General Electric," *New York Times*, February 3, 1925, 34; "Power Inquiry Referred," *New York Times*, January 21, 1925, 32.

38. Federal Trade Commission, *Electric Power Industry, Control of Power Companies*.

39. Ibid., 68.

40. Philip J. Funigiello, *Toward a National Power Policy: The New Deal and the Electric Utility Industry, 1933–1941* (Pittsburgh: University of Pittsburgh Press, 1973), 3–31. Funigiello provides a detailed chronology of this first FTC study of the power industry, the second study initiated in 1929, and the political maneuvering surrounding congressional intervention in the affairs of the electric power industry.

41. Federal Trade Commission, *Electric Power Industry, Control of Power Companies*, 77–92.

42. William Eugene Mosher et al., *Electrical Utilities* (New York: Harper & Brothers, 1929).

43. For a brief description of municipal regulation and the power industry before 1907, see Patrick McGuire, "Instrumental Class Power and the Origin of Class-Based State Regulation in the U.S. Electric Utility Industry," *Critical Sociology* 16, no. 2 (2004): 186–188.

44. For a discussion of the effect of local control on early electrification in Chicago and the role of both Samuel Insull and NELA in promoting state-level regulation, see Harold L. Platt, *The Electric City: Energy and the Growth of the Chicago Area, 1880–1930* (Chicago: University of Chicago Press, 1991).

45. For a listing of state-by-state establishment of utility commissions, see "Regulatory Commissions," National Association of Regulatory Utility Commissioners website, accessed January 9, 2017, https://www.naruc.org/about-naruc/regulatory-commissions/.

46. Richard F. Hirsh, *Power Loss: The Origins of Deregulation and Restructuring in the American Electric Utility System* (Cambridge, MA: MIT Press, 1999).

47. For a sampling of the discussion about the American approach to power system regulation, see Robert L. Bradley Jr., "The Origins of Political Electricity: Market Failure or Political Opportunism," *Energy Law Journal* 17, no. 59 (1996): 59–102; William M. Emmons III, "Private and Public Responses to Market Failure in the U.S. Electric Power Industry, 1882–1942," *Journal of Economic History* 51, no. 2 (1991): 452–454; McGuire, "Instrumental Class Power and the Origin of Class-Based State Regulation in the U.S. Electric Utility Industry," 181–203; Paul L. Joskow and Roger G. Noll, "The Bell Doctrine: Applications in Telecommunications, Electricity, and Other Network Industries," *Stanford Law Review* 51, no. 5 (1999): 1249–1315; Marc Schneiberg, "Resisting and Regulating Corporations through Ecologies of Alternative Enterprise: Insurance and Electricity in the US Case," in *The Corporation: A Critical, Interdisciplinary Handbook*, ed. André Spicer and Grietje Baars (Cambridge: Cambridge University Press, 2015); Jane Summerton and Ted K. Bradshaw, "Towards a Dispersed Electrical System: Challenges to the Grid," *Energy Policy* 19, no. 1 (1991): 24–34. For international comparisons of power system regulation, see Richard J. Gilbert and Edward Kahn, *International Comparisons of Electricity Regulation* (New York: Cambridge University Press, 1996).

48. When system operators *closed* ties, electricity flowed between systems. When they *opened* the ties, electricity no longer flowed across the links between the systems.

49. With very few exceptions, the system operators identified in the technical publications of the late nineteenth century and early twentieth century were men.

50. Nathan Cohn, "Power Flow Control—Basic Concepts for Interconnected Systems," *Electric Light and Power* 28, no. 8–9 (1950): 82–88, 93-94, 100-107, 131.

51. George P. Roux, "Load Dispatching System of the Philadelphia Electric Company," *Electric Journal* 16, no. 11 (1919): 470–474; E. C. Stone, "The Function of the Load Dispatcher," *Electric Journal* 16, no. 11 (1919): 469; J. P. Jollyman, "Operation of Interconnected Systems," *Electrical World* 71, no. 20 (1918): 1020–1022.

52. L. L. Elden, "Notes on Operation of Large Interconnected Systems" (paper presented at the 37th Annual and 10th Pacific Coast Convention of the American Institute of Electrical Engineers, Salt Lake City, Utah, June 22, 1921).

53. Hughes, *Networks of Power*, 372.

54. Ibid., 14.

55. Ibid., 372.

56. Henry E. Warren, "Utilizing the Time Characteristics of Alternating Current," *Proceedings of the American Institute of Electrical Engineers* 38, no. 5 (1919): 629–643.

57. Henry E. Warren, "Modern Electric Clocks," ClockHistory.com, accessed January 24, 2011, http://clockhistory.com/telechron/company/documents/warren_1937/. Alexander Bain patented the first electric clock in England in 1840, although he failed to put this expensive timepiece into wide production. In 1862, engineer Mathias Hipp installed a networked system in Geneva, with a master clock controlling fifteen secondary clocks connected by low-voltage wires. Western Union later used similar systems in numerous cities, as it made accurate time both accessible and a necessity.

58. F. Hope Jones, *Electric Clocks* (London: NAG Press, 1931), 261; "Obituary: Dr. M. Hipp," *The Electrical Engineer* 15, no. 268 (1893): 609; Warren, "Modern Electric Clocks."

59. Warren, "Modern Electric Clocks." In 1917, General Electric acquired a major interest in the Warren Clock Company. The name changed to Warren Telechron in 1926, and eventually GE absorbed the company entirely.

60. Harry S. Holcombe and Robert Webb, "The Warren Telechron Master Clock Type A," *NAWCC Bulletin* 27, no. 1 (1985): 35–37; Nathan Cohn, "The Way We Were," *IEEE Computer Applications in Power Magazine* 1, no. 1 (1988): 4–8.

61. H. E. Warren, "Synchronous Electric Time Service," *Electrical Engineering* 51, no. 4 (1932): 228–232; J. U. Benziger and J. T. Johnson Jr., "Automatic Frequency Control at Mitchell Dam," *Electrical World* 93, no. 26 (1929): 1332–1334; "Correct Time, a New Central-Station Service," *Electrical World* 87, no. 8 (1926): 399–410. Charles Steinmetz, in 1918, noted that improved parallel operations between stations could result in "a saving of many millions of tons of coal." Steinmetz, "America's Energy Supply," *Transactions of the American Institute of Electrical Engineers* 37, no. 6 (1918): 161; quote in Nathan Cohn, "Historical Perspectives" (paper presented at the Professional Workshop on Power Systems Control, San Luis Obispo, California, April 28–29, 1977), 4; W. R. Hamilton, "A Discussion of Frequency and Load Control Presented at Meeting of System Operating Committee, C. D. Cordes, Chairman, Pennsylvania Electric Association," in Leeds & Northrup Company, W. Spencer Bloor Collection courtesy of North American Electric Reliability Corporation; now located at Hagley Library, Manuscripts and Archives Department, Accession Number 2612, Wilmington, Delaware (hereafter cited as Electrical Power Systems Records, Hagley).

62. Warren, "Utilizing the Time Characteristics of Alternating Current," 769.

63. Ibid., 770. In the ensuing years, the efficacy of Warren's design received global affirmation. Booker, "A New Design for an Electrically Driven Clock," *Model Engineer and Electrician* 44, no. 1039 (1921): 234–236; P. Schubert, "Electrically Operated Timekeepers," *Engineering Progress* 3, no. 8 (1922): 177–179; Alex Steuart, "An Electric Clock with Detached Pendulum and Continuous Motion," *Royal Society of Edinburgh—Proceedings* 43, no. 2 (1923): 154–159; H. Voigt, "Electrical Clocks," *Engineering Progress* 4, no. 8 (1923): 161–163; "Clocks and Timing Devices," *Electrical West* 60, no. 6 (1928): 457–458.

64. This is the terminology the utilities used.

65. "Correct Time, a New Central-Station Service"; Warren, "Modern Electric Clocks," 8; quote in E. Whitehorne, "Correct Time for Public Relations," *Electrical World* 92, no. 4 (1928): 171–173.

66. Warren, "Utilizing the Time Characteristics of Alternating Current."

67. G. E. Moore, "Synchronous Motor Clocks," *Engineer* 148, no. 3859 (1929): 682–684.

68. Robert Brandt, "Historical Approach to Speed and Tie-Line Control," *Transactions of the American Institute of Electrical Engineers. Part III: Power Apparatus and Systems*, 72, no. 2 (1953): 7–9. Robert Brandt spent his career at New England Power Company. He worked in the Operating Department in the late 1920s, was head of power planning in the 1950s and 1960s, and eventually served as vice president of the company. Brandt was involved in some of the earliest tests of automatic control instruments and made major contributions to the art and practice of control of power on interconnected systems throughout his career.

69. "Frequency Control," *Electrical Review* 105, no. 2713 (1929): 920–921; "Speed-Time—a Method of Time Control," *Electrical West* 62, no. 6 (1929): 397–399.

70. George S. Humphrey, "The Interconnection of Power Systems Surrounding the Pittsburgh District," *Electric Journal* 24, no. 6 (1927): 254.

71. Nathan Cohn, "Recollections of the Evolution of Realtime Control Applications to Power Systems," *Automatica* 20, no. 2 (1984): 148.

72. C. L. Edgar, "Discussion," *Transactions of the American Institute of Electrical Engineers* 47, no. 2 (1928): 408–422.

73. "Look at Load through the Dispatcher's Eyes," *Modern Precision* 8, no. 1 (1948): 7.

74. *Research and Development Center, North Wales, Pennsylvania* (Philadelphia: Leeds & Northrup Company, 1960); William P. Vogel Jr., *Precision, People and Progress: A Business Philosophy at Work* (Philadelphia: Leeds & Northrup Company, 1949).

75. Experimental Committee, Reports and Minutes, December, 1911, Acc. 1110, Reel 1, Leeds & Northrup Company, Hagley Library, Wilmington, Delaware (hereafter cited as L&N, Hagley).

76. Vogel Jr., *Precision, People and Progress*; Experimental Committee, Reports and Minutes, June 17, 1912, Acc. 1110, Reel 1, L&N, Hagley; C. E. Kenneth Mees, "The Organization of Industrial Scientific Research," *Science* 43, no. 1118 (1916): 763–773; Experimental Committee, Reports and Minutes, Development Department Annual Report, May 31, 1916, July 5, 1916, Acc. 1110, Reel 1, L&N, Hagley.

77. Executive Committee Minutes, Leeds & Northrup Company, December 5, 1918, Acc. 1110, Reel 1, L&N, Hagley.

78. "Speech: 'Our Company—Leeds & Northrup' A Unit of General Signal (Company History and Growth)," March 6, 1989, L&N, Hagley.

79. Executive Committee Minutes, Leeds & Northrup Company, August 19, 1919, Acc. 1110, Reel 2, L&N, Hagley; "Speech: 'Our Company—Leeds & Northrup' A Unit of General Signal (Company History and Growth)"; Experimental Committee, Reports and Minutes, May 31, 1920, Acc. 1110, Reel 1, L&N, Hagley; Cohn, "The Way We Were," 6.

80. Development Committee, January 30, 1923, Acc. 1110, Reel 1, L&N, Hagley.

81. Ibid.

82. Development Committee, June 10, 1924, Acc. 1110, Reel 3, L&N, Hagley.

83. Vogel Jr., *Precision, People and Progress*, 99–100; "Speech: 'Our Company—Leeds & Northrup,' A Unit of General Signal (Company History and Growth)," 6; Federal Trade Commission, *Control of Power Companies*, xxxviii; "Development Committee."

84. Development Committee, June 10, 1924; Felix Wunsch, "System of Frequency or Speed Measurement and Control," US Patent 1751538, filed March 27, 1925, and issued March 25, 1930; Cohn, "The Way We Were," 6.

85. Wunsch, "System of Frequency and Speed Measurement and Control."

86. Philip Sporn, Reprint of "Progress in Power" speech given to the Edison Electric Institute, Los Angeles, California, June 16, 1955, in *Vistas in Electric Power*, 1st ed. (Oxford: Pergamon Press, 1968), 325. Philip Sporn (1896–1978) obtained a degree in electrical engineering in 1918 and began his career in the power industry at American Gas and Electric Company, later the American Electric Power Company, in 1920. He held the positions of chief electrical engineer (1927), chief engineer (1933), vice president in charge of engineering activities (1934), director (1943), executive vice president (1947), and president and chief executive officer (1947). Sporn was a pioneer in the development and use of a variety of advanced power system technologies and also actively participated in exchanges with other power system experts. He was widely respected nationally and internationally. "Philip Sporn Biography," *Engineering and Technology Wiki* (ETHW), last modified February 16, 2016, http://ethw.org/Philip_Sporn.

87. Development Committee, October 19, 1926, Acc. 1110, Reel 7, L&N, Hagley.

88. *Frequency Measurement and Control, Bulletin No. 985, No. 3 in Power Plant Series* (Philadelphia: Leeds & Northrup Company, 1927); Development Committee Misc. Report 180, 2-6-28, February 6, 1928, Acc. 1110, Reel 7, L&N, Hagley.

89. Development Committee Misc. Report 180, 2-6-28.

90. Ibid.

91. Roux, "Load Dispatching System of the Philadelphia Electric Company," 473.

92. Ibid.

93. Jollyman, "Operation of Interconnected Systems."

94. Nevin E. Funk, "The Economic Value of Major System Interconnections," *Journal of the Franklin Institute* 212, no. 2 (1931): 171–208; Roux, "Load Dispatching System of the Philadelphia Electric Company"; Stone, "The Function of the Load Dispatcher," 469.

95. Elden, "Notes on Operation of Large Interconnected Systems," 1126.

96. Humphrey, "The Interconnection of Power Systems Surrounding the Pittsburgh District," 255.

97. W. E. Mitchell, "Progress and Problems from Interconnection in Southeastern States," *Transactions of the American Institute of Electrical Engineers* 47, no. 2 (1928): 382–392.

98. Edgar, "Discussion," 414.

99. Thank you to Joseph Pratt for suggesting the phrasing "shared management and divided authority."

100. Singer, "Power to the People."

101. W. C. L. Eglin, "Symposium on Interconnection Conowingo Hydroelectric Project with Particular Reference to Interconnection," *Transactions of the American Institute of Electrical Engineers* 47, no. 2 (1928): 373–374; Singer, "Power to the People," 106–117.

102. Eglin, "Symposium on Interconnection Conowingo Hydroelectric Project with Particular Reference to Interconnection," 376–377.

103. Funk, "The Economic Value of Major System Interconnections," 206.

104. Eglin, "Symposium on Interconnection Conowingo Hydroelectric Project with Particular Reference to Interconnection," 376–377.

105. Ibid. The PNJ did not adopt automatic load and frequency control until much later than other interconnected groups.

106. H. B. Gear, "Interconnection and Power Development in Chicago and the Middle West," *Transactions of the American Institute of Electrical Engineers* 47, no. 2 (1928): 402.

107. Humphrey, "The Interconnection of Power Systems Surrounding the Pittsburgh District," 254.

108. Gear, "Interconnection and Power Development in Chicago and the Middle West," 401.

109. Ibid., 404.

110. Edgar, "Discussion," 419.

111. Mitchell, "Progress and Problems from Interconnection in Southeastern States," 384.

112. Gear, "Interconnection and Power Development in Chicago and the Middle West"; Mitchell, "Progress and Problems from Interconnection in Southeastern States"; Edgar, "Discussion."

113. Nathan Cohn, "Developments in Computer Control of Interconnected Power Systems: Exercises in Cooperation and Coordination among Independent Entitites, from Genesis to Columbus" (paper presented to the Measurement, Computation, and Control Section of the South African Institute of Electrical Engineers, 75th Anniversary Year, Johannesburg, Durban, and Capetown, South Africa, 1984), 2.

114. Humphrey, "The Interconnection of Power Systems Surrounding the Pittsburgh District"; H. S. Fitch, "Some Phases of Operation of Interconnected System,"

N.A.C.A. Bulletin 19, no. 5 (1932): 283–290; quote in S. L. Kerr, "Frequency and Load Control on Electric System," *Power Plant Engineering* 34, no. 12 (1930): 682.

115. Hamilton, "A Discussion of Frequency and Load Control Presented at Meeting of System Operating Committee, C. D. Cordes, Chairman, Pennsylvania Electric Association."

116. "Frequency Control."

117. Edgar, "Discussion," 408.

118. Ibid., 411.

119. G. M. Keenan, "Interconnection Development and Operation," *Electrical Engineering* 51, no. 3 (1932): 177–179.

120. "Statement by the Leeds & Northrup Company for the Annual Report of the Operating Committee of the Empire State Gas & Electric Associaton" (Empire State Gas & Electric Association, 1930).

121. Robert Brandt, "Automatic Frequency Control," *Electrical World* 93, no. 8 (1929): 387.

122. Ibid., 388.

123. Development Committee Misc. Report 192, 7-24-28, July 24, 1928, Acc. 1110, Reel 8, L&N, Hagley.

124. Development Committee, December 10, 1929, Acc. 1110, Reel 13, L&N, Hagley.

125. Development Committee Misc. Report 268, August 15, 1929, Acc. 1110, Reel 11, L&N, Hagley.

126. Ibid.

127. Development Committee Misc. Report 287, December 10, 1929, Acc. 1110, Reel 12, L&N, Hagley.

128. Ibid.

129. Ibid.

130. Development Committee Misc. Report 288, February 25, 1930, Acc. 1110, Reel 12, L&N, Hagley.

131. Leslie O. Heath, "Apparatus for Speed Control, United States," filed February 14, 1933; Edgar D. Doyle and Leslie O. Heath, "Method and Apparatus for Controlling Alternating Current Generating Units, United States," filed December 18, 1934.

132. Quote in Development Committee Misc. Report 287; Development Committee Misc. Report 288, December 10, 1929, Acc. 1110, Reel 12, L&N, Hagley.

133. Development Committee Misc. Report 287.

134. Ibid.; quote in Development Committee, July 8, 1930, Acc. 1110, Reel 13, L&N, Hagley.

135. Cohn, "Recollections of the Evolution of Realtime Control Applications to Power Systems," 146.

136. Lloyd F. Hunt and Hydraulic Power Committee Subcommittee on Automatic Frequency Control, "Automatic Frequency Control in Hydroelectric Plants," *Electrical West* 64, no. 6 (1930): 337–354; R. Bailey, "Fundamental Plan of Power Supply in the Philadelphia Area," *Transactions of the American Institute of Electrical Engineers* 49, no. 2 (1930): 605–620; Philip Sporn and W. M. Marquis, "Frequency, Time and Load Control in Interconnected Systems," *Electrical World* (1932): 495, 618; Brandt, "Automatic Frequency Control"; Development Committee Technical Report #333, July 8, 1930, Acc. 1110, Reel 12, L&N, Hagley. Utilities conducting field tests and mentioned in the minutes of the Development Committee include Philadelphia Electric Company, Pennsylvania Power & Light Company, New England Power Company, and Commonwealth Edison.

137. Development Committee Technical Report #333.

138. Development Committee Misc. Report 287; Development Committee Misc. Report 180, 2-6-28; Development Committee Technical Report #337, September 2, 1930, Acc. 1110, Reel 8, L&N, Hagley; Development Committee, July 8, 1930; "Development Committee Miscellaneous Report No. 340"; quote in "Development Committee Technical Report #333." This reflects a characteristic of the industry dating back to the multitude of lawsuits between General Electric and Westinghouse regarding early lighting systems. After years of litigation, the companies decided to pool patents in 1896 and carry on with their work. Harold C. Passer, *The Electrical Manufacturers, 1875–1900: A Study in Competition, Entrepreneurship, Technical Change, and Economic Growth*, Technology and Society (New York: Arno Press, 1972).

139. Memorandum: Automatic Frequency Control, March 20, 1930, Box 40, Nathan Cohn Papers, MIT.

140. Sporn, Reprint of "Progress in Power Transmission" (paper presented to meeting of the Edison Electric Institute, Los Angeles, California, June 16, 1955), 325.

141. Cohn, "Historical Perspectives."

142. Ibid.; Factors Effecting the Utilization of Hydroelectric Plants for Automatic Control, November 30, 1931, Box 44, Nathan Cohn Papers, MIT; Kerr, "Frequency and Load Control on Electric System."

143. Hamilton, "A Discussion of Frequency and Load Control Presented at Meeting of System Operating Committee, C. D. Cordes, Chairman, Pennsylvania Electric Association," 2.

144. NELA Subcommittee Frequency Control Status, October 2, 1930, Electrical Power Systems Records, Hagley.

145. H. S. Fitch, "The Pennsylvania-Ohio-West Virginia Interconnection," *Transactions of the American Institute of Electrical Engineers* 50, no. 4 (1931): 1264–1274; Merriam, Memorandum to R. L. Thomas: New England Trip General Summary, August 28, 1931, Electrical Power Systems Records, Hagley; Heath, "N.E.L.A. Subcommittee Frequency Control Status."

146. Memorandum to Development Committee: Your Note of March 3, 1931 on Frequency Control for Crawford Station of Commonwealth Edison Company, October 2, 1931, Electrical Power Systems Records, Hagley.

147. Handwritten Letter from Morehouse to Heath: Frequency Control Philo, March 30, 1932, 1931, Electrical Power Systems Records, Hagley.

148. Memorandum to Leslie Heath: Pennsylvania-Ohio Frequency Control Tests, December 7, 1931, Electrical Power Systems Records, Hagley.

149. Memorandum to Heath: Interconnection Frequency Control Test, November 30, 1931, Electrical Power Systems Records, Hagley.

150. Memorandum to Heath: Field Tests of Frequency Control Equipment on Penn-Ohio Interconnection, December 29, 1931, Electrical Power Systems Records, Hagley.

151. Report on System Operations Meeting at Windsor Station, West Pennsylvania Power Pool, July 16, 1932, 1932, Electrical Power Systems Records, Hagley; Resume of Events, March 30, 1932, Electrical Power Systems Records, Hagley; S. B. Morehouse, Report of Parallel Tests of Crawford Station, June 6, 1934, Electrical Power Systems Records, Hagley; Report, May 1932, Electrical Power Systems Records, Hagley.

152. Quote in Sporn and Marquis, "Frequency, Time and Load Control in Interconnected Systems," 623; K. N. Reardon, Letter to Langstaff, H. A. P., Copy to Moat, December 7, 1931, Electrical Power Systems Records, Hagley.

153. Sporn and Marquis, "Frequency, Time and Load Control in Interconnected Systems," 624.

154. Cohn, "Recollections of the Evolution of Realtime Control Applications to Power Systems," 148; Hunt and Subcommittee on Automatic Frequency Control, "Automatic Frequency Control in Hydroelectric Plants," 337, 348–350.

155. Hunt and Subcommittee on Automatic Frequency Control, "Automatic Frequency Control in Hydroelectric Plants."

156. History and Highlights, Undated, Box 40, Nathan Cohn Papers, MIT.

157. G. M. Keenan, "Interconnection Development and Operation," *Transactions of the American Institute of Electrical Engineers* 50, no. 4 (1931): 1275, 1277.

158. Robert Brandt, "To Control System Frequency," *Electrical World* 104, no. 10 (1934): 71.

159. Ibid.

160. Reardon, Letter to Langstaff, H.A.P., Copy to Moat; Merriam, Memorandum to R. L. Thomas: New England Trip General Summary; Morehouse, Report of Parallel Tests of Crawford Station.

4 Balancing Reliability and Economy, 1930–1940

1. *Report on the Status of Interconnections and Pooling of Electric Utility Systems in the United States* (Washington, DC: Edison Electric Institute, 1962).

2. Philip Sporn, "Interconnected Electric Power Systems," *Electrical Engineering* 57, no. 1 (1938): 16–25.

3. Tennessee Valley Authority Act of 1933, 16 U.S.C. § 831 (1933); National Industrial Recovery Act of 1933, 15 U.S.C. § 703 (1933); Rural Electrification Act of 1936, 7 U.S.C. § 901 (1936); Bonneville Project Act of 1937, 16 U.S.C. § 832 (1937).

4. Public Utility Holding Company Act of 1935, 15 U.S.C. § 79 (1935); Federal Power Act of 1935, 16 U.S.C. § 791 (1935).

5. *National Power Survey Interim Report*, Federal Power Commission, 1935, accessed April 1, 2015, http://babel.hathitrust.org/cgi/pt?id=mdp.39015021243368;view=1up;seq=93.

6. *National Power Survey Interim Report*, part 1, 54.

7. Federal Water Power Act of 1920, 16 U.S.C. § 791 (1920); Jerome G. Kerwin, *Federal Water-Power Legislation*, Studies in History, Economics, and Public Law (New York: Columbia University Press, 1926).

8. Federal Water Power Act of 1920; Kerwin, *Federal Water-Power Legislation*; Walter H. Voskuil, "Water-Power Situation in the United States," *Journal of Land & Public Utility Economics* 1, no. 1 (1925): 89–101.

9. *Annual Report of the Federal Power Commission* (Washington, DC: Government Printing Office, 1921, 1922, 1923, 1924).

10. Public Utilities Commission of Rhode Island v. Attleboro Steam and Electric Company, 273 US 83 (1927).

11. Ibid.; F. G. Crawford, "Control of Interstate Transmission of Electricity," *Journal of Land & Public Utility Economics* 5, no. 3 (1929): 229–234.

12. Federal Power Act of 1930, 16 U.S.C. § 791 (1930).

13. Harold L. Platt, *The Electric City: Energy and the Growth of the Chicago Area, 1880–1930* (Chicago: University of Chicago Press, 1991), 271–273.

14. Franklin D. Roosevelt, "The 'Portland Speech': A Campaign Address on Public Utilities and Development of Hydro-Electric Power, Portland Oregon, September 21, 1932," the New Deal Network, accessed November 7, 2007, http://newdeal.feri.org/speech/1932a.htm.

15. Philip J. Funigiello, *Toward a National Power Policy: The New Deal and the Electric Utility Industry, 1933–1941* (Pittsburgh: University of Pittsburgh Press, 1973).

16. Franklin D. Roosevelt, "The Portland Speech."

17. Ibid.

18. *Historical Statistics of the United States, Colonial Times to 1970, Bicentennial Edition* (Washington, DC: Government Printing Office, 1975).

19. Edward Eyre Hunt, *The Power Industry and the Public Interest, a Summary of a Survey of the Relations between the Government and the Electric Power Industry* (New York: Twentieth Century Fund, 1944), 190; Thomas K. McCraw, *TVA and the Power Fight, 1933–1939* (Philadelphia: Lippincott, 1971), 85.

20. National Industrial Recovery Act of 1933, Title II, Section 202, 15 U.S.C. § 703.

21. Arthur D. Gayer, *Public Works in Prosperity and Depression* (New York: National Bureau of Economic Research, 1935), 106.

22. Jay L. Brigham, *Empowering the West: Electrical Politics before FDR*, Development of Western Resources (Lawrence: University Press of Kansas, 1998); Martin V. Melosi, *Coping with Abundance: Energy and Environment in Industrial America*, 1st ed. (Philadelphia: Temple University Press, 1985), 130, 132–133. Congress also considered, and rejected, proposals to create seven river valley authorities similar to the TVA. Ronald C. Tobey, *Technology as Freedom: The New Deal and the Electrical Modernization of the American Home* (Berkeley: University of California Press, 1996).

23. Tobey, *Technology as Freedom.*

24. The name of this project changed from Boulder Dam to Hoover Dam in 1930, back to Boulder Dam in 1933, and finally back to Hoover Dam in 1947. Because the changes took place during the years covered in this chapter, the name will appear as Boulder/Hoover Dam in the text. "The Controversial Naming of the Dam," Public Broadcasting System, accessed January 9, 2017, http://www.pbs.org/wgbh/americanexperience/features/general-article/hoover-controversy/.

25. Tobey, *Technology as Freedom*, 113–115; Brigham, *Empowering the West*, 141.

26. For a detailed discussion of Roosevelt's policy efforts, see Funigiello, *Toward a National Power Policy*; Ellis Wayne Hawley, *The New Deal and the Problem of Monopoly: A Study in Economic Ambivalence* (New York: Fordham University Press, 1995).

27. McCraw, *TVA and the Power Fight, 1933–1939.*

28. *National Power Survey Interim Report*, 261.

29. Robert L. Bradley Jr., "The Origins of Political Electricity: Market Failure or Political Opportunism," *Energy Law Journal* 17, no. 59 (1996): 85.

30. Energy Information Administration, *Public Utility Holding Company Act of 1935: 1935–1992*, reprint (Washington, DC: Government Printing Office, 1993); Funigiello, *Toward a National Power Policy*; William Lasser, *Benjamin V. Cohen: Architect of the New Deal* (New Haven: Yale University Press, 2002).

31. Congressional acts affecting the work of the FPC included the Tennessee Valley Authority Act of 1933, the Federal Power Act of 1935, the Bonneville Project Act of 1937, the Natural Gas Act of 1938, the Flood Control Act of 1938, and the Fort Peck Act of 1938. Clyde L. Seavey, "Functions of the Federal Power Commission," *The Annals of the American Academy of Political and Social Science* 201 (1939): 73–81. For more on regulation of natural gas pipelines, see Christopher James Castaneda, *Regulated Enterprise: Natural Gas Pipelines and Northeastern Markets, 1938–1954*, Historical Perspectives on Business Enterprise Series (Columbus: Ohio State University Press, 1993).

32. *Twentieth Annual Report of the Federal Power Commission: Fiscal Year Ended June 30, 1940 with Additional Activities to December 1940* (Washington, DC: Government Printing Office, 1941), 4.

33. As quoted in McCraw, *TVA and the Power Fight, 1933–1939*, 52.

34. Ibid., 109.

35. Federal Power Commission, *Ninth Annual Report of the Federal Power Commission* (Washington, DC: Government Printing Office, 1929), 3.

36. For more on the negotiations for electricity and water from Boulder/Hoover Dam, see Sarah Elkind, *How Local Politics Shape Federal Policy: Business, Power, and the Environment in Twentieth Century Los Angeles*, ed. William H. Becker, the Luther H. Hodges Jr. and Luther H. Hodges Sr. Series on Business, Society, and the State (Chapel Hill: University of North Carolina Press, 2011); Andrew Needham, *Power Lines: Phoenix and the Making of the Modern Southwest*, ed. William Chafe et al., Politics and Society in Twentieth-Century America (Princeton, NJ: Princeton University Press, 2014).

37. S. B. Morehouse, Letter to United States Department of the Interior, Bureau of Reclamation: Subject: Automatic Frequency-Time-Load Control for Grand Coulee Power Plant, October 10, 1939, attachment titled "Summary of Regulating Methods on Principal Interconnected Power Systems and Recommendations on Automatic Load-Frequency Control for Grand Coulee Dam of the Bureau of Reclamation," 1, Box 43, Nathan Cohn Papers, MIT.

38. *National Power Survey Interim Report*, xi.

39. S. B. Morehouse, "Some Historical Highlights in the Operation of Electric Utility Systems on an Interconnected and Coordinated Basis" (paper presented at the Annual Joint Meeting of the Northeast Regional Committee of the Interconnected Systems Group and the System Operation Committee of the Pennsylvania Electric Association, Pittsburgh, Pennsylvania, 1965), 8.

40. Ibid.

41. E. C. Stone, "Some Problems in the Operation of Power Plants in Parallel," *Transactions of the American Institute of Electrical Engineers* 38, no. 2 (1919): 1651–1674.

42. Nathan Cohn, "Recollections of the Evolution of Realtime Control Applications to Power Systems," *Automatica* 20, no. 2 (1984): 147.

43. C. H. Linder et al., "Telemetering," *Transactions of the American Institute of Electrical Engineers* 48, no. 3 (1929): 766–772.

44. Morehouse, "Some Historical Highlights in the Operation of Electric Utility Systems on an Interconnected and Coordinated Basis," 8.

45. Robert Brandt, "Historical Approach to Speed and Tie-Line Control," *Transactions of the American Institute of Electrical Engineers. Part III: Power Apparatus and Systems*, 72, no. 2 (1953): 8.

46. *Compendex*, s.v. "telemeter," Engineering Village, accessed December 12, 2016, https://www.engineeringvillage.com/. A search for the term "telemeter" on *Compendex* for the years 1900–1945 produced 176 results. Of these, only 13 appeared in publications before 1928. The rest of the articles were evenly spread across the remaining years, with a low of only three published in 1943, a high of fifteen published in 1945, and an average of nine per year.

47. *Telemetering and Totalizing Station Loads, Bulletin No. 874, No. 4 Power Plant Series* (Philadelphia: Leeds & Northrup Company, 1933).

48. W. J. Lank, "Interconnection Economies through Telemeter Totalizing," *Electrical World* 103, no. 26 (1934): 948–949.

49. F. Zogbaum, "Load Totalizing in the New York Area," *Transactions of the American Institute of Electrical Engineers* 53, no. 6 (1934): 886–889.

50. A. Trenner, "Automatic Telemetering and Supervisory Control," *AEG Progress (English Edition)*, no. 2 (1935): 21–24; *A-C Network Operation 1936–1937* (New York: Edison Electric Institute, 1939); Iwane Schigyo and Takashi Hioki, "Power Line Carrier Telemeter," *Shibaura Review* 16, no. 4 (1937): 139–145; S. Jimbo and T. Ito, "Carrier-Current Telemeter," *Electrotechnical Journal—Japan* 2, no. 2 (1938): 32–35; B. O. Mongain, "Electrical Devices for Distant Supervision and Control of Engineering Plant," *Transactions of the Institution of Civil Engineers of Ireland* 64 (1938): 221–236; F. Jaggi, "Telemetering Equipments for Transmission of Any Desired Measurements over Long Distances," *Brown Boveri Review* 32, no. 4 (1945): 147–148.

51. R. L. Wegel and C. R. Moore, "An Electrical Frequency Analyzer," *Transactions of the American Institute of Electrical Engineers* 43 (1924): 465.

52. H. L. Hazen, O. R. Schurig, and M. F. Gardner, "The M.I.T. Network Analyzer Design and Application to Power System Problems," *Transactions of the American Institute of Electrical Engineers* 49, no. 3 (1930): 1108.

53. H. L. Hazen and M. F. Gardner, "Solving System Problems by Means of the Power Network Analyser," *Power Plant Engineering* 33, no. 22 (1929): 1220–1222; Hazen, Schurig, and Gardner, "The M.I.T. Network Analyzer Design and Application to Power System Problems," 1102; "Solution of Commercial Power-System Problems on M.I.T. Network Analyzer," *Massachusetts Institute of Technology—Department of Electrical Engineering* (1931): 9; "M.I.T. Network Analyser," *Electricien* 106, no. 2770 (1931): 22–23.

54. F. Preston, "Vannevar Bush's Network Analyzer at the Massachusetts Institute of Technology," *Annals of the History of Computing, IEEE* 25, no. 1 (2003): 77.

55. Hazen, Schurig, and Gardner, "The M.I.T. Network Analyzer Design and Application to Power System Problems," 1104.

56. John Casazza, *The Development of Electric Power Transmission: The Role Played by Technology, Institutions, and People,* IEEE Case Histories of Achievement in Science and Technology (New York: Institute of Electrical and Elecronics Engineers, 1993), 80–84; Preston, "Vannevar Bush's Network Analyzer at the Massachusetts Institute of Technology."

57. While best known for his role in the Manhattan Project during World War II, Vannevar Bush's work at MIT on the Network Analyzer laid the theoretical groundwork for digital circuit design. Ioan James, "Claude Elwood Shannon 30 April 1916–24 February 2001," *Biographical Memoirs of Fellows of the Royal Society* 55 (2009): 257–265.

58. H. P. Kuehni and R. G. Lorraine, "A New A-C Network Analyzer," *Transactions of the American Institute of Electrical Engineers* 57, no. 2 (1938): 67–73; Karl L. Wildes and Nilo A. Lindgren, *A Century of Electrical Engineering and Computer Science at MIT, 1882–1982* (Cambridge, MA: MIT Press, 1985), 103; Edith Clarke and S. B. Crary, "Stability Limitations of Long-Distance A-C Power-Transmission Systems," *Transactions of the American Institute of Electrical Engineers* 60, no. 12 (1941): 1051–1059; Preston, "Vannevar Bush's Network Analyzer at the Massachusetts Institute of Technology."

59. For a discussion of the field test of automatic frequency control equipment at Windsor station, see chapter 3.

60. Development Committee Miscellaneous Report No. 371, July 17, 1934, Acc. 1110, Reel 10, L&N, Hagley; Albert J. Williams and Stephen B. Morehouse, "Electrical Generating System," US Patent 2124724, filed July 8, 1936, and issued July 26, 1938.

61. Development Committee, November 14, 1934, Acc. 1110, Reel 3, L&N, Hagley; Remote Metering Demand and Integration, 1930, Box 43, Nathan Cohn Papers, MIT; *General Electric Torque Balance Telemetering Bulletin* (General Electric Company, 1931); Memorandum to Cohn, Emerich, Moat, Robinson, Wyeth, Greer, Cleeland, Keene, April 16, 1931, Box 43, Nathan Cohn Papers, MIT; Memorandum to Cohn, Emerich, Robinson, Moat, Morehouse, Wyeth, June 1, 1931, Box 43, Nathan Cohn Papers, MIT; Memorandum to Technical Division Salesmen, March 20, 1933, Box 43, Nathan Cohn Papers, MIT.

62. Memorandum to N. Cohn Regarding Lincoln Thermal Converters, January 22, 1934, Box 43, Nathan Cohn Papers, MIT; Memorandum to L. O. Heath Regarding Load Recording, September 17, 1934, Box 43, Nathan Cohn Papers, MIT; Memorandum to L. Heath Regarding Lincoln Thermal Converters, January 18, 1934, Box 43, Nathan Cohn Papers, MIT; Letter to Leeds & Northrup Company, Attn.: L. O. Heath, September 5, 1934, Box 43, Nathan Cohn Papers, MIT; Letter to Leeds & Northrup Company, Attn.: L. O. Heath, August 23, 1934, Box 43, Nathan Cohn Papers, MIT; Cohn, "Historical Perspectives" (paper presented to the Professional Workshop on Power Systems Control, Electric Power Research Institute, April 28–29, 1977), 7; David P. Billington and Donald C. Jackson, *Big Dams of the New Deal Era: A Confluence of Engineering and Politics* (Norman: University of Oklahoma Press, 2006), 143.

63. *The Hoover Dam Power and Water Contracts and Related Data* (Washington, DC: Government Printing Office, 1933).

64. Nathan Cohn, Memorandum to Emerich Re: Boulder Dam Generating Station, April 17, 1934, Box 4, Nathan Cohn Papers, MIT.

65. Nathan Cohn, Memorandum to Emerich Re: Boulder Dam Frequency Load Control, March 20, 1935, Box 43, Nathan Cohn Papers, MIT.

66. Billington and Jackson, *Big Dams of the New Deal Era,* 125; Various correspondence, Box 43, Nathan Cohn Papers, MIT.

67. Cohn, Memorandum to Emerich Re: Boulder Dam Frequency Load Control.

68. Ibid.

69. Cohn, "Historical Perspectives," 8.

70. Ibid.

71. Ibid.

72. Cohn, "Recollections of the Evolution of Realtime Control Applications to Power Systems," 153.

73. Ibid., 154.

74. Ibid.

75. Ibid.

76. Ibid.

77. Morehouse, Letter to United States Department of the Interior, Bureau of Reclamation: Subject: Automatic Frequency-Time-Load Control for Grand Coulee Power Plant, Cover Letter.

78. Ibid., attachment titled "Summary of Regulating Methods on Principal Interconnected Power Systems and Recommendations on Automatic Load-Frequency Control for Grand Coulee Dam of the Bureau of Reclamation."

79. Ibid., 11.

80. Ibid.

81. Ibid.

82. Ibid.

83. Earle Wild, "Methods of System Control in a Large Interconnection," *Transactions of the American Institute of Electrical Engineers* 60, no. 5 (1941): 234–235.

84. Nathan Cohn, "Developments in Computer Control of Interconnected Power Systems: Exercises in Cooperation and Coordination among Independent Entitites, from Genesis to Columbus" (paper presented to the Measurement, Computation, and Control Section of the South African Institute of Electrical Engineers, 75th Anniversary Year, Johannesburg, Durban, Capetown, South Africa, 1984), 3.

5 Power Transformations on the Home Front, 1935–1950

1. *Historical Statistics of the United States, Colonial Times to 1970, Bicentennial Edition* (Washington, DC: Government Printing Office, 1975).

2. *Directory of Electric Utilities in the United States, 1941* (Washington, DC: Federal Power Commission, 1941); *Electric Power Statistics, 1920–1940* (Washington, DC: Federal Power Commission, 1941).

3. *Sixteenth Annual Report of the Federal Power Commission* (Washington, DC: Government Printing Office, 1936).

4. See chapter 3 for a discussion of the effect of the 1935 Public Utility Holding Company Act on the number of holding companies and their subsidiaries.

5. *Historical Statistics of the United States, 1789–1945; a Supplement to the Statistical Abstract of the United States* (Washington, DC: Government Printing Office, 1949).

6. *Electric Power Statistics, 1920–1940.*

7. H. S. Bennion, "Electric Power in American Industry," *Military Engineer* 32, no. 186 (1940): 393–396.

8. *Nineteenth Annual Report of the Federal Power Commission* (Washington, DC: Government Printing Office, 1940); *National Power Survey: A Report by the Federal Power Commission, 1964* (Washington, DC: Government Printing Office, 1964).

9. "The 150,000-Volt Big Creek Development—I," *Electrical World* 63, no. 1 (1914): 33–38; "The World's Largest Transmission Line," *Electrical World* 63, no. 2 (1914): 71; "The Big Creek Transmission System," *Electrical World* 64, no. 14 (1914): 646.

10. "Trade Publications," *Electrical World* 64, no. 12 (1914): 591.

11. *National Power Survey: A Report by the Federal Power Commission, 1964*, 149.

12. See chapter 3 for related discussions of the expanded federal role in both regulation of the power industry and investment in power infrastructure.

13. Franklin D. Roosevelt, "Executive Order 6251 Designating the Federal Power Commission an Agency of the Public Works Administration," August 19, 1933, available online at Gerhard Peters and John T. Woolley, *The American Presidency Project*, accessed May 13, 2015, http://www.presidency.ucsb.edu/ws/?pid=14504.

14. Letter of Transmittal, *National Power Survey, Interim Report, Power Series No. 1* (Washington, DC: Government Printing Office, 1935).

15. Ibid., xi.

16. Ibid.

17. Ibid., 7.

18. Douglas D. Anderson, *Regulatory Politics and Electric Utilities: A Case Study in Political Economy* (Boston: Auburn House, 1981); Philip J. Funigiello, *Toward a National Power Policy: The New Deal and the Electric Utility Industry, 1933–1941* (Pittsburgh: University of Pittsburgh Press, 1973); Ellis Wayne Hawley, *The New Deal and the Problem of Monopoly: A Study in Economic Ambivalence* (New York: Fordham University Press, 1995); Thomas K. McCraw, *TVA and the Power Fight, 1933–1939* (Philadelphia: Lippincott, 1971); Ronald C. Tobey, *Technology as Freedom: The New Deal and the Electrical Modernization of the American Home* (Berkeley: University of California Press, 1996).

19. Funigiello, *Toward a National Power Policy*; McCraw, *TVA and the Power Fight, 1933–1939*.

20. *Decisions of the Comptroller General of the United States* (Washington, DC: Government Printing Office, 1939).

21. "Sees Utilities Set for Defense Task," *New York Times*, August 10, 1940, 39; "Power for Arms Is Called Ample," *New York Times*, August 30, 1941, 35.

22. Philip J. Funigiello, "Kilowatts for Defense: The New Deal and the Coming of the Second World War," *Journal of American History* 56, no. 3 (1969): 604–620.

23. Franklin D. Roosevelt, "Letter Merging the National Defense Power Committee and the National Power Policy Committee," October 13, 1939, available online at Gerhard Peters and John T. Woolley, *The American Presidency Project*, accessed May 13, 2015, http://www.presidency.ucsb.edu/ws/?pid=15824.

24. *Decisions of the Comptroller General of the United States*; Funigiello, "Kilowatts for Defense," 604–620; Funigiello, *Toward a National Power Policy*; McCraw, *TVA and the Power Fight, 1933–1939*.

25. Leland Olds, "Forecasting Defense Power," *Electrical World* 114 (1940): 55–58.

26. Ibid., 56.

27. *Twentieth Annual Report of the Federal Power Commission* (Washington, DC: Government Printing Office, 1941), 6.

28. Bennion, "Electric Power in American Industry," 393–396; M. W. Smith, "War Emergency Power from Present Systems," *Electrical World* 112, no. 3 (1939): 45–94; Thomas P. Swift, "Utilities Geared to Aid Defense," *New York Times*, June 5, 1940, 41; "Utilities Pledge Aid for Defense," *New York Times*, June 6, 1940, 43, 48; "Defense Is Theme of Utility Session," *New York Times*, June 2, 1940, 57.

29. "Sees Utilities Set for Defense Task," 39.

30. Ibid.

31. Ibid.

32. Edward Eyre Hunt, *The Power Industry and the Public Interest, a Summary of a Survey of the Relations between the Government and the Electric Power Industry* (New York: Twentieth Century Fund, 1944), 147.

33. "Southeast to Pool Power for Defense," *New York Times*, May 26, 1941, 17; "Rationing of Power Faces Southeast Due to Wide, Long Extended Drought," *Christian Science Monitor*, May 27, 1941, 17; "Power Rationing Plan Ready for Southeast," *Wall Street Journal*, June 7, 1941, 3; *Opinions and Decisions of the Federal Power Commission, with Appendix of Selected Orders in the Nature of Opinions*, ed. Federal Power Commission, vol. 2 (Washington, DC: Government Printing Office, 1943), 990–999; Raymond Daniell, "Drought Dims Lights in Dixie to Save Power," *Washington Post*, June 15, 1941, 16.

34. "Utilities Are Seen Facing More Curbs," *New York Times*, December 26, 1942, 19.

35. "Parleys to Weigh Power Emergency," *New York Times*, June 2, 1941, 29.

36. W. H. Lawrence, "Power Unit Set Up to Spur Defense," *New York Times*, July 22, 1941, 9.

37. *Nineteenth Annual Report of the Federal Power Commission*, No. 496; *Production of Electric Energy and Capacity of Generating Plants* (Washington, DC: Government Printing Office, 1941); *Twentieth Annual Report of the Federal Power Commission*, 21; quote in Lawrence, "Power Unit Set Up to Spur Defense."

38. Thomas P. Swift, "Pooling Speeded in Electric Field," *New York Times*, August 3, 1941, 82, 84; Hunt, *The Power Industry and the Public Interest*; "Utility Program Speeded by SEC," *New York Times*, November 9, 1941, F1, F5; "Power Emergency Seen," *New York Times*, December 17, 1941, 45.

39. Exec. Order No. 9024, 7 FR 302, January 16, 1942; "Power Authority Received by WPB," *New York Times*, April 30, 1942, 38.

40. "F.P.C. Waives Policy on Power Connections," *New York Times*, October 17, 1942, 20; *Opinions and Decisions of the Federal Power Commission, with Appendix of Selected Orders in the Nature of Opinions*, vol. 2, 990–998, 1021–1026, 1059–1060, 1095–1096; *Opinions and Decisions of the Federal Power Commission, with Appendix of Selected Orders in the Nature of Opinions*, vol. 3 (Washington, DC: Government Printing Office, 1944), 668–669, 678–679, 712–716, 750–751, 765–766, 777–780, 795–799, 834–836, 846–851, 859–862, 869–873, 889–891, 912–916, 920–922, 930–932, 934–935, 1072–1077; *Opinions and Decisions of the Federal Power Commission, with Appendix of Selected Orders in the Nature of Opinions*, vol. 4 (Washington, DC: Government Printing Office, 1946), 406–408, 420–425, 431–434, 438–444, 506–508, 515–517, 528–533, 615–618, 713–715.

41. "Defense of Nation in War Discussed," *New York Times*, January 16, 1941, 15. The *New York Times* report quoted H. S. Bennion, who read a paper by Philip Sporn to the American Society of Civil Engineers at their 88th annual meeting.

42. P. W. Swain, "Power Teamwork for Victory," *Power* 86, no. 9 (1942): 63, 66, 67; W. C. Heston, "Kilowatt-Hours Pooled for War," *Electrical West* 92, no. 3 (1944): 51; "Defense of Nation in War Discussed"; "Utilities Ready for Raids," *New York Times*, March 5, 1942, 25; C. S. Lynch, "Southwest Power Pool," *Electric Light and Power* 20, no. 8 (1942): 41.

43. Israel Putnam, "Wartime Prices That Fell—Gas and Electricity," *Public Utilities Fortnightly* 40, no. 3 (1947): 151–162; "Why Power Curtailment Is Necessary," *The Tuscaloosa News*, November 5, 1941, 6; "Black-Outs Must Continue So Factories Can Run Full-Time," *The Tuscaloosa News*, November 26, 1941, 6.

44. Lynch, "Southwest Power Pool."

45. Electric Bond and Share Company was one of the holding companies that restructured following enactment of the PUHCA in 1935. Ebasco Services, Inc., provided engineering and consulting services after this but did not act as a holding company.

46. Thomas P. Swift, "Plan Is Devised to Increase Power," *New York Times*, August 9, 1942. F1, F3; "Idle Electric Generating Capacity," *Edison Electric Institute—Bulletin* 10,

no. 7 (1942): 249–259; L. Elliott, "Meeting Power Demand during War," *Mechanical Engineering* 64, no. 12 (1942): 872–876; "New Power Line Set up in West," *New York Times*, September 8, 1942, 32.

47. *Electric Power Requirements and Supply in the United States, 1940–1945: War Impact on Electric Utility Industry* (Washington, DC: Federal Power Commission, 1945).

48. *Twenty-Sixth Annual Report of the Federal Power Commission* (Washington, DC: Government Printing Office, 1947), 1; Philip Sporn, reprint of "A Plan for Maintaining Power Supply in a Destructive Emergency" (paper presented to the Industrial College of the Armed Forces, Washington, DC, January 21, 1948), in *Vistas in Electric Power* (New York: Pergamon Press, 1968), 955.

49. *Annual Report of the Administrator of the Bonneville Power Administration to the Secretary of the Interior* (Washington, DC: Government Printing Office, 1943).

50. Heston, "Kilowatt-Hours Pooled for War," 51.

51. Lynch, "Southwest Power Pool," 52.

52. Ibid., 41–45.

53. Ibid.; S. B. Morehouse, "Inter-System Power Coordination in Southwest Region," *Electric Light and Power* 23, no. 12 (1945): 62–68; Lynch, "Southwest Power Pool"; quote in Edward Falck, "Power Pooling during War," *Power Plant Engineering* 49, no. 10 (1945): 85.

54. Falck, "Power Pooling during the War."

55. Ibid., 86.

56. Ibid.

57. Morehouse, "Inter-System Power Coordination in Southwest Region"; Heston, "Kilowatt-Hours Pooled for War"; Lynch, "Southwest Power Pool." The East central region included Indiana, Ohio, West Virginia, Kentucky, western Pennsylvania, and Virginia.

58. C. K. Duff, "Control of Load, Frequency, and Time of Interconnected Systems," *Electrical Engineering* 64, no. 11 (1945): 778.

59. Ibid.

60. "Transcripts from Leeds & Northrup Company's Presentation to American Electric Power Service Corporation," in Electrical Power Systems Records, Hagley, 22.

61. See, for example, Nathan Cohn, "Inadvertent Interchange and Time Error Accumulations—Causes and Coordinated Techniques for Correction" (paper presented to Engineering for the Conservation of Mankind, IEEE Region Six Conference, Sacramento, California, May 13, 1971), reprint by Leeds & Northrup Company, 1971, 1–6; Nathan Cohn, "Hidden Costs of Inadvertent Interchange" (paper presented to

Costs Associatied with AGC, Session No. 12, INSIGHTS '86 Engineering and Operating Forum, Edison Electric Institute, Washington, DC, September 7–10, 1986).

62. Nathan Cohn, "Recollections of the Evolution of Realtime Control Applications to Power Systems," *Automatica* 20, no. 2 (1984): 154. See chapter 3 for additional details about the experiments on the Indiana and Michigan system.

63. Morehouse, "Inter-System Power Coordination in Southwest Region," 65.

64. Ibid., 67.

65. Lynch, "Southwest Power Pool"; quote in Morehouse, "Inter-System Power Coordination in Southwest Region," 105.

66. Heston, "Kilowatt-Hours Pooled for War," 51.

67. S. B. Smith and M. L. Blair, "Automatic Load and Frequency Control in Northwest Power Pool," *Electrical World* 124, no. 25 (1945): 56.

68. Heston, "Kilowatt-Hours Pooled for War," 59.

69. Ibid.

70. Morehouse, "Inter-System Power Coordination in Southwest Region," 68.

71. Ibid.; quote in Robert Brandt, "Historical Approach to Speed and Tie-Line Control," *Transactions of the American Institute of Electrical Engineers. Part III: Power Apparatus and Systems*, 72, no. 2 (1953): 8.

72. "1944 Electricity Production Measured Country's War Effort," *Edison Electric Institute—Bulletin* 13, no. 5 (1945): 125.

73. Swain, "Power Teamwork for Victory," 63.

74. Falck, "Power Pooling during War," 84.

75. J. M. Gaylord, "Integration of Power Systems," *Engineering and Science Monthly* 8, no. 6 (1945): 3.

76. Falck, "Power Pooling during War"; John P. Callahan, "Office Ends Today for War Utilities," *New York Times*, September 30, 1945, F1, F5; quote in Gaylord, "Integration of Power Systems," 3.

77. Elliott, "Meeting Power Demand during War," 873.

78. Ibid.

79. Philip Sporn, reprint of "Realism in Post War Planning," *Electric Light and Power* 22, no. 6 (1944), in *Vistas in Electric Power*, 295.

80. Heston, "Kilowatt-Hours Pooled for War."

81. "1944 Electricity Production Measured Country's War Effort," 126.

82. Sporn, reprint of "A Plan for Maintaining Power Supply in a Destructive Emergency," 955; Walker L. Cisler, "Electric Power and National Defense," *Electrical Engineering* 67, no. 4 (1948): 319–324. *Electric Power Requirements and Supply in the United States, 1940–1945: War Impact on Electric Utility Industry*; "Shortages of Reserve Capacity Tax Systems' Capabilities," *Electrical World* 128, no. 21 (1947): 74–77; W. J. Lyman, "Determination of Required Reserve Generation Capacity," *Electrical World* 127, no. 19 (1947): 92–95; H. W. Phillips, "Determination of Reserve Requirements for Interconnected System," *Edison Electric Institute—Bulletin* 14, no. 4 (1946): 117–120.

83. *Twenty-Sixth Annual Report of the Federal Power Commission*; Richard F. Hirsh, *Technology and Transformation in the American Electric Utility Industry* (Cambridge: Cambridge University Press, 1989), 48–51; David E. Nye, *Consuming Power: A Social History of American Energies* (Cambridge, MA: MIT Press, 1998), 204; Mark H. Rose, *Cities of Light and Heat: Domesticating Gas and Electricity in Urban America* (University Park: Pennsylvania State University Press, 1995), 173–174.

84. Donald M. Nelson, *Arsenal of Democracy: The Story of American War Production*, ed. Frank Frieidel, Da Capo Press Reprint Series (New York: Da Capo Press, 1973), 365. Donald M. Nelson served as director of priorities for the Office of Production Management from 1941 to 1942 and then as chairman of the War Production Board from 1942 to 1944 until it was replaced by the Office of Production Management.

85. Horace M. Gray, "The Integration of the Electric Power Industry," *American Economic Review* 41, no. 2 (1951): 545.

6 Nuances of Control in an Increasingly Interconnected World, 1945–1965

1. John Casazza, *The Development of Electric Power Transmission: The Role Played by Technology, Institutions, and People*, IEEE Case Histories of Achievement in Science and Technology (New York: Institute of Electrical and Electronics Engineers, 1993); Leonard S. Hyman, Andrew S. Hyman, and Robert C. Hyman, *America's Electric Utilities: Past, Present and Future*, 8th ed. (Vienna, VA: Public Utilities Reports, 2005), 151.

2. Nathan Cohn, "Power Flow Control—Basic Concepts for Interconnected Systems," *Electric Light and Power* 28, no. 8–9 (1950): 82.

3. R. O. Usry, "Interconnected Systems Group Test Committee Meeting, Commonwealth Edison Building—Chicago, Illinois, November 19–20, 1959, Minutes," Electrical Power Systems Records, Hagley.

4. Notably, the Edison Electric Institute, which constructed the maps in figures 5.1 and 5.2, began to identify interconnections, not just high-tension transmission lines, after 1946.

5. *National Power Survey: A Report by the Federal Power Commission, 1964* (Washington, DC: Government Printing Office, 1964), 1:14, 200.

6. Ibid., 14.

7. *1970 National Power Survey of the Federal Power Commission* (Washington, DC: Government Printing Office, 1970), I-13-4.

8. Ibid.

9. *Prevention of Power Failures: An Analysis and Recommendations Pertaining to the Northeast Failure and the Reliability of U.S. Power Systems: A Report to the President by the Federal Power Commission* (Washington, DC: Government Printing Office, 1967), 35.

10. *National Power Survey: A Report by the Federal Power Commission, 1964,* 1:14.

11. According to the NERC *Glossary of Terms, spinning reserve* is "unloaded generation that is synchronized and ready to serve additional demand." In other words, the power plants are capable of generating extra power that is not being used at that moment, but since it is already at synchronous speed and connected to the grid, the unit is quickly available for unanticipated demand. *Glossary of Terms Used in NERC Reliability Standards,* North American Electric Reliability Corporation, published online, updated May 19, 2015, http://www.nerc.com/files/glossary_of_terms.pdf.

12. This section is largely drawn from the Edison Electric Institute's (EEI) *Report on the Status of Interconnections and Pooling of Electric Utility Systems in the United States* (New York: Edison Electric Institute, 1962). The EEI acknowledged that the report primarily addressed investor-owned systems but included discussion of government projects when they were part of interconnected systems. Before World War I, engineers sometimes described economy scheduling as a "conservation" practice as well as an economy practice. After the war, "conservation" all but disappeared from the lexicon of power systems operation.

13. Ibid., 123.

14. Ibid., 31.

15. Ibid., 107.

16. Nathan Cohn, "Developments in Computer Control of Interconnected Power Systems: Exercises in Cooperation and Coordination among Independent Entitites, from Genesis to Columbus" (paper presented to the Measurement, Computation, and Control Section of the South African Institute of Electrical Engineers, 75th Anniversary Year, October 25–November 1, 1984, Johannesburg, Durban, Capetown, South Africa).

17. Co-ordination of Desired Generation Computer with Area Control, Discussion by: F. H. Light, Senior Engineer, Economy Division, Philadelphia Electric Company, 1959, Box 3, Nathan Cohn Papers, MIT.

18. J. B. Ward, "Analogue Computer for Use in Design of Servo Systems," *Proceedings of the Institution of Electrical Engineers—Part II, Power Engineering* 99, no. 72 (1952): 521–532.

19. For a discussion of this episode in data technologies and power systems growth that addresses the process of making a transition from analog to digital computing, see Julie Cohn, "Transitions from Analog to Digital Computing in Electric Power Systems," *IEEE Annals of the History of Computing* 37, no. 3 (2015): 32–43. © [2015] IEEE. Some of this material reappears in the following sections by permission.

20. H. H. Johnson and M. S. Umbenhauer, "Effective Load Dividing Device," *Edison Electric Institute Bulletin* 7, no. 8 (1939): 386.

21. E. E. George, "Intrasystem Transmission Losses," *Electrical Engineering* 62, no. 3 (1943): 153–158; Max Jacob Steinberg and Theodore Hunter Smith, *Economy Loading of Power Plants and Electric Systems* (New York: J. Wiley & Sons, 1943).

22. E. E. George, H. W. Page, and J. B. Ward, "Co-ordination of Fuel Cost and Transmission Loss by Use of the Network Analyzer to Determine Plant Loading Schedules," *Transactions of the American Institute of Electrical Engineers* 68, no. 2 (1949): 1163.

23. Ibid., 1162.

24. L. K. Kirchmayer and G. W. Stagg, "Evaluation of Methods of Co-Ordinating Incremental Fuel Costs and Incremental Transmission Losses [Includes Discussion]," *Transactions of the American Institute of Electrical Engineers. Part III: Power Apparatus and Systems* 71, no. 1 (1952): 513–521; C. A. Imburgia, L. K. Kirchmayer, and G. W. Stagg, "Transmission-Loss Penalty Factor Computer," *Transactions of the American Institute of Electrical Engineers. Part III: Power Apparatus and Systems* 73, no. 12 (1954): 567–571.

25. E. D. Early, "Central Power Coordination Control for Maximum System Economy," *Electric Light and Power* 31, no. 14 (1953): 96–98; E. D. Early, "Power Co. Sees $200,000 Saved Yearly by Using Power Cost Computer," *Modern Precision* 15, no. 2 (1955): 5; A. H. Willennar and G. W. Stagg, "Penalty Factor Computer Teams with Slide Rule," *Electrical World* 143, no. 16 (1955): 120–122.

26. George, Page, and Ward, "Co-ordination of Fuel Cost and Transmission Loss," 1162.

27. C. D. Morrill and J. A. Blake, "Computer for Economic Scheduling and Control of Power Systems," *Transactions of the American Institute of Electrical Engineers. Part III: Power Apparatus and Systems* 74, no. 21 (1955): 1136–1141; E. D. Early, W. E. Phillips, and W. T. Shreve, "Incremental Cost of Power-Delivered Computer," *Transactions of the American Institute of Electrical Engineers. Part III: Power Apparatus and Systems* 74, no. 18 (1955): 529–535.

28. Entire issue, *Electrical World* 143, no. 16 (1955); "Utility Promotes Computer to a Top Job," *Business Week*, September 10, 1955, 112–118.

29. Memorandum from Kuhl (L&N Cleveland Office) to Balbirnie (L&N Philadelphia Office): Cleveland Electric Illuminating Co. Load Control Negotiations, October 1, 1958; Letter to Mr. James R. Guy, the Cleveland Electric Illuminating Company, October 3, 1958; Memorandum from Balbirnie to Hissey: Cleveland Electric Illuminating Company, November 28, 1958; Memorandum from Hissey to Mackay, Patent Department: Competitive Activities—Load Frequency Control, Westinghouse Electric Corporation Entry into Field, November 19, 1958; Memorandum from Nichols to W. G. Amey, Research: Simulation of Control Schemes Associated with Load Control Recommendations for Cleveland Electric Illuminating Company, July 17, 1958; all in Box 3, Nathan Cohn Papers, MIT.

30. D. H. Cameron and W. S. Burt, "Load Scheduling Goes Automatic," *Electrical World* 143, no. 16 (1955): 125–126; R. H. Travers et al., "Loss Evaluation; Part III. Economic Dispatch Studies of Steam-Electric Generating Systems [Includes Discussion]," *Transactions of the American Institute of Electrical Engineers. Part III: Power Apparatus and Systems* 73, no. 2 (1954): 1091–1104; Morrill and Blake, "A Computer for Economic Scheduling and Control of Power Systems"; "Computer for Economic Scheduling and Control of Power Systems," *Transactions of the American Institute of Electrical Engineers. Part III: Power Apparatus and Systems*, 74, no. 21 (1955): 1136–1141; J. E. Van Ness and W. C. Peterson, "The Use of Analogue Computers in Power System Studies [Includes Discussion]," *Transactions of the American Institute of Electrical Engineers. Part III: Power Apparatus and Systems* 75, no. 3 (1956): 238–242; J. B. Ward and H. W. Hale, "Digital Computer Solution of Power-Flow Problems [Includes Discussion]," *Transactions of the American Institute of Electrical Engineers. Part III: Power Apparatus and Systems* 75, no. 3 (1956): 398–404.

31. Van Ness and Peterson, "Use of Analogue Computers in Power System Studies," 23.

32. John M. Undrill et al., "Interactive Computation in Power System Analysis," *Proceedings of the IEEE* 62, no. 7 (1974): 1009–1018. John M. Undrill began working for General Electric in 1966, making important contributions to computer-based analysis and control of power system operations. Undrill joined Power Technologies, Inc., in 1972 and joined in founding Electric Power Consultants, Inc., in 1986. General Electric acquired Electric Power Consultants in 1996, and Undrill retired from GE in 2006. He is still active in the industry.

33. Ward and Hale, "Digital Computer Solution of Power-Flow Problems"; S. B. Morehouse, "Automatic Economic Loading Practices on Interconnected Power Systems in U.S.A." (paper presented at the International Conference on Large Electrical Systems, June 8–18, 1966, Paris, France); G. W. Bills, "Digital Computers to Speed Studies," *Electrical World* 143, no. 16 (1955): 115–117; *National Power Survey: A Report by the Federal Power Commission, 1964*, 1:165.

34. Proceedings of the Symposium on Scheduling and Billing of Economy Interchange on Interconnected Power Systems, American Power Conference, Chicago, Illinois, March 26, 27, 28, 1958; J. W. Lamont and J. R. Tudor, "Survey of Operating Computer Applications," *Proceedings of the American Power Conference, 21–23 April 1970, Chicago, IL, 1970* (Chicago: Illinois Institute of Technology, 1970), 1142–1148; William Aspray, "Edwin L. Harder and the Anacom: Analog Computing at Westinghouse," *IEEE Annals of the History of Computing* 15, no. 2 (1993): 35–52.

35. The different techniques for automatic frequency and load control are delineated in chapter 3 and further explained in chapter 4. See figure 3.10.

36. See chapter 4 for a description of this experiment.

37. See chapter 5 for descriptions of the control approaches used by these two power pools during World War II.

38. Nathan Cohn, "Recollections of the Evolution of Realtime Control Applications to Power Systems," *Automatica* 20, no. 2 (1984): 154–155.

39. Cohn, "Power Flow Control—Basic Concepts for Interconnected Systems."

40. Ibid., 84.

41. Cohn, "Recollections of the Evolution of Realtime Control Applications to Power Systems," 155.

42. Robert Brandt, "Theoretical Approach to Speed and Tie Line Control," *Transactions of the American Institute of Electrical Engineers* 66, no. 1 (1947): 24–30.

43. Cohn, "Power Flow Control—Basic Concepts for Interconnected Systems."

44. Adapted from Nathan Cohn, "Power-System Interconnections: Control of Generation and Power Flow," in *Standard Handbook for Electrical Engineers*, 10th ed. (New York: McGraw Hill, 1968), section 15-10.

45. NERC defines *Area Control Error* as "the instantaneous difference between a Balancing Authority's net actual and scheduled interchange, taking into account the effects of Frequency Bias, correction for meter error, and Automatic Time Error Correction (ATEC), if operating in the ATEC mode. ATEC is applicable only to Balancing Authorities in the Western Interconnection." *Glossary of Terms Used in NERC Reliability Standards*, North American Electric Reliability Council, updated December 12, 2016, http://www.nerc.com/files/glossary_of_terms.pdf.

46. This explanation per John Adams, principal engineer, Electric Reliability Council of Texas, Inc., personal communication with author, March 4, 2013.

47. Adams, personal communication with author, March 4, 2013.

48. Portions of the episode described in this chapter appear in Julie Cohn, "Bias in Electrical Power Systems: A Technological Fine Point at the Intersection of

Commodity and Service," in *Electric Worlds/Mondes Electriques: Creations Circulations, Tensions, Transitions (19th–21st C.)*, ed. Alain Beltran, Léonard Laborie, Pierre Lanthier, and Stéphanie Le Gallic (New York: Peter Lang, 2016), 271–293.

49. NERC defines *frequency bias setting* as "a number, either fixed or variable, usually expressed in MW/0.1 Hz, included in a Balancing Authority's Area Control Error equation to account for the Balancing Authority's inverse Frequency Response contribution to the Interconnection, and discourage response withdrawal through secondary control systems." *Glossary of Terms Used in NERC Reliability Standards*, North American Electric Reliability Council, updated December 12, 2016, http://www.nerc.com/files/glossary_of_terms.pdf. *Balancing Authorities* are the entities that now operate designated segments of the grid in the United States and Canada.

50. This chapter will use the term "natural characteristic," as that was the term that appeared most frequently in the papers published in the 1950s. Today, all large generators are obligated to have automatic governor controllers as a condition of connecting to the grid, and the industry is in general agreement for the bias setting. Walt Stadlin, personal communication, October 20, 2012; Nathan Cohn, *Control of Generation and Power Flow on Interconnected Power Systems*, 2nd ed. (New York: J. Wiley, 1967), 19; Nathan Cohn, "Some Aspects of Tie-Line Bias Control on Interconnected Power Systems," *Transactions of the American Institute of Electrical Engineers. Part III: Power Apparatus and Systems* 75, no. 3 (1957): 1415–1436.

51. Letter to Cohn, March 28, 1972, Box 1, Nathan Cohn Papers, MIT.

52. Cohn, "Some Aspects of Tie-Line Bias Control on Interconnected Power Systems," 2.

53. *Report on the Status of Interconnections and Pooling of Electric Utility Systems in the United States*, 58.

54. Ibid.; Miscellaneous Minutes, Standards, Newsletters, Reports, Correspondence of the Interconnected Systems Group, 1933–1938, Electrical Power Systems Records, Hagley.

55. Memorandum from J. R. Smith, Chairman, to All Members of the Northwest Regional Committee of the Interconnected Systems Committee, October 21, 1952, Electrical Power Systems Records, Hagley. In other words, the bias would be set to 1 percent of an area's natural characteristic.

56. Ibid.

57. "Miscellaneous Minutes, Standards, Newsletters, Reports, Correspondence of the Interconnected Systems Group."

58. Cohn, "Power Flow Control—Basic Concepts for Interconnected Systems," 102.

59. Letter to Members of Northwest Regional Committee, June 24, 1955, Electrical Power Systems Records, Hagley.

60. Memorandum to S. B. Morehouse Regarding System Operations, September 22, 1955, Box 44, Nathan Cohn Papers, MIT.

61. Report from the Test Committee to the Interconnected Systems Committee, 1955, Box 4, Nathan Cohn Papers, MIT.

62. Letter to W. T. Pavely, March 21, 1955, Electrical Power Systems Records, Hagley.

63. Report from the Test Committee to the Interconnected Systems Committee; Letter to Nathan Cohn, November 28, 1955, Box 4, Nathan Cohn Papers, MIT; Information on the Interconnected Systems Group, Electrical Power Systems Records, Hagley.

64. Some Aspects of Bias Control—Getting the Most from It during and Following Periods of System Disturbance, April, 1955, Box 4, Nathan Cohn Papers, MIT; Conference Minutes, April 9, 1955, Box 4, Nathan Cohn Papers, MIT.

65. AIEE System Control Subcommittee, "Report on the Current Status of Load—Frequency Control Methods and Equipment by the System Controls Subcommittee of the Committee on System Engineering" (paper presented to the AIEE Fall General Meeting, Chicago, Illinois, October 1–5, 1956), 2, Electrical Power Systems Records, Hagley.

66. Letter to L. V. Leonard, Chairman, Interconnected Systems Test Committee, June 24, 1955, Electrical Power Systems Records, Hagley.

67. Memorandum to Members of the Test Committee, July 7, 1955, Electrical Power Systems Records, Hagley.

68. E. S. Miller, Letter to Members of Northwest Regional Committee; L. V. Leonard, Memorandum to Members of the Test Committee. Utilities represented on the Test Committee and the states in which they were headquartered included Arkansas Power & Light Company, Arkansas; Duke Power Company, North Carolina; Tennessee Valley Authority, Tennessee; Indiana & Michigan Electric Company, Indiana; Louisiana Power & Light Company, Louisiana; The Ohio Power Company, Ohio; Georgia Power Company, Georgia; Illinois Power Company, Illinois; Cleveland Electric Illuminating Company, Ohio; and Appalachian Electric Power Company, Virginia. Leonard, "Letter to Nathan Cohn"; Letter to Test Committee Members, Interconnected Systems, January 6, 1956, Box 4, Nathan Cohn Papers, MIT.

69. Report from the Test Committee to the Interconnected Systems Committee, 3.

70. Leonard, Letter to Test Committee Members, Interconnected Systems.

71. Ibid.

72. Leonard, Letter to Nathan Cohn.

73. Inter-Office Memorandum to N. Cohn Re: Bias-T.V.A.-Almond, February 7, 1956, Box 4, Nathan Cohn Papers, MIT.

74. Minutes of the Test Committee Meeting, February 15–16, 1956, Cincinnati, Ohio, February 15–16, 1956, Box 4, Nathan Cohn Papers, MIT.

75. Letter to Nathan Cohn, April 9, 1956, Box 4, Nathan Cohn Papers, MIT.

76. Letter to Russ L. Purdy, April 23, 1956, Box 4, Nathan Cohn Papers, MIT.

77. Letter to A. L. Richmond, March 1, 1956, Box 4, Nathan Cohn Papers, MIT; Letter to Nathan Cohn, March 2, 1956, Box 4, Nathan Cohn Papers, MIT.

78. Interconnected Systems Committee General Meeting Attendance List, April 26–27, 1956, Box 40, Nathan Cohn Papers, MIT; Interconnected Systems Committee General Meeting Agenda, April 26–27, 1956, Box 40, Nathan Cohn Papers, MIT; quote in Report of Bias Analysis Survey for June 21, 1955, April 26, 1956, Box 4, Nathan Cohn Papers, MIT.

79. Nathan Cohn, "A Step-by-Step Analysis of Load Frequency Control Showing the System Regulating Response Associated with Frequency Bias" (paper presented to a meeting of the Interconnected Systems Committee, Des Moines, Iowa, 1956), 1, reprinted with permission by Leeds & Northrup Company, 1956.

80. Ibid., 24.

81. Letter to Nathan Cohn, May 10, 1956, Box 40, Nathan Cohn Papers, MIT; Letter to Nathan Cohn, June 5, 1956, Box 40, Nathan Cohn Papers, MIT.

82. Information on the Interconnected Systems Group.

83. Letter to R. T. Purdy Re: Prize Paper, August 15, 1957, Box 41, Nathan Cohn Papers, MIT.

84. Cohn, "Some Aspects of Tie-Line Bias Control on Interconnected Power Systems," 16.

85. Ibid., 15–20.

86. Ibid., 21.

87. Ibid.

88. Memorandum to Nathan Cohn, Subject: Your Paper on Tie-Line Bias Control, July 26, 1956, Box 41, Nathan Cohn Papers, MIT (photocopy of page from *Electrical West* 117, no. 1 attached); Memorandum to D. E. Moat, Subject: AIEE Paper 56–670; "Some Aspects of Tie-Line Bias Control on Interconnected Power Systems," January 14, 1957, Box 41, Nathan Cohn Papers, MIT; Cohn, "A Step-by-Step Analysis of Load Frequency Control Showing the System Regulating Response Associated with Frequency Bias"; Cohn, "Some Aspects of Tie-Line Bias Control on Interconnected Power Systems."

89. Memorandum from Mollman to Howell, Subject: Test Committee, May 22, 1957, Electrical Power Systems Records, Hagley.

90. O. A. Demuth, Letter to The Lamme Medal Committee of the American Institute of Electrical Engineers, Box 3, Nathan Cohn Papers, MIT.

91. Operating Recommendations for the Interconnected Systems Sponsored by the Test Committee and Approved by the Main Committee, May 1, 1957, Electrical Power Systems Records, Hagley.

92. Letter to Cohn, April 4, 1972, Box 1, Nathan Cohn Papers, MIT; Operating Recommendations for the Interconnected Systems Sponsored by the Test Committee and Approved by the Main Committee, April 22, 1960, Electrical Power Systems Records, Hagley; North American Power Systems Interconnection Committee Minutes of Meeting January 15–16, 1963—New Orleans, La., February 18, 1963, Electrical Power Systems Records, Hagley.

93. Memorandum from Mollman to Howell, Subject: Test Committee.

94. Stefan Timmermans and Steven Epstein, "A World of Standards but Not a Standard World: Toward a Sociology of Standards and Standardization," *Annual Review of Sociology* 36 (2010): 69–89; Andrew L. Russell, *Open Standards and the Digital Age: History, Ideology, and Networks*, ed. Louis Galambos and Geoffrey Jones, Cambridge Studies in the Emergence of Global Enterprise (New York: Cambridge University Press, 2014). The bias setting debate reflected the four key aspects of standards enumerated by Timmermans and Epstein: creation, resistance, implementation, and outcomes.

95. Interconnected Systems Group Test Committee Meeting, Sheraton-Jefferson Hotel—St. Louis, Missouri, February 18–19, 1960, Minutes, February 29, 1960, 2, Electrical Power Systems Records, Hagley.

96. Report on Progress on Interconnections and Summary of Meeting Held in Omaha on April 25, 1962 to Discuss the Future Operations of Systems, April 26, 1962, Box 1, Nathan Cohn Papers, MIT; Memorandum to System Representatives Re: Interconnection Coordination Committee, May 4, 1962, Box 1, Nathan Cohn Papers, MIT.

97. Kleinbach, Report on Progress on Interconnections and Summary of Meeting Held in Omaha on April 25, 1962 to Discuss the Future Operations of Systems, Box 1, Nathan Cohn Papers, MIT.

98. Ibid.

99. Ibid.

100. Interconnection Coordination Committee, Minutes of Meeting, August 28–29, 1962—Denver, Colorado, September 12, 1962, Electrical Power Systems Records, Hagley; Interconnection Coordination Committee, Minutes of Meeting, June 12–13, 1962—Chicago, Illinois, June 22, 1962, Electrical Power Systems Records, Hagley. As mentioned in a previous note, *unintentional* and *inadvertant* both refer to electricity

that crosses from one network to another outside of any planned power exchanges. In this case, Kleinbach used the term "unintentional" in his meeting report.

101. Canady, North American Power Systems Interconnection Committee Minutes of Meeting January 15–16, 1963—New Orleans, La., Electrical Power Systems Records, Hagley.

102. NERC went through several additional transformations, eventually becoming the North American Electric Reliability Corporation. This name change reflects NERC's responsibility to power systems in the United States, Canada, and parts of Mexico. The change from Council to Corporation further reflects NERC's status as an independent organization.

103. Peter Braestrup, "National Power Survey Is Urged as Step to Cut Consumer Costs," *New York Times*, January 24, 1962, 16.

104. "Swidler Asks Nation-Wide Power Tie-In," *Washington Post, Times Herald*, January 24, 1962, A12.

105. In fact, the average price for electricity in 1962 (adjusted for inflation) was 9.4 cents per kilowatt-hour, and in 1980, it was 9.8 cents per kilowatt-hour—an increase of about 4 percent. "Total Energy: Annual Energy Review, September 2012," US Energy Information Agency website, released September 27, 2012, http://www.eia.gov/totalenergy/data/annual/showtext.cfm?t=ptb0810; *National Power Survey: A Report by the Federal Power Commission, 1964*, 279–288.

106. *National Power Survey: A Report by the Federal Power Commission, 1964*, 4.

107. Ibid., 1.

108. The survey specifically addressed questions related to power supply and demand in the target year of 1980. While the Canadian public did not fall directly within the intended audience for the survey, the authors assumed the grid would also be international in scope, accessing the vast water resources of the northern part of the continent.

109. "Swidler Asks Nation-Wide Power Tie-In."

110. Braestrup, "National Power Survey Is Urged as Step to Cut Consumer Costs"; *Report on the Status of Interconnections and Pooling of Electric Utility Systems in the United States*.

111. Braestrup, "National Power Survey Is Urged as Step to Cut Consumer Costs."

112. "How to Save the Taxpayers, $380,000," *Wall Street Journal*, February 6, 1962, 16.

113. "F.P.C. Establishes New Advisory Unit," *New York Times*, March 9, 1962, 3; quote in Joseph Swidler, Letter to the President, March 20, 1962, White House Central Files, Federal Power Commission, Executive, FG 232, Box 171, John F. Kennedy Presidential Library. Kennedy appointed Swidler to the FPC in June 1961 and named

him chair of the commission in August. Swidler took up this post on September 1, 1961. "Swidler Nominated to Head F.P.C.," *New York Times*, August 19, 1961.

114. *National Power Survey: A Report by the Federal Power Commission, 1964*, vol. 1, 293–296; Letter to Clyde T. Ellis, General Manager, National Rural Electric Cooperative Association, December 11, 1962, UT 2 Electricity, Box 993, White House Central Files, General, John F. Kennedy Presidential Library; Julius Duscha, "Swidler Tells Electric Leaders to Generate New Power Uses," *Washington Post, Times Herald*, June 5, 1963, C8; Julius Duscha, "Power Commission Split by Licensing Proposal," *Washington Post, Times Herald*, January 19, 1963, A2; Clyde Ellis, Letter to the President from the National Rural Electric Cooperative Association, March 19, 1963, White House Central Files, UT 2-1, Box 993, John F. Kennedy Presidential Library; Alex Radin, Letter to Lee C. White, with Attachments, September 23, 1963, White House Central Files, Public Power, UT 2-1, Box 993, John F. Kennedy Presidential Library; Charles A. Robinson Jr., Letter to Stewart Udall in Re: Proposed Regulations Governing the Granting of Rights-of-Way for Electric Transmission Lines across Public Lands, May 15, 1963, White House Central Files, UT 2-1, Box 993, John F. Kennedy Presidential Library.

115. Gene Smith, "Dec. 14 Is 'D-Day' for U.S. Utilities," *New York Times*, November 16, 1964, 49; Gene Smith, "Utilities Fear Power Policies New Government Might Adopt," *The New York Times*, November 29, 1964, F1, F13.

116. Gene Smith, "F.P.C. Seeks Cut in Power Costs," *New York Times*, December 13, 1964, 1, 16.

117. "FPC Power Suggestion Finds Mixed Reactions," *Washington Post, Times Herald*, December 15, 1964, D6.

118. Smith, "F.P.C. Seeks Cut in Power Costs."

119. Ibid.

120. "Planning for Power," *Wall Street Journal*, December 18, 1964, 14.

121. Ibid.

122. Smith, "F.P.C. Seeks Cut in Power Costs"; Eric Wentworth, "Federal Power Agency to Seek Controls over Growth of Electric-Utility Industry," *Wall Street Journal*, December 14, 1964, 4.

123. "U.S. Survey Sets Goal for Electricity Price Cut," *Washington Post, Times Herald*, December 13, 1964, C9; Julius Duscha, "Projection to 1980 Says Electric Rates Can Be Cut by 27%," *Washington Post, Times Herald*, December 13, 1964, A7; Smith, "F.P.C. Seeks Cut in Power Costs."

124. Gene Smith, "Utilities Study National Power Survey and Relax," *New York Times*, December 20, 1964, F1, F9.

125. Smith, "F.P.C. Seeks Cut in Power Costs"; Smith, "Utilities Study National Power Survey and Relax."

126. Gene Smith, "A Powerful Year in U.S. Electricity," *New York Times*, January 11, 1965, 67.

127. Wentworth, "Federal Power Agency to Seek Controls over Growth of Electric-Utility Industry."

7 Drifting "Lazily" into Synchrony: From Blackout to Grid, 1965–1967

1. Many thanks to engineer James Robinson for this technical clarification: "Simply said, electric energy generated on the west coast can be scheduled to supply customers on the east coast, but it will never fully arrive. Due to the long length of the transmission wires, a large portion of the energy would just add heat to the numerous transmission wires, which is then lost to the surrounding air." With electrical engineering degrees from Drexel University and Lehigh University, Mr. Robinson began his career as a cooperative student at Delmarva Power and Light Company in 1968. He helped plan the peninsula's transmission system on a Network Analyzer at the Franklin Institute. Following many years of service in multiple positions, Mr. Robinson retired from Pennsylvania Power & Light Electric Utilities as a transmission asset manager in 2007. He founded Relion Associates in 2008 and continues to consult for the industry. In 2003, he served as team leader of the Sequence of Events Team investigating the August 14, 2003, Northeast blackout for NERC.

2. Numerous scholars have described and analyzed the 1965 blackout, offering analyses of the social, technical, economic, and political implications of this event. For detailed discussions, see David E. Nye, *When the Lights Went Out: A History of Blackouts in America* (Cambridge, MA: MIT Press, 2010), 67–103; Joseph A. Pratt, *A Managerial History of Consolidated Edison*, 1936–1981 (New York: Consolidated Edison Company of New York, 1988), 147–152; Phillip F. Schewe, *The Grid: A Journey through the Heart of Our Electrified World* (Washington, DC: J. Henry Press, 2007), 115–156. The website of "The Blackout History Project," accessed January 10, 2017, http://blackout.gmu.edu, provides a useful compendium of reports, essays, personal accounts, data, and images for both the 1965 blackout and later blackouts.

3. The North American Electric Reliability Corporation (NERC) defines *cascading* as "the uncontrolled successive loss of System Elements triggered by an incident at any location. Cascading results in widespread electric service interruption that cannot be restrained from sequentially spreading beyond an area predetermined by studies." *Glossary of Terms Used in NERC Reliability Standards*, North American Electric Reliability Corporation, updated December 12, 2016, http://www.nerc.com/files/glossary_of_terms.pdf.

4. *National Power Survey: A Report of the Federal Power Commission, 1964* (Washington, DC: Government Printing Office, 1964).

5. Phillip G. Harris, then president and CEO of PJM, described the industry thus at a Senate hearing following the 2003 Northeast blackout: "Yet, this industry was built, financed, and operated for over 80 years by a *gaggle* of over 4,000 different entities providing varying aspects of the service of generation and delivery of electricity" (emphasis added). "Blackout in the Northeast and Midwest," Hearing before Committee on Energy and Natural Resources, US Senate, February 24, 2004, Federal Document Clearing House Congressional Testimony, *Proquest Congressional*, accessed May 22, 2017, http://www.proquest.com/. Also quoted in Nye, *When the Lights Went Out*, 164.

6. The New England Pool included New England Electric Systems (NEES), Boston Edison, Eastern Utilities Associates, Connecticut Valley Electric Exchange (CONVEX), and Vermont Electric Power Company. The New York Pool covered three smaller pools: the Upstate Interconnected Systems, the Southeastern New York Power Pool, and the Michigan-Canadian Group, with connections into New York. Consolidated Edison Company of New York was the major power provider in Southeastern New York Pool.

7. The utilities in the northeast used primarily imported energy resources—natural gas and coal—to generate electricity, which contributed in part to the higher electricity rates. *Table 8.4b Consumption for Electricity Generation by Energy Source: Electric Power Sector, 1949–2011*, Energy Information Administration, US Department of Energy, Washington, DC, 2011, accessed March 5, 2013, http://www.eia.gov/totalenergy/data/annual/pdf/sec8_18.pdf.

8. Energy Information Administration, *Average Retail Price of Electricity to Ultimate Customers: Total by End-Use Sector*, US Department of Energy, Washington, DC, 2011, accessed March 5, 2013, http://www.eia.gov/electricity/monthly/epm_table_grapher.cfm?t=epmt_5_3.

9. "N.Y. Utilities Form Group for State-Wide Studies," *Electrical World* 153, no. 2 (1960): 40–43; "EEI Task Force Study Reveals 1970 Pooling Plans," *Electrical World* 158, no. 5 (1962): 30; Public Service Commission, *Annual Report* (Albany: State of New York Public Service Commission, 1962); James F. Fairman, "Hard-Head Engineering" (paper presented to the 28th Annual Convention of the Edison Electric Institute, Atlantic City, New Jersey, June 7, 1960).

10. G. S. Vassell, "Northeast Blackout of 1965," *IEEE Power Engineering Review* 11, no. 1 (1991): 4.

11. When a relay on a power line "trips," this means it operates to open the power line—in other words, to stop the flow of electricity at that location.

12. The four separate areas were (1) the Ontario system, (2) an area around Niagara served by the Power Authority of the State of New York (PASNY), (3) a second area

around Niagara with excess generation, and (4) the remainder of the pool extending east to Boston and south to New York City.

13. Nathan Cohn, "L & N and the Control of Electric Power Systems" (paper presented to the Leeds & Northrup Shareholders Meeting, Philadelphia, Pennsylvania, September 14, 1966); "Oral-History: Jack Casazza," an oral history conducted in 1994 by Loren J. Butler, IEEE History Center, Hoboken, New Jersey, accessed January 10, 2017, http://ethw.org/Oral-History:Jack_Casazza.

14. Letter from Jack. K. Busby, President, Pennsylvania Power & Light Company to Customers, November 10, 1965, Box 39, Nathan Cohn Papers, MIT.

15. *Report to the President by the Federal Power Commission on the Power Failure in the Northeastern United States, and the Province of Ontario on November 9–10, 1965* (Washington, DC: Government Printing Office, 1965).

16. Schewe, *The Grid*; *Report to the President by the Federal Power Commission on the Northeast Power Failure in the Northeastern United States and the Province of Ontario on November 9 and 10, 1965*. Schewe offers an entertaining and suspense-filled account of the events unfolding in the Consolidated Edison control room.

17. *Report to the President by the Federal Power Commission on the Power Failure in the Northeastern United States, and the Province of Ontario on November 9–10, 1965*; J. J. O'Connor, "Northeast Blackout Triggers Plans for . . . Firm Power Supplies," *Power* 110, no. 1 (1966): 141–148; Pratt, *A Managerial History of Consolidated Edison, 1936–1981*, 138–152.

18. James Doyle, "The Blackout All Started in a Little Ontario Relay," *Boston Globe* November 16, 1965, 1; John M. Lee, "Ontario Accepts Blame for Blackout in Northeast," *New York Times*, November 16, 1965, 58. A Google News search for the term "blackout" between the dates of November 9, 1965, and November 16, 1965, returned hundreds of articles from around the world documenting the power failure, most of which repeated wire service stories. For example, there were ninety-seven articles calling the blackout a "mystery," eighty-two headlines asking for the cause of the blackout, twenty regarding President Johnson's plan for the Federal Power Commission to probe the blackout, and seventeen explaining that the grid itself spread the blackout. *Google News*, s.v. "blackout," Google, accessed October 29, 2012, https://news.google.com.

19. McCandlish Phillips, "Behind the Light Switch Lies Complex Power Network Covering Entire Northeast," *New York Times*, November 15, 1965, 42.

20. Lawrence J. Hollander, "The Big Blackout: Whooping Cranes & Power Failures," *The Nation* 202 (1966): 33.

21. "Large-Scale Federal Sale of Electricity in Northeast to Be Sought by President," *Wall Street Journal*, July 12, 1965, 28.

22. "The Great Blackout—It's Still a Big Mystery," *St. Petersburg (FL) Evening Independent*, November 10, 1965, 1.

23. "Did Blackout Tarnish Utility Image?," *Electrical World* 164, no. 21 (1965): 31–34.

24. Ibid.

25. Pratt, *A Managerial History of Consolidated Edison, 1936–1981*, 148.

26. "Did Blackout Tarnish Utility Image?," 31–34.

27. "Paralysis of Power," *New York Times*, November 11, 1965, 46.

28. Murray Illson, "Blackout Is News All over World," *New York Times*, November 11, 1965, 38.

29. "Blackout Is Traced to Canadian Plant," *Hartford (CT) Courant*, November 16, 1965, 1; Illson, "Blackout Is News All over World," 38; Gene Smith, "Utilities Failed Major Test for Grid," *New York Times*, November 14, 1965, F1, F12.

30. "Utilities Agree on a Prediction: Statewide Failures Can Recur," *New York Times*, November 11, 1965, 38.

31. "Trouble Cascades over Six-State Area," *Electrical World* Special News Summary on Massive Northeast Power Failure (1965): 44A–44D.

32. Eileen Shanahan, "Experts Still Unsure of Failure's Cause," *New York Times*, November 11, 1965, 38.

33. Ibid.

34. *Report to the President by the Federal Power Commission on the Power Failure in the Northeastern United States, and the Province of Ontario on November 9–10, 1965.*

35. Ironically, systems in Texas and New Mexico experienced a brief power outage on December 3, just days before President Johnson, then at his "Texas White House," expected to receive the report. The following paragraphs summarize key points covered in the report.

36. *Report to the President by the Federal Power Commission on the Power Failure in the Northeastern United States, and the Province of Ontario on November 9–10, 1965*, 79.

37. Ibid.

38. *Load shedding* refers to the practice of disconnecting a customer or collection of customers from the generating system in order to reduce the total load on the network.

39. *Report to the President by the Federal Power Commission on the Power Failure in the Northeastern United States, and the Province of Ontario on November 9–10, 1965*, 24.

40. Ibid., 41.

41. "Northeast Power Failure, November 9, 10, 1965, Hearings before the Special Subcommittee to Investigate Power Failures of the Committee on Interstate and

Foreign Commerce, House of Representatives, Eighty-Ninth Congress, First and Second Sessions . . . December 15, 1965; February 24, 25, 1966," Serial No. 89–40 (Washington, DC: Government Printing Office, 1966).

42. Examples of headlines include "Chairman Urges More Authority for FPC," *St. Petersburg (FL) Times*, December 16, 1965; "Laws Called Answer to Power Losses," *Toledo (OH) Blade*, December 16, 1965; "Legislation Held Need in Power Problem," *Lexington (NC) Dispatch*, December 16, 1965; Eileen Shanahan, "F.P.C. Asks Right to Set Electric Power Standards," *New York Times*, December 16, 1965, 36.

43. *Google News*, s.v. "Federal Power Commission," Google, accessed November 3, 2012, https://news.google.com. A Google News search for the term "Federal Power Commission" produced fifty-eight news articles published in early December 1965. Representative headlines from December 6 and 7 include the "Blackout Study Asks New Regulations" (*Milwaukee Journal*), "Change Asked in Power Act" (*Spokane Daily Chronicle*), and "FPC Indicates New Legislation" (*New York Times*).

44. Eileen Shanahan, "F.P.C. Criticizes Power Systems in Nov. 9 Failure," *New York Times*, December 7, 1965, 1, 41.

45. Ibid.

46. Shanahan, "F.P.C. Asks Right to Set Electric Power Standards," 36.

47. Response from Rural Electrification Administration administrator Norman Clapp, "Responses to Inquiries about the Northeast Power Failure November 9 and 10, 1965; Interim Report of the Committee on Commerce, United States Senate on the Northeast Power Failure, March 22 (Legislative Day, March 21), 1966" (Washington, DC: Government Printing Office, 1966), 3.

48. Ibid., 161.

49. Ibid., 180.

50. Ibid.

51. Ibid., response from A. H. McDowell Jr., Virginia Electric & Power Co., 197.

52. Ibid., response from Donald C. Cook, President, American Electric Power Co., Inc., 147.

53. Ibid., response from James L. Grahl, Basin Electric Power Cooperative, 153.

54. "Northeast Power Failure, November 9, 10, 1965, Hearings before the Special Subcommittee to Investigate Power Failures of the Committee on Interstate and Foreign Commerce, House of Representatives, Eighty-Ninth Congress, First and Second Sessions . . . December 15, 1965; February 24, 25, 1966."

55. Ibid., 189.

56. Ibid., 195.

57. Ibid., 198–199.

58. Ibid., 386.

59. Ibid., 464.

60. Ibid., 376.

61. Smith, "Utilities Failed Major Test for Grid," F1, F12.

62. Gordon D. Friedlander, "Prevention of Power Failures: The FPC Report of 1967," *Spectrum, IEEE* 5, no. 2 (1968): 53–61; Clarence P. E. Paulus, "Questions Engineers' Courage," *Electrical World* 165, no. 6 (1966): 6; Pratt, *A Managerial History of Consolidated Edison, 1936–1981.*

63. Friedlander, "Prevention of Power Failures: The FPC Report of 1967," 54.

64. State of New York Public Service Commission, *Annual Report,* 1966, Albany, New York; Consolidated Edison, *Annual Report* (New York: Consolidated Edison, 1968); C. Girard Davidson, *Report to the City of New York's Consumer Council on Reliability of Service, Adequacy of Future Power Supply, and Rates to Consumers Provided by Consolidated Edison Company* (Washington, DC: Davidson, Sharkey & Cummings, 1968); Friedlander, "Prevention of Power Failures: The FPC Report of 1967."

65. Federal Power Commission, *Prevention of Power Failures: An Analysis and Recommendations Pertaining to the Northeast Failure and the Reliability of U.S. Power Systems; a Report to the President,* 3 vols. (Washington, DC: Government Printing Office, 1967).

66. *Hearing before the Committee on Commerce on S. 1934 Amending the Federal Power Act and Related Bills, S. 683, S. 1834, and S. 2227, August 22, 1967* (Washington, DC: Government Printing Office, 1967).

67. Ibid.; *Hearing before the Committee on Commerce on S. 1934 Amending the Federal Power Act and Related Bills, S. 683, S. 1834, and S. 2227, Part 2, December 20 and 21, 1967* (Washington, DC: Government Printing Office, 1968); *Hearing before the Committee on Commerce on S. 1934 Amending the Federal Power Act and Related Bills, S. 683, S. 1834, and S. 2227, Part 3, April 26 and 29, 1968* (Washington, DC: Government Printing Office, 1968).

68. *Hearing before the Committee on Commerce on S. 1934 Amending the Federal Power Act and Related Bills, S. 683, S. 1834, and S. 2227, August 22, 1967,* 113.

69. Ibid., 135; Memorandum of Lelan F. Sillin Jr., President and Chief Executive Officer, Central Hudson Gas & Electric Corporation, concerning the Federal Power Commission's Proposed Electric Reliability Act of 1967.

70. The National Electric Reliability Council changed the name to North American Electric Reliability Council in 1981 to reflect Canadian participation and then to the North American Electric Reliability Corporation in 2007. The organization continued to use the acronym NERC throughout.

71. Gene Smith, "Electric Utilities Form Group," *New York Times*, June 12, 1968, 61.

72. *Federal Power Commission Oversight, Hearing before the Subcommittee on Energy, Natural Resources, and the Environment, January 30, 1970* (Washington, DC: Government Printing Office, 1970).

73. For some examples, see Karl Boyd Brooks, *Public Power, Private Dams: The Hells Canyon High Dam Controversy* (Seattle: University of Washington Press, 2006); Robert Lifset, *Power on the Hudson: Storm King Mountain and the Emergence of Modern American Environmentalism* (Pittsburgh: University of Pittsburgh Press, 2014); Daniel Pope, *Nuclear Implosions: The Rise and Fall of the Washington Public Power Supply System* (New York: Cambridge University Press, 2008); Thomas Raymond Wellock, *Critical Masses: Opposition to Nuclear Power in California, 1958–1978* (Madison: University of Wisconsin Press, 1998); Paul David Wellstone and Barry M. Casper, *Powerline: The First Battle of America's Energy War*, 1st ed. (Minneapolis: University of Minnesota Press, 2003).

74. Richard N. L. Andrews, *Managing the Environment, Managing Ourselves: A History of American Environmental Policy*, 2nd ed. (New Haven: Yale University Press, 2006). The Wilderness Act of 1964 authorized permanent protection of wilderness lands. The 1965 Water Quality Act provided federal standards to prevent water pollution. The Air Quality Act of 1967 introduced explicit federal authority over stationary sources of air pollution. In 1967, the *Storm King v. Federal Power Commission* decision affirmed the right of environmental groups to have standing in environmental lawsuits. In 1970, President Richard Nixon signed the National Environmental Policy Act into law, launching a series of follow-on legislation that strengthened federal control over environmental policy across the country. The 1970 Clean Air and Water Quality Acts followed.

75. *The 1970 National Power Survey of the Federal Power Commission* (Washington, DC: Government Printing Office, 1970), I-1-15.

76. Ibid., I-2-11.

77. "Expansion Set by Big Utility," *New York Times*, October 13, 1968, F17; "T.V.A. and Southern Map Power Protection Accord," *New York Times*, March 11, 1968, 57; "Electric Utilities Form Five Regional Councils," *New York Times*, October 31, 1969, 76; "T.V.A., Utilities Set Power Deal," *New York Times*, September 14, 1969, F13; Claude Koprowski, "Pepco Signs Regional Pact to Help Prevent Blackouts," *Washington Post, Times Herald (1959–1973)*, 1968, C8; Gene Smith, "Utility Goal: A Shoehorn for Volts," *New York Times*, February 23, 1969, 2.

78. U.S.-Canada Power System Outage Task Force and Energy United States, "Final Report on the August 14, 2003 Blackout in the United States and Canada Causes and Recommendations," US Department of Energy, accessed October 5, 2015, http://purl.access.gpo.gov/GPO/LPS47061; "List of Power Outages," *Wikipedia*, accessed October 5, 2015, http://en.wikipedia.org/wiki/List_of_power_outages.

79. "Special Message to Congress on Conservation, March 1, 1962," Papers of John F. Kennedy, Presidential Papers, President's Office Files, Subjects: Conservation.

80. President John F. Kennedy, "Special Message on Natural Resources, 1961," Theodore C. Sorensen Papers, JFK Speech Files 1961–1963, Box 67, John F. Kennedy Presidential Library; *National Power Survey: A Report of the Federal Power Commission, 1964*; "East-West Closure Will Parallel 94% of US Capacity," *Electrical World* 166, no. 20 (1966): 100–103; "East-West Power Intertie Closure Test Scheduled February 7," *United States Department of the Interior News Release*, January 26, 1967.

81. Quote in "Mr. Udall's Empire Grows," *Chicago Tribune*, November 16, 1966, 20; "Giant Power Intertie for U.S., Canada to Be Tested Early in 1967," *Wall Street Journal*, November 14, 1966, 8; "U.S.-Canada Power Grid Trial Set," *Chicago Tribune*, November 13, 1966, 1, C13; "Coast-Coast Power Link Due in 1967," *Washington Post, Times Herald*, November 13, 1966, 1, A6.

82. "East-West Closure Will Parallel 94% of US Capacity"; "Editorial—Thoughts about an East-West Closure," *Electrical World* 166, no. 20 (1966): 63; "Pooling Changes Planning and Operating Patterns," *Electrical World* 166, no. 20 (1966): 98–99.

83. "Editorial—Thoughts about an East-West Closure." Task force members represented the Bureau of Reclamation offices in North Dakota, South Dakota, and Colorado; the Consumers Public Power District, Nebraska; Pacific Power & Light, Oregon; Public Service Company, Colorado; Idaho Power, Idaho; Utah Power & Light Company, Utah; Iowa Power & Light Company, Iowa; Montana Power Company, Montana; and Iowa Public Service Company, Iowa. East-West Tie Closure Task Force Meeting Agenda, July 27, 1967, Record Series 1206–1213, Box 3, Folder 10, Seattle City Light Regional Power Management Records, Seattle Municipal Archives, Seattle, Washington.

84. "East-West Power Intertie Closure Test Scheduled February 7," *Chicago Tribune*, January 26, 1967, 22; "U. S.-Canadian Power Hookup Set for Feb. 7," *Chicago Tribune*, January 26, 1967, 22; "Huge Power Grid to Be Formed," *Hartford Courant*, January 26, 1967, 15; "Huge Power Grid to Get Test Feb. 7," *New York Times*, January 26, 1967, 23; "Power Pool Planned for February 7," *Baltimore Sun*, January 26, 1967, 1.

85. Quote in "Huge Power Grid to Get Test Feb. 7." Lee C. White replaced Joseph Swidler as chair of the Federal Power Commission 1966. In his memoirs, Swidler explains that he had planned to enter private law practice at the end of his FPC term in June 1965 but then elected to stay in office until December 1, 1965, in order to complete work on a major natural gas rate-fixing case (the Permian Basin Case). Following

the blackout in November, President Lyndon Johnson asked Swidler to continue as FPC chair through the end of the calendar year, which he did. Pratt, *A Managerial History of Consolidated Edison, 1936–1981*, 147–152.

86. Frank W. Lachicotte, "Emergency Action after Automatic Separation and Normal Opening Points for Prolonged Separation" (unpublished: Bureau of Reclamation, 1966), Electrical Power Systems Records, Hagley; Walt Stadlin, personal communication, December 6, 2012; "East-West Power Intertie Closure Test Scheduled February 7."

87. Frank W. Lachicotte, The East-West Tie Closure, Staff Information Letter, February 27, 1967, attached to letter from Nathan Cohn to Frank Lachicotte, and attached documents, April 24, 1967, Box 38, Nathan Cohn Papers, MIT.

88. Ibid.

89. Walt Stadlin, personal communication, December 7, 2012. This paragraph paraphrases comments provided by Mr. Stadlin.

90. Lachicotte, "The East-West Tie Closure, Staff Information Letter, February 27, 1967."

91. "Electrical Week: Intertie," *Electrical World* 167, no. 7 (1967): 11; "East-West Ties Hold; US Systems in Phase," *Electrical World* 167, no. 8 (1967): 49–51; "U.S.-Canada Power Grid Passes Test," *New York Times*, February 8, 1967, 61; "North American Grid Put Together to Test Blackout Prevention," *Wall Street Journal*, February 8, 1967, 11; "Power System Is Tested for Blackout Guard," *Washington Post, Times Herald*, February 8, 1967, 1, D7; "Closing Circuits," *Christian Science Monitor*, November 13, 1967, 1; Neal Stanford, "Nationwide Power Net Nears," *Christian Science Monitor*, February 7, 1967, 3; "East West Tie," *The Lamplighter Newsletter, Black Hills Power and Light Company* 17, no. 3 (1967).

92. Letter from Nathan Cohn to Frank Lachicotte and Attached Documents, April 24, 1967, Box 38, Nathan Cohn Papers, MIT; Minutes of East-West Tie Closure Task Force Meeting, July 27, 1967, Record Series 1206–1213, Box 3, Folder 10, Seattle City Light Regional Power Management Records, Seattle Municipal Archives, Seattle, Washington; Letter from Frank W. Lachicotte, Chairman, East-West Task Force to R. P. Marean, Chairman, Western Operations Committee, August 31, 1967, Record Series 1206–1213, Box 3, Folder 10, Seattle City Light Regional Power Management Records, Seattle Municipal Archives, Seattle, Washington.

93. *Serving the West: Western Area Power Administration's First 25 Years as a Power Marketing Agency* (Lakewood, CO: Western Area Power Administration, US Department of Energy, 2002), 33, available online, accessed January 10, 2017, https://www.wapa.gov/newsroom/Publications/Documents/25yr-history_2.pdf.

94. Stadlin, personal communication, December 7, 2012.

95. Walt Stadlin, James Resek, and David Nevius, personal communication, June 29, 2012.

8 Reaching Maturity: Integration, Security, and Advanced Technologies, 1965–1990

1. Note that in the twenty-first century, *system security* has taken on a new meaning in terms of physical and cybersecurity. Thank you to David Nevius for underscoring this expanded definition.

2. Richard Hirsh provides a detailed analysis of how economic, technical, and social trends created a crisis environment for the investor-owned utilities by the 1970s, exacerbated by physical limits on generator size and moribund technical departments within the companies. Richard F. Hirsh, *Technology and Transformation in the American Electric Utility Industry* (Cambridge: Cambridge University Press, 1989).

3. Lizette Cintrón, ed., *Historical Statistics of the Electric Utility Industry through 1992* (Washington, DC: Edison Electric Institute, 1995), 243–244.

4. Ibid., 319.

5. For recent discussions of the US energy crises of the 1970s, see Robert Lifset, ed., *American Energy Policy in the 1970s* (Norman: University of Oklahoma Press, 2014).

6. For effects of these oil-producing country actions on the US power industry, see Hirsh, *Technology and Transformation in the American Electric Utility Industry*.

7. "Domestic Crude Prices First Purchase Prices by Area," *Petroleum and other Liquids Data*, Energy Information Administration webpage, released March 1, 2016, http://www.eia.gov/dnav/pet/pet_pri_dfp1_k_a.htm.

8. *CPI Detailed Report: Data for December 2015* (Washington, DC: US Bureau of Labor Statistics, 2016), 72, 74.

9. "Table 8.10 Average Retail Prices of Electricity, 1960–2011," *Annual Energy Review*, September 2012, Energy Information Administration website, accessed May 22, 2017, http://www.eia.gov/totalenergy/data/annual/showtext.cfm?t=ptb0810.

10. For example, see Karl Boyd Brooks, *Public Power, Private Dams: The Hells Canyon High Dam Controversy* (Seattle: University of Washington Press, 2006); Thomas Raymond Wellock, *Critical Masses: Opposition to Nuclear Power in California, 1958–1978* (Madison: University of Wisconsin Press, 1998).

11. For a few examples, see Elizabeth D. Blum, *Love Canal Revisited: Race, Class, and Gender in Environmental Activism* (Lawrence: University Press of Kansas, 2008); Robert Lifset, *Power on the Hudson: Storm King Mountain and the Emergence of Modern American Environmentalism* (Pittsburgh: University of Pittsburgh Press, 2014); Daniel Pope, *Nuclear Implosions: The Rise and Fall of the Washington Public Power Supply System* (New York: Cambridge University Press, 2008); Joseph A. Pratt, *A Managerial History of Consolidated Edison, 1936–1981* (New York: Consolidated Edison Company of New York, 1988).

12. Congress passed the Clean Air Act (1963), the Wilderness Act (1964), the Solid Waste Disposal Act (1965), the National Historic Preservation Act (1966), the Water Quality Act (1965), the Air Quality Act (1967), and NEPA (1969). For more on the modern environmental movements, see Richard N. L. Andrews, *Managing the Environment, Managing Ourselves: A History of American Environmental Policy*, 2nd ed. (New Haven: Yale University Press, 2006); Robert Gottlieb, *Forcing the Spring: The Transformation of the American Environmental Movement*, rev. and updated ed. (Washington, DC: Island Press, 2005); Lifset, *Power on the Hudson: Storm King Mountain and the Emergence of Modern American Environmentalism*; Martin V. Melosi, "Environmental Policy," in *A Companion Guide to Lyndon B. Johnson*, ed. Mitchell Lerner (New York: Blackwell, 2012), 187–209; Adam Rome, "Conservation, Preservation, and Environmental Activism: A Survey of the Historical Literature," National Park Service, US Department of the Interior, accessed January 8, 2017, http://www.cr.nps.gov/history/hisnps/NPSThinking/nps-oah.htm; Adam Rome, "'Give Earth a Chance': The Environmental Movement and the Sixties," *Journal of American History* 90, no. 2 (2003): 525–554; Thomas Raymond Wellock, *Preserving the Nation: The Conservation and Environmental Movements, 1870–2000*, American History Series (Wheeling, IL: Harlan Davidson, 2007).

13. Andrews, *Managing the Environment, Managing Ourselves.*

14. Federal Power Commission, *Annual Report* (Washington, DC: Government Printing Office, 1966–1971 and 1972–1976), available online at *Hathitrust Digital Library*, https://www.hathitrust.org/.

15. Hirsh, *Technology and Transformation in the American Electric Utility Industry.*

16. For examples, see Henry F. Bedford, *Seabrook Station: Citizen Politics and Nuclear Power* (Amherst: University of Massachusetts Press, 1990); H. Craig Miner, *Wolf Creek Station: Kansas Gas and Electric Company in the Nuclear Era*, Historical Perspectives on Business Enterprise Series (Columbus: Ohio State University Press, 1993); Pope, *Nuclear Implosions*; Pratt, *A Managerial History of Consolidated Edison, 1936–1981*; Joseph A. Stromberg, "Atomic Cowboys: The South Texas Project and the Decline of Nuclear Power" (PhD diss., University of Houston, 2012); Wellock, *Critical Masses.*

17. *The 1970 National Power Survey of the Federal Power Commission* (Washington, DC: Government Printing Office, 1970), I-13-4.

18. Cintrón, *Historical Statistics of the Electric Utility Industry through 1992*, 566–567.

19. *National Power Survey: A Report of the Federal Power Commission, 1964* (Washington, DC: Government Printing Office, 1964), 1.

20. *The 1970 National Power Survey of the Federal Power Commission*, I-1-15.

21. Federal Power Commission, *Annual Report: Fiscal Year 1967* (Washington, DC: Government Printing Office, 1968), 4.

22. Federal Power Commission, *1977 Final Annual Report* (Washington, DC: Government Printing Office, 1978), 44.

23. The reorganization of federal energy agencies took place in two steps during the 1970s. Congress passed the Energy Reorganization Act of 1974, which split the Atomic Energy Commission into two agencies—the Nuclear Regulatory Commission and the Energy Research and Development Administration (ERDA)—the latter taking over research activities focused on nuclear energy but also incorporating some general energy research (42 U.S.C.A. § 5801). In 1977, Congress enacted the Department of Energy Organization Act, which consolidated ERDA and other federal energy activities into a single department, the Department of Energy (DOE). But the Federal Power Commission, renamed the Federal Energy Regulatory Commission (FERC), remained independent. FERC then took on the new responsibility of hearing appeals to decisions for the DOE in addition to its traditional regulatory duties (42 U.S.C. § 7134).

24. C. Sulzberger, "History—When the Lights Went out, Remembering 9 November 1965," *Power and Energy Magazine, IEEE* 4, no. 5 (2006): 95; G. S. Vassell, "Northeast Blackout of 1965," *IEEE Power Engineering Review* 11, no. 1 (1991): 4; quote in *The 1970 National Power Survey of the Federal Power Commission*, I-1-15.

25. (Untitled) Draft Newsletter, November 10, 1969, Box 39, Nathan Cohn Papers, MIT.

26. Letter from J. P. Newbauer, Manager, System Operation Department, Consolidated Edison to S. B. Morehouse, Leeds & Northrup Co., Box 39, Nathan Cohn Papers, MIT.

27. Memorandum from K. W. Conners to Mr. R. E. Hill, Box 39, Nathan Cohn Papers, MIT.

28. 11/9/65 Frequency Charts for Three Impedance Bridge Recorders Including Vepco and AEP, Box 39, Nathan Cohn Papers, MIT.

29. Handwritten Notes Regarding Call from New York Post, November 11, 1965, Box 39, Nathan Cohn Papers, MIT; Typed Note Regarding Call from Mr. Harner of Electronic Design Magazine, November 12, 1965, Box 39, Nathan Cohn Papers, MIT; Typed Note Regarding Call from Gordon Friedlander of IEEE, November 12, 1965, Box 39, Nathan Cohn Papers, MIT; L&N Management Bulletin No. 1048, November 12, 1965, Signed by George E. Beggs Jr., Box 39, Nathan Cohn Papers, MIT.

30. Nathan Cohn, "The Automatic Control of Electric Power in the United States," *IEEE Spectrum* 2, no. 11 (1965): 67–77.

31. Typed Note Regarding Call from Jack Kinn of the IEEE Staff, November 11, 1965, Box 39, Nathan Cohn Papers, MIT.

32. Memorandum from N. Cohn to Messers. S. B. Morehouse, W. E. Phillips, E. F. Peterson, F. M. Hamilton, F. B. Davis, November 19, 1965, Box 39, Nathan Cohn Papers, MIT.

33. Nathan Cohn, *L&N and the Control of Electric Power Systems* (Philadelphia: Leeds & Northrup Company, 1966), 12.

34. Ibid., 12–13.

35. Ibid., 15.

36. S. Bennett, *A History of Control Engineering, 1930–1955* (London: Peter Peregrinus, 1993); Thomas Parke Hughes and Agatha C. Hughes, eds., *Systems, Experts, and Computers: The Systems Approach in Management and Engineering, World War II and After*, Dibner Institute Studies in the History of Science and Technology (Cambridge, MA: MIT Press, 2000); Ronald R. Kline, *The Cybernetics Moment: Or Why We Call Our Age the Information Age* (Baltimore: Johns Hopkins University Press, 2015).

37. Cohn, *L&N and the Control of Electric Power Systems*, 15.

38. A. C. Hartranft and F. H. Light, "A Survey of the Application of Automatic Devices for Electric Power Generation," *IRE Transactions on Industrial Electronics* PGIE, no. 7 (1958): 55–62.

39. Nathan Cohn, "Methods of Controlling Generation on Interconnected Power Systems," *Transactions of the American Institute of Electrical Engineers. Part III: Power Apparatus and Systems* 80, no. 3 (1961): 270–279; Cohn, "The Automatic Control of Electric Power in the United States."

40. *IEEExplore*, s.v. "automatic generation control," IEEExplore Digital Library, accessed January 21, 2016, http://ieeexplore.ieee.org; *Compendex*, s.v. "automatic generation control," Engineering Village, accessed January 21, 2016, http://www.engineeringvillage.com. A search of both the IEEExplore and the *Compendex* databases revealed that the first usages of the complete phrase "automatic generation control" occurred in 1961 and 1962, respectively.

41. "IEEE Standard Definitions of Terms for Automatic Generation Control on Electric Power Systems," *IEEE Transactions on Power Apparatus and Systems* PAS-89, no. 6 (1970): 1356–1364. In 1963, the American Institute of Electrical Engineers (AIEE) and the Institute of Radio Engineers combined to form the Institute of Electrical and Electronics Engineers (IEEE).

42. Ibid., 1359. Most recently, NERC has defined *automatic generation control* to mean "equipment that automatically adjusts generation in a Balancing Authority Area from a central location to maintain the Balancing Authority's interchange schedule plus Frequency Bias. AGC may also accommodate automatic inadvertent payback and time error correction." *Balancing Authority Area, Balancing Authority*, and *Frequency Bias* are also defined. *Glossary of Terms Used in NERC Reliability*

Standards, North American Electric Reliability Corporation website, updated December 12, 2016, http://www.nerc.com/files/glossary_of_terms.pdf.

43. Federal Power Commission, *Prevention of Power Failures: An Analysis and Recommendations Pertaining to the Northeast Failure and the Reliability of U.S. Power Systems; a Report to the President by the Federal Power Commission* (Washington, DC: Government Printing Office, 1967), II.i.

44. Ibid., xiii.

45. "Advisory Group Probes Bulk Power Reliability: An Interview with J. J. Busby, C. P. Almon, and T. J. Nagel," *Electrical World*, May 1,1967, 69.

46. *Prevention of Power Failures: An Analysis and Recommendations Pertaining to the Northeast Failure and the Reliability of U.S. Power Systems; a Report to the President by the Federal Power Commission*, II.1.

47. K. W. James, "Progress in Power-System Automatic Control," *Electronics and Power* 11, no. 8 (1965): 278–280.

48. Ibid., 280.

49. K. Fruhauf et al., "The Application of a Database System to On-line Security Assessment in An EHV-Network," *IEEE Conference Proceedings, Power Industry Computer Applications Conference, 1979 (PICA-79)*, 165–169; J. W. Klein et al., "Dynamic Stability Assessment Model of a Parallel AC-DC Power System," *IEEE Transactions on Power Apparatus and Systems* PAS-96, no. 4 (1977): 1296–1304; Hasan Modir and R. A. Schlueter, "A Dynamic State Estimator for Dynamic Security Assessment," *IEEE Power Engineering Review* PER-1, no. 11 (1981): 43; K. N. Pragnell and B. J. Cory, "Security Assessment in Distribution Systems," *Proceedings of the Institution of Electrical Engineers* 117, no. 1 (1970): 161–164; H. Rudnick and A. Brameller, "Transient Security Assessment Methods," *Proceedings of the Institution of Electrical Engineers* 125, no. 2 (1978): 135–140; F. G. Vervloet and A. Brameller, "A.C. Security Assessment," *Proceedings of the Institution of Electrical Engineers* 122, no. 9 (1975): 897–902; R. C. Wilson and A. E. Di Marco, "The Intermediate-Term Security Assessment of a Power Generating System," *IEEE Transactions on Power Apparatus and Systems* PAS-94, no. 4 (1975): 1417–1424.

50. Clarence F. Paulus, "Keep Generators Running: Improve Reliability at Little Cost," *IEEE Transactions on Power Apparatus and Systems* PAS-92, no. 1 (1973): 243–247.

51. Tomas E. Dy Liacco, "The Adaptive Reliability Control System," *IEEE Transactions on Power Apparatus and Systems* PAS-86, no. 5 (1967): 517–533.

52. Ibid., 517.

53. Ibid.

54. Ibid., 528.

55. Ibid., 530.

56. Ibid., 529.

57. Ibid., 517–533. One measure of the influence of one publication on an industry is the frequency with which other engineers refer to the work. Dy Liacco's 1967 paper had been cited more than 250 times in related technical literature as of January 2016 and four times as recently as 2015: 257 citations in *Google Scholar,* https://scholar.google.com/; 134 citations recorded in *Scopus,* http://www.scopus.com/; 110 citations in *IEEExplore,* http://ieeexplore.ieee.org/Xplore/home.jsp. All accessed January 21, 2016.

58. Tomas E. Dy Liacco, "The Emerging Concept of Security Control," in *Purdue 1970 Symposium on Power Systems: Power Systems in the Seventies* (Lafayette, IN: Purdue University, 1970).

59. Ibid., 1.

60. See chapter 7 for a description of the 1965 Northeast power failure and the actions taken by the Consolidated Edison system operator.

61. Tomas E. Dy Liacco, "Real-Time Computer Control of Power Systems," *Proceedings of the IEEE* 62, no. 7 (1974): 884–885.

62. *IEEExplore,* s.v. "system security" and "power," but not "bio" and "ether," accessed January 21, 2016, http://ieeexplore.ieee.org/Xplore/home.jsp; *Compendex,* s.v. "system security" and "power," but not "bio" and "ether," Engineering Village, accessed January 21, 2016, http://www.engineeringvillage.com. Searches for the terms "system security" and "power" on both IEEExplore and *Compendex* illustrate the change in usage and concern following the 1965 blackout. Before the appearance of "system security" in the 1967 FPC report *Prevention of Power Failures,* the separate terms "system" and "security" occurred together in earlier professional papers but not generally in reference to the overall reliability and stability of a power network. The combined term "system security" began to appear in the late 1960s, with a sharp rise in usage in the 1970s.

63. *Prevention of Power Failures: An Analysis and Recommendations Pertaining to the Northeast Failure and the Reliability of U.S. Power Systems; a Report to the President,* 44–46. As James Robinson explained, in practice, engineers described the states as "steady state," "dynamic state," and "transient state with several seconds duration" as a subcategory of the longer dynamic state. Personal communication with the author, December 19, 2016.

64. *Prevention of Power Failures: An Analysis and Recommendations Pertaining to the Northeast Failure and the Reliability of U.S. Power Systems; a Report to the President,* 44.

65. Dy Liacco, "The Adaptive Reliability Control System," 518.

66. Dy Liacco, "The Emerging Concept of Security Control," 3. In the early twenty-first century, engineers continue to use Dy Liacco's characterizations of normal, emergency, and restorative states, with the added "alert" state that represents a power system departing from normal but not yet in a state of emergency. Engineers are also grappling with the challenge of cyberattacks on power systems, which may represent a fifth state. K. R. Davis et al., "A Cyber-Physical Modeling and Assessment Framework for Power Grid Infrastructures," *IEEE Transactions on Smart Grid* 6, no. 5 (2015): 2464–2475.

67. Dy Liacco, "The Adaptive Reliability Control System," 519; quote in Dy Liacco, "The Emerging Concept of Security Control," 1.

68. R. O. Hinkel, F. J. Keglovitz, and M. F. Daumer, "Security Monitoring and Security Analysis for Pennsylvania Power and Light Company's Power Control Center" (paper presented to the Power Industry Computer Applications Conference, PICA '77, Toronto, Ontario, May 24–27, 1977), 83–86, in proceedings.

69. James Robinson, personal communication with the author, May 23, 2016.

70. Hinkel, Keglovitz, and Daumer, "Security Monitoring and Security Analysis for Pennsylvania Power and Light Company's Power Control Center," 86.

71. Robinson, personal communication with the author, May 23, 2016.

72. Ibid.

73. Hirsh, *Technology and Transformation in the American Electric Utility Industry*, 131–138. Hirsh offers a brief analysis of industry research and development activities in the 1960s and 1970s, with a particular focus on power generation technologies.

74. Ibid., 137–139.

75. Lester H. Fink and Kjell Carlsen, eds. *Proceedings of Systems Engineering for Power: Status and Prospects: Henniker, New Hampshire, 17–22 August, 1975* (Washington, DC: US Energy Research and Development Administration, 1975).

76. A. S. Debs and A. R. Benson, "Security Assessment of Power Systems," in *Proceedings of Systems Engineering for Power*, 144–172; Harvey H. Happ, "Optimal Power Dispatch," in *Proceedings of Systems Engineering for Power*, 36–49; J. Zaborszky, A. K. Subramanian, and K. M. Lu, "Control Interfaces in Generation Allocation," in *Proceedings of Systems Engineering for Power*, 52–71.

77. Happ, "Optimal Power Dispatch," 36–49; M. Ruane, "Economic—Environmental Operation of Power Systems," in *Proceedings of Systems Engineering for Power*, 87–100; Zaborszky, Subramanian, and Lu, "Control Interfaces in Generation Allocation," 52–71.

78. Debs and Benson, "Security Assessment of Power Systems," 144–172.

79. See Charles P. Steinmetz, "Power Control and Stability of Electric Generating Stations," *Transactions of the American Institute of Electrical Engineers* 39, no. 2 (1920): 1215–1287.

80. Debs and Benson, "Security Assessment of Power Systems," 144.

81. John L. Scheidt, "A Survey of Power System Control Center Justifications," *IEEE Transactions on Power Apparatus and Systems* PAS-98, no. 1 (1979): 135–140.

82. Hirsh, *Technology and Transformation in the American Electric Utility Industry.*

83. P. M. Anderson, "Reliability Criteria for System Dynamic Performance," *IEEE Transactions on Power Apparatus and Systems* PAS-96, no. 6 (1977): 1815–1818.

84. Ibid., 1817.

85. Federal Energy Regulatory Commission, *The Con Edison Power Failure of July 13 and 14, 1977: Final Staff Report, June 1978* (Washington, DC: Government Printing Office, 1978).

86. Donald Christiansen, "Spectral Lines: Blackout Prevention," *Spectrum, IEEE* 14, no. 10 (1977): 21–22.

87. Tomas E. Dy Liacco and D. L. Rosa, "Survey of System Control Centers for Generation-Transmission Systems" (Cleveland: Cleveland Electric Illuminating Company, 1986); Nathan Cohn, Handwritten Table of Dy Liacco Control Center Data and List of Abbreviations and Definitions, in Nathan Cohn Papers (Houston, TX: Author's Personal Collection, 1985).

88. John M. Undrill, "Where We Stand (Power System Computer Applications)," *IEEE Computer Applications in Power* 1, no. 1 (1988): 9–10.

89. G. W. Stagg et al., "Thirty Years of Power Industry Computer Applications," *IEEE Computer Applications in Power* 7, no. 2 (1994): 43.

90. Scheidt, "A Survey of Power System Control Center Justifications," 135–140; B. F. Wollenberg and Walt Stadlin, "A Real Time Optimizer for Security Dispatch," *IEEE Transactions on Power Apparatus and Systems* PAS-93, no. 5 (1974): 1640–1649.

91. J. Dillow, U. G. Knight, and J. Hewson, "System Control in the Central Electricity Generating Board" (paper presented at the Sixth Triennial World Congress of the International Federation of Automatic Control, Boston/Cambridge, Massachusetts, August 24–30, 1975).

92. C. W. Taylor and R. L. Cresap, "Real-Time Power System Simulation for Automatic Generation Control," *IEEE Transactions on Power Apparatus and Systems* PAS-95, no. 1 (1976): 375–384.

93. J. Dillow and D. E. Plews, "Design Features of Graphical Displays Used in the Monitoring of Electricity Distribution Systems" (paper presented at the Sixth

Triennial World Congress of the International Federation of Automatic Control, Boston/Cambridge, Massachusetts, August 24–30, 1975).

94. From papers presented at the Sixth Triennial World Congress of the International Federation of Automatic Control, Boston/Cambridge, Massachusetts, August 24–30, 1975; Lars Gustafsson and John Lindqvist, "The New Information System for the National Dispatching Centre at the Swedish State Power Board"; Vivian Saminaden, "Electrical Dispatching in France: Status and Trends"; Yasuo Tamura et al., "Energy Control Centers in Japan—Evaluation of Design Principles and Functions from Security Veiwpoints"; V. A. Venikov, B. I. Golovitsyn, and M. S. Liseyev, "Power System Cybernetics and Modern Energy Control."

95. M. A. Laughton and M. W. Humphrey Davies, "Numerical Techniques in Solution of Power-System Load-Flow Problems," *Proceedings of the Institution of Electrical Engineers* 111, no. 9 (1964): 1575–1588; Harvey H. Happ and John M. Undrill, "Multicomputer Configurations and Diakoptics: Real Power Flow in Power Pools," *IEEE Transactions on Power Apparatus and Systems* PAS-88, no. 6 (1969): 789–796; William F. Tinney and C. E. Hart, "Power Flow Solution by Newton's Method," *IEEE Transactions on Power Apparatus and Systems* PAS-86, no. 11 (1967): 1449–1460. General Electric engineers Harvey H. Happ and John M. Undrill reported research on the use of distributed computing to work around the problem of unwieldy data processing on a single computer.

96. James Robinson provided a description of the burdensome direct method: "The network impedance model was large since each transmission facility was modeled, but individual substation MW loads had to be predicted by hand and entered into the model prior to solving the network for predicted MW flow on each transmission facility. Each model dataset was large, stored on magnetic tape, and was cumbersome to prepare for each key generation and customer peak load pattern. Typically, the original input and changes to the data were submitted to the computer on punched cards." Personal communication with the author, May 21, 2016.

97. Laughton and Davies, "Numerical Techniques in Solution of Power-System Load-Flow Problems," 1575–1588; Tinney and Hart, Tinney and Hart, "Power Flow Solution by Newton's Method," 1449–1460.

98. "Power Flow Solution by Newton's Method," 1449–1460.

99. "Long-Retired BPA Engineer's Achievement Lasts a Lifetime," *Newsroom*, Bonneville Power Administration website, November 17, 2011, accessed March 1, 2016, https://www.bpa.gov/news/newsroom/Pages/Long-retired-BPA-engineers-achievement-lasts-a-lifetime.aspx.

100. Hermann W. Dommel and William F. Tinney, "Optimal Power Flow Solutions," *IEEE Transactions on Power Apparatus and Systems* PAS-87, no. 10 (1968): 1866–1876. In 2000, the North American Power Symposium listed this paper as one of the five most

influential papers of the twentieth century for electric power engineering. G. T. Heydt, S. S. Venkata, and Nagaraj Balijepalli, "High Impact Papers in Power Engineering, 1900–1999," *Proceedings 2000 North American Power Symposium* 1 (2000): 1–7.

101. *Compendex*, s.v. "computer" and "power," Engineering Village, accessed January 25, 2016, http://www.engineeringvillage.com.

102. H. P. St. Clair and G. W. Stagg, "Experience in Computation of Load-Flow Studies Using High-Speed Computers," *Transactions of the American Institute of Electrical Engineers. Part III, Power Apparatus and Systems*, 77, no. 3 (1958): 1275–1282.

103. "History of PSCC," Power Systems Computation Conference website, updated October 16, 2014, http://www.pscc-central.org/en/about-pscc/history-of-pscc.html.

104. Mark Enns and William F. Tinney, "Scanning the Issue: Computers in the Power Industry," *Proceedings of the IEEE* 62, no. 7 (1974): 868–871.

105. Some images indicated a third small system called the Southwest Public Service System in the Texas panhandle area.

106. *National Power Survey: A Report of the Federal Power Commission, 1964*, 14.

107. *The 1970 National Power Survey of the Federal Power Commission*, I-13-4.

108. Ibid.

109. Nathan Cohn, "Automatic Control of Power Systems" (paper presented at the IEEE International Convention Record, Part 12—Large Interconnected Power Systems, New York, New York, 1966), 1–8; Nathan Cohn, *Control of Generation and Power Flow on Interconnected Power Systems* (New York: J. Wiley, 1966); Cohn, "L & N and the Control of Electric Power Systems."

110. Nathan Cohn, "Power-System Interconnections: Control of Generation and Power Flow," in *Standard Handbook for Electrical Engineers*, 10th ed., ed. Donald G. Fink and John M. Carroll (New York: McGraw-Hill, 1968), sections 15-3–15-4.

111. Nathan Cohn, "Power-System Interconnections: Control of Generation and Power Flow," in *Standard Handbook for Electrical Engineers*, 11th ed., ed. Donald G. Fink and H. Wayne Beaty (New York: McGraw-Hill, 1978), section 16-6.

112. Anderson, "Reliability Criteria for System Dynamic Performance," 6.

113. For examples, see the US Department of Energy web page titled "Top 9 Things You Didn't Know about America's Power Grid," accessed February 25, 2016, http://energy.gov/articles/top-9-things-you-didnt-know-about-americas-power-grid; a sample online journal article by Loren Thompson, "Five Reasons The U.S. Power Grid is Overdue For A Cyber Catastrophe," *Forbes.com*, August 19, 2015, http://www.forbes.com/

home_usa/; and a book by Phillip F. Schewe, *The Grid: A Journey through the Heart of Our Electrified World* (Washington, DC: J. Henry Press, 2007).

114. Ferber R. Schlief et al., "A Swing Relay for the East-West Intertie," *IEEE Transactions on Power Apparatus and Systems* PAS-88, no. 6 (1969): 821–825.

115. Ibid., 824.

116. Ibid.

117. Quote from Walter D. Brown, former president of NERC, "Whatever Happened To," *IEEE Spectrum* 22, no. 3 (1985): 27.

118. Michael Hamilton, "Bulk Power Supply Reliability and the Proposed National Grid System: Signposts Pointing toward What Destination?," *Policy Studies Journal* 7, no. 1 (1978): 97.

119. "The Early HVDC Development: The Key Challenge in the HVDC Technique," ABB AB Grid Systems—HVDC, Asea Brown Boveri website, accessed January 10, 2017, https://library.e.abb.com/public/93e7f5ea0e800b7cc1257ac3003f4955/HVDC_50years.pdf.

120. "Existing HVDC Projects Listing, Prepared for the HVDC and Flexible Transmission Subcommittee of the IEEE Transmission and Distribution Committee," University of Idaho Electrical and Computer Engineering Department website, March 2012, accessed March 1, 2016, http://www.ece.uidaho.edu/hvdcfacts/Projects/HVDCProjectsListingMarch2012-existing.pdf.

121. F. J. Ellert and N. G. Hingorani, "HVDC for the Long Run," *IEEE Spectrum* 13, no. 8 (1976): 37.

122. Ibid., 42.

123. Mark Dowling and J. Barry Winter, *Tri-State Generations of Power* (Westminster, CO: Tri-State Generation and Transmission Association, 2002)

124. "Tri-State Milestones," Tri-State Generation and Transmission Association webpage, updated February 11, 2013, http://www.tristategt.org/AboutUs/milestones.cfm.

125. By the end of the twentieth century, there were six HVDC ties between the Eastern and Western Interconnected Systems, four between the Eastern Interconnected System and the Quebec Interconnection, and two between the Eastern Interconnected System and Texas, which were used for scheduled or emergency trades of electricity. Dennis A. Woodford and Wang Xuegong, "Synchronous Operation of Adjacent Power Systems," *Proceedings of POWERCON '98 International Conference of Power System Technology* (1998): 914–917.

126. Paul A. David and Julie Ann Bunn, "The Economics of Gateway Technologies and Network Evolution: Lessons from Electricity Supply History," *Information Economics and Policy* 3, no. 2 (1988): 165–202.

9 Deregulation and Disaggregation: A Brief Overview, 1980–2015

1. For further reading, see Severin Borenstein and James Bushnell, "The U.S. Electricity Industry after 20 Years of Restructuring," *NBER Working Papers* 21113, available online, National Bureau of Economic Research, accessed January 5, 2017, http://www.nber.org/papers/w21113 (2015), 1–31; G. Bruce Doern, *Canadian Energy Policy and the Struggle for Sustainable Development* (Toronto: University of Toronto Press, 2005); William J. Hausman, Peter Hertner, and Mira Wilkins, *Global Electrification: Multinational Enterprise and International Finance in the History of Light and Power, 1878–2007*, ed. Louis Galambos and Geoffrey Jones, Cambridge Studies in the Emergence of Global Enterprise (New York: Cambridge University Press, 2008), 64–72, 262–275; Richard F. Hirsh, *Power Loss: The Origins of Deregulation and Restructuring in the American Electric Utility System* (Cambridge, MA: MIT Press, 1999); Paul L. Joskow, "Markets for Power in the United States: An Interim Assessment," *Energy Journal* 27, no. 1 (2006): 1–36; Paul L. Joskow, "Lessons Learned from Electricity Market Liberalization," in "The Future of Electricity: Papers in Honor of David Newbery," special issue, *Energy Journal* (2008): 9–42; John Kwoka, "Restructuring the U.S. Electric Power Sector: A Review of Recent Studies," *Review of Industrial Organization* 32, no. 3–4 (2008): 165–195; Jeremiah D. Lambert, *The Power Brokers: The Struggle to Shape and Control the Electric Power Industry* (Cambridge, MA: MIT Press, 2015), 131–205; Paul W. MacAvoy, *The Unsustainable Costs of Partial Deregulation* (New Haven: Yale University Press, 2007); David E. Nye, *When the Lights Went Out: A History of Blackouts in America* (Cambridge, MA: MIT Press, 2010), 137–171; Rebecca Slayton, "Efficient, Secure Green: Digital Utopianism and the Challenge of Making the Electrical Grid 'Smart,'" *Information & Culture* 48, no. 4 (2013): 448–478.

2. Public Utility Regulatory Policies Act of 1978, 16 U.S.C. § 46 (1978).

3. Hirsh, *Power Loss*, 95–99.

4. The Energy Policy Act of 1992, 42 U.S.C. § 134.

5. Federal Energy Regulatory Commission, "Promoting Wholesale Competition through Open Access Non-Discriminatory Transmission Services by Public Utilities and Recovery of Stranded Costs by Public Utilities and Transmitting Utilities," Order No. 888.

6. Federal Energy Regulatory Commission, "Regional Transmission Organizations," Order No. 2000, 14–16. The following summarizes the more detailed description in FERC Order No. 2000.

7. According to the FERC website, the agency suggested the establishment of Independent System Operators following the issuance of Order No. 888. "Regional

Transmission Organizations (RTO)/Independent System Operators (ISO)," FERC, accessed January 7, 2017, https://www.ferc.gov/industries/electric/indus-act/rto.asp. Today, the agency defines the ISO as "an independent, Federally regulated entity established to coordinate regional transmission in a non-discriminatory manner and ensure the safety and reliability of the electric system." "Glossary," FERC, accessed January 7, 2017, https://www.ferc.gov/resources/glossary.asp#I.

8. FERC Order No. 2000.

9. The eleven principles, as first set out in FERC Order 888, Section IV.F.4. Bi-lateral Coordination Arrangements: ISO Principles, 279–286, are summarized in the appendix.

10. "Today in Energy," April 4, 2011, *Energy Information Agency*, accessed January 7, 2017, http://www.eia.gov/todayinenergy/detail.php?id=790.

11. The Energy Policy Act of 2005, 42 U.S.C. § 149.

12. David Nevius, "Challenges and Opportunities for Improving Bulk Power System Reliability in North America," *European Review of Energy Markets*, no. 4 (2007): 1–19, accessed June 21, 2016, https://www.eeinstitute.org/european-review-of-energy-market/erem-4-article-d.-nevius.

13. Ibid., 8.

14. Federal Energy Regulatory Commission, "The Western Energy Crisis, the Enron Bankruptcy, and FERC's Response," report issued April 13, 2005, accessed May 22, 2017, https://www.ferc.gov/industries/electric/indus-act/wec/chron/chronology.pdf. For some additional reading, see James B. Bushnell, Erin T. Mansur, and Celeste Saravia, "Vertical Arrangements, Market Structure, and Competition: An Analysis of Restructured US Electricity Markets," *American Economic Review* 98, no. 1 (2008): 237–266; Martin Chick and H. V. Nelles, "Nationalisation and Privatisation: Ownership, Markets and the Scope for Introducing Competition into the Electricity Supply Industry," *Revue Économique* 58, no. 1 (2007): 277–293; Hirsh, *Power Loss*; MacAvoy, *The Unsustainable Costs of Partial Deregulation*; Richard Munson, *From Edison to Enron: The Business of Power and What It Means for the Future of Electricity* (Westport, CT: Praeger, 2005); James L. Sweeney, *The California Electricity Crisis* (Stanford: Hoover Institution Press, 2002); Susan E. Taylor, "Energy Emergency Powers in a National Interconnected System," *Natural Resources & Environment* 21, no. 1 (2006): 43–69.

15. U.S.-Canada Power System Outage Task Force, "Final Report on the August 14, 2003 Blackout in the United States and Canada: Causes and Recommendations," North American Electric Reliability Corporation website, April 2004, accessed December 17, 2016, http://www.nerc.com/pa/rrm/ea/Pages/Blackout-August-2003.aspx.

16. "Fact Sheet: Energy Policy Act of 2005," Federal Energy Regulatory Commission, August 8, 2006, accessed December 17, 2016, https://www.ferc.gov/legal/fed-sta/epact-fact-sheet.pdf.

17. "NERC certified as Electric Reliability Organization; Western Region Reliability Advisory Body Accepted," News Release, *Federal Energy Regulatory Commission,* July 20, 2006, available online at Federal Energy Regulatory Commission website, accessed January 10, 2017, https://www.ferc.gov/media/news-releases/2006/2006-3/07-20-06-E-5.asp.

18. Within Texas, the majority of power companies do not participate in trades across state lines.

19. *Reliability Functional Model: Function Definitions and Functional Entities, Version 5* (Princeton, NJ: North American Electric Reliability Corporation, 2009), available online at North American Electric Reliability Corporation website, accessed January 10, 2017, http://www.nerc.com/pa/Stand/Pages/FunctionalModel.aspx.

20. *NERC Active Compliance Registry Matrix as of 1/15/2016* (Atlanta: North American Electric Reliability Corporation, 2016).

21. *Appendix 5a: Organization Registration and Certification Manual* (Atlanta: North American Electric Reliability Corporation, 2015), available online at North American Electric Reliability Corporation website, accessed January 10, 2017, http://www.nerc.com/FilingsOrders/us/RuleOfProcedureDL/Appendix_5A_OrganizationRegistration_20151015.pdf.

10 Conclusion

1. Engineers at ERCOT met with the author on two occasions: March 1, 2013, and January 8, 2016. They provided a tour of the central control room for the Texas network and offered valuable explanations, insights, and commentary on how grids work in the twenty-first century. Meeting participants on March 1, 2013, included Joel Mickey, B. J. Behroon, and John Adams. Meeting participants on January 8, 2016, included Joel Mickey, Bill Blevins, Dan Jones, Julia Matevosyan, N. D. R. Sarma Nuthalipati, John Adams, Sai Morty, and Dave Maggio.

2. System operator's personal communication with the author, January 8, 2016. This individual prefers to remain anonymous.

3. Conversation between ERCOT engineers and system operators and the author, January 8, 2016.

4. Kristian M. Koellner et al., "Synchrophasors across Texas: The Deployment of Phasor Measurement Technology in the ERCOT Region," *IEEE Power & Energy Magazine* 13, no. 5 (2015): 36–40.

5. Phil Overholt, David Ortiz, and Alison Silverstein, "Synchrophasor Technology and the DOE," *IEEE Power & Energy Magazine* 13, no. 5 (2015): 14–17.

6. Nathan Cohn, "Power Flow Control—Basic Concepts for Interconnected Systems," *Electric Light and Power* 28, no. 8-9 (1950): 84.

Selected Bibliography

Abbate, Janet. *Inventing the Internet*. Cambridge, MA: MIT Press, 1999.

Anderson, Douglas D. *Regulatory Politics and Electric Utilities: A Case Study in Political Economy*. Boston: Auburn House, 1981.

Andrews, Richard N. L. *Managing the Environment, Managing Ourselves: A History of American Environmental Policy*. 2nd ed. New Haven: Yale University Press, 2006.

Armitage, Kevin C. *The Nature Study Movement: The Forgotten Popularizer of America's Conservation Ethic*. Lawrence: University Press of Kansas, 2009.

Armstrong, Christopher, and H. V. Nelles. *Monopoly's Moment: The Organization and Regulation of Canadian Utilities, 1830–1930*. Technology and Urban Growth. Philadelphia: Temple University Press, 1986.

Bakke, Gretchen. *The Grid: The Fraying Wires between Americans and Our Energy Future*. New York: Bloomsbury, 2016.

Balogh, Brian. *Chain Reaction: Expert Debate and Public Participation in American Commercial Nuclear Power, 1945–1975*. New York: Cambridge University Press, 1991.

Beck, Bill. *At Your Service: An Illustrated History of Houston Lighting and Power*. Houston: Houston Lighting and Power Company, 1990.

Beck, Bill. *Interconnections: The History of the Mid-continent Area Power Pool*. 1st ed. Minneapolis: The Pool, 1988.

Bedford, Henry F. *Seabrook Station: Citizen Politics and Nuclear Power*. Amherst: University of Massachusetts Press, 1990.

Belfield, Robert Blake. "The Niagara Frontier: The Evolution of Electric Power Systems in New York and Ontario, 1880–1935." PhD diss., University of Pennsylvania, 1981.

Bennett, S. *A History of Control Engineering, 1930–1955*. London: Peter Peregrinus, 1993.

Berghoff, Hartmut and Adam Rome, *Green Capitalism: Business and the Environment in the Twentieth Century*. Hagley Perspectives on Business and Culture, edited by Roger Horowitz. Philadelphia: University of Pennsylvania Press, 2017.

Bijker, Wiebe E., Thomas Parke Hughes, and T. J. Pinch. *The Social Construction of Technological Systems: New Directions in the Sociology and History of Technology*. Cambridge, MA: MIT Press, 1987.

Billington, David P., and Donald C. Jackson. *Big Dams of the New Deal Era: A Confluence of Engineering and Politics*. Norman: University of Oklahoma Press, 2006.

Bix, Amy Sue. *Girls Coming to Tech!: A History of American Engineering Education for Women*. Cambridge, MA: MIT Press, 2013.

Blum, Elizabeth D. *Love Canal Revisited: Race, Class, and Gender in Environmental Activism*. Lawrence: University Press of Kansas, 2008.

Bowker, Geoffrey C., Karen Baker, Florence Millerand, and David Ribes. "Toward Information Infrastructure Studies: Ways of Knowing in a Networked Environment." In *International Handbook of Internet Research*, edited by Jeremy Hunsinger, Lisbeth Klastrup, and Matthew Allen, 97–117. London: Springer, 2010.

Breyer, Stephen G. *Regulation and Its Reform*. Cambridge, MA: Harvard University Press, 1982.

Brigham, Jay L. *Empowering the West: Electrical Politics before FDR*. Development of Western Resources. Lawrence: University Press of Kansas, 1998.

Brooks, Karl Boyd. *Public Power, Private Dams: The Hells Canyon High Dam Controversy*. Seattle: University of Washington Press, 2006.

Carlson, W. Bernard. *Tesla: Inventor of the Electrical Age*. Princeton, NJ: Princeton University Press, 2013.

Casazza, John. *The Development of Electric Power Transmission: The Role Played by Technology, Institutions, and People*. IEEE Case Histories of Achievement in Science and Technology. New York: Institute of Electrical and Electronics Engineers, 1993.

Castaneda, Christopher James. *Regulated Enterprise: Natural Gas Pipelines and Northeastern Markets, 1938–1954*. Historical Perspectives on Business Enterprise Series. Columbus: Ohio State University Press, 1993.

Chandler, Alfred D. *The Visible Hand: The Managerial Revolution in American Business*. Cambridge, MA: Belknap Press of Harvard University Press, 1977.

Charpak, Georges, and Richard L. Garwin. *Megawatts and Megatons: A Turning Point in the Nuclear Age?* New York: Alfred A. Knopf, 2001.

Coutard, Olivier. *The Governance of Large Technical Systems*. New York: Routledge, 1999.

Coutard, Olivier, Richard E. Hanley, and Rae Zimmerman. *Sustaining Urban Networks: The Social Diffusion of Large Technical Systems*. The Networked Cities Series. London: Routledge, 2005.

Doern, G. Bruce. *Canadian Energy Policy and the Struggle for Sustainable Development.* Toronto: University of Toronto Press, 2005.

Durden, Robert F. *Electrifying the Piedmont Carolinas: The Duke Power Company, 1904–1997.* Durham, NC: Carolina Academic Press, 2001.

Elkind, Sarah. *How Local Politics Shape Federal Policy: Business, Power, and the Environment in Twentieth-Century Los Angeles.* The Luther H. Hodges Jr. and Luther H. Hodges Sr. Series on Business, Society, and the State, edited by William H. Becker. Chapel Hill: University of North Carolina Press, 2011.

Evenden, Matthew. *Allied Power: Mobilizing Hydro-Electricity during Canada's Second World War.* Toronto: University of Toronto Press, 2015.

Farmer, Jared. *Glen Canyon Dammed: Inventing Lake Powell and the Canyon Country.* Tucson: University of Arizona Press, 1999.

Finn, Bernard S. *The History of Electrical Technology: An Annotated Bibliography.* Garland Reference Library of the Humanities. New York: Garland, 1991.

Friedricks, William B. *Henry E. Huntington and the Creation of Southern California.* Historical Perspectives on Business Enterprise Series, edited by Mansel G. Blackford and K. Austin Kerr. Columbus: Ohio State University Press, 1992.

Frost, Robert L. *Alternating Currents: Nationalized Power in France, 1946–1970.* Ithaca, NY: Cornell University Press, 1991.

Funigiello, Philip J. *Toward a National Power Policy: The New Deal and the Electric Utility Industry, 1933–1941.* Pittsburgh: University of Pittsburgh Press, 1973.

Gilbert, Richard J., and Edward Kahn. *International Comparisons of Electricity Regulation.* Cambridge: Cambridge University Press, 1996.

Gottlieb, Robert. *Forcing the Spring: The Transformation of the American Environmental Movement.* Rev. and updated ed. Washington, DC: Island Press, 2005.

Grose, Peter. *Power to People: The Inside Story of AES and the Globalization of Electricity.* Washington, DC: Island Press, 2007.

Hannah, Leslie. *Electricity before Nationalisation: A Study of the Development of the Electricity Supply Industry in Britain to 1948.* Johns Hopkins Studies in the History of Technology. Baltimore: Johns Hopkins University Press, 1979.

Hargrove, Erwin C. *Prisoners of Myth: The Leadership of the Tennessee Valley Authority, 1933–1990.* 1st ed. Knoxville: University of Tennessee Press, 2001.

Harrison, Robert. *Congress, Progressive Reform, and the New American State.* New York: Cambridge University Press, 2004.

Hausman, William J., Peter Hertner, and Mira Wilkins. *Global Electrification: Multinational Enterprise and International Finance in the History of Light and Power, 1878–2007.*

Cambridge Studies in the Emergence of Global Enterprise, edited by Louis Galambos and Geoffrey Jones. New York: Cambridge University Press, 2008.

Hawkey, David, and Janette Webb. "Social Studies of Technology, Energy Systems and Modern Societies." In *Sustainable Urban Energy Policy: Heat and the City*, edited by David Hawkey, Janette Webb, Heather Lovell, David McCrone, Margaret Tingey, and Mark Winskel, 21–43. New York: Routledge, 2016.

Hawley, Ellis Wayne. *The New Deal and the Problem of Monopoly: A Study in Economic Ambivalence*. New York: Fordham University Press, 1995.

Hays, Samuel P. *Conservation and the Gospel of Efficiency: The Progressive Conservation Movement, 1890–1920*. Pittsburgh: University of Pittsburgh Press, 1959.

Hecht, Gabrielle. *The Radiance of France: Nuclear Power and National Identity after World War II*. Inside Technology. Cambridge, MA: MIT Press, 1998.

Hirsh, Richard F. *Power Loss: The Origins of Deregulation and Restructuring in the American Electric Utility System*. Cambridge, MA: MIT Press, 1999.

Hirsh, Richard F. *Technology and Transformation in the American Electric Utility Industry*. Cambridge: Cambridge University Press, 1989.

Hirt, Paul W. *The Wired Northwest: The History of Electric Power, 1870s–1970s*. Lawrence: University Press of Kansas, 2012.

Hornig, James F. *Social and Environmental Impacts of the James Bay Hydroelectric Project*. Montreal: McGill-Queen's Press, 1999.

Hughes, Thomas Parke. *Networks of Power: Electrification in Western Society, 1880–1930*. Baltimore: Johns Hopkins University Press, 1983.

Hughes, Thomas Parke, and Agatha C. Hughes, eds. *Systems, Experts, and Computers: The Systems Approach in Management and Engineering, World War II and After*. Dibner Institute Studies in the History of Science and Technology, edited by Jed Buchwald and Evelyn Simha. Cambridge, MA: MIT Press, 2000

Hunt, Edward Eyre. *The Power Industry and the Public Interest: A Summary of a Survey of the Relations between the Government and the Electric Power Industry*. New York: Twentieth Century Fund, 1944.

Hunter, Louis C., and Lynwood Bryant. *A History of Industrial Power in the United States, 1780–1930*. Vol. 3, *The Transmission of Power*. Cambridge, MA: MIT Press, 1991.

Hyman, Leonard S., Andrew S. Hyman, and Robert C. Hyman. *America's Electric Utilities: Past, Present and Future*. 8th ed. Vienna, VA: Public Utilities Reports, 2005.

Israel, Paul. *Edison: A Life of Invention*. New York: John Wiley, 1998.

John, Richard R. *Network Nation: Inventing American Telecommunications*. Cambridge, MA: Belknap Press of Harvard University Press, 2010.

Jones, Christopher F. *Routes of Power: Energy and Modern America*. Cambridge, MA: Harvard University Press, 2014.

Jonnes, Jill. *Empires of Light: Edison, Tesla, Westinghouse, and the Race to Electrify the World*. 1st ed. New York: Random House, 2003.

Kerwin, Jerome G. *Federal Water-Power Legislation*. Studies in History, Economics, and Public Law. New York: Columbia University Press, 1926.

Kline, Ronald R. *The Cybernetics Moment: Or Why We Call Our Age the Information Age*. Baltimore: Johns Hopkins University Press, 2015.

Koppel, Ted. *Lights Out: A Cyberattack, a Nation Unprepared, Surviving the Aftermath*. New York: Crown, 2015.

Lambert, Jeremiah D. *The Power Brokers: The Struggle to Shape and Control the Electric Power Industry*. Cambridge, MA: MIT Press, 2015.

Lasser, William. *Benjamin V. Cohen: Architect of the New Deal*. New Haven: Yale University Press, 2002.

Layton, Edwin T. *The Revolt of the Engineers: Social Responsibility and the American Engineering Profession*. Baltimore: Johns Hopkins University Press, 1986.

LeCain, Timothy J. *Mass Destruction: The Men and Giant Mines That Wired America and Scarred the Planet*. New Brunswick, NJ: Rutgers University Press, 2009.

Lifset, Robert, ed. *American Energy Policy in the 1970s*. Norman: University of Oklahoma Press, 2014.

Lifset, Robert. *Power on the Hudson: Storm King Mountain and the Emergence of Modern American Environmentalism*. Pittsburgh: University of Pittsburgh Press, 2014.

MacAvoy, Paul W. *The Unsustainable Costs of Partial Deregulation*. New Haven: Yale University Press, 2007.

Manore, Jean. *Cross-Currents: Hydroelectricity and the Engineering of Northern Ontario*. Waterloo, ON: Wilfrid Laurier University Press, 1999.

Mazur, Allen. *Energy and Electricity in Industrial Nations: The Sociology and Technology of Energy*. New York: Routledge, 2013.

McCraw, Thomas K. *Prophets of Regulation: Charles Francis Adams, Louis D. Brandeis, James M. Landis, Alfred E. Kahn*. Cambridge, MA: Belknap Press of Harvard University Press, 1984.

McCraw, Thomas K. *TVA and the Power Fight, 1933–1939*. Philadelphia: Lippincott, 1971.

Melosi, Martin V. *Atomic Age America*. Boston: Pearson, 2013.

Melosi, Martin V. *Coping with Abundance: Energy and Environment in Industrial America*. 1st ed. Philadelphia: Temple University Press, 1985.

Miller, Char. *Gifford Pinchot and the Making of Modern Environmentalism*. Washington, DC: Island Press/Shearwater Books, 2001.

Miner, H. Craig. *Wolf Creek Station: Kansas Gas and Electric Company in the Nuclear Era*. Historical Perspectives on Business Enterprise Series. Columbus: Ohio State University Press, 1993.

Minteer, Ben A., and Robert E. Manning. *Reconstructing Conservation: Finding Common Ground*. Washington, DC: Island Press, 2003.

Munson, Richard. *From Edison to Enron: The Business of Power and What It Means for the Future of Electricity*. Westport, CT: Praeger, 2005.

Murphy, Craig N., and Joanne Yates. *The International Organization for Standardization (ISO): Global Governance through Voluntary Consensus*. Routledge Global Institutions, edited by Thomas G. Weiss and Rorden Wilkinson. New York: Routledge Taylor and Francis Group, 2009.

Needham, Andrew. *Power Lines: Phoenix and the Making of the Modern Southwest*. Politics and Society in Twentieth-Century America, edited by William Chafe, Gary Gerstle, Linda Gordon, and Julian Zelizer. Princeton, NJ: Princeton University Press, 2014.

Nelles, H. V. *The Politics of Development: Forests, Mines and Hydro-Electric Power in Ontario, 1849–1941*. Toronto: Macmillan of Canada, 1974.

Noble, David F. *America by Design: Science, Technology, and the Rise of Corporate Capitalism*. 1st ed. New York: Knopf, 1977.

Norwood, Gus. *Columbia River Power for the People: A History of Policies of the Bonneville Power Administration*. Portland, OR: US Department of Energy, Bonneville Power Administration, 1981.

Nye, David E. *Consuming Power: A Social History of American Energies*. Cambridge, MA: MIT Press, 1998.

Nye, David E. *Electrifying America: Social Meanings of a New Technology, 1880–1940*. Cambridge, MA: MIT Press, 1990.

Nye, David E. *When the Lights Went Out: A History of Blackouts in America*. Cambridge, MA: MIT Press, 2010.

Palfrey, John G., and Urs Gasser. *Interop: The Promise and Perils of Highly Interconnected Systems*. New York: Basic Books, 2012.

Passer, Harold C. *The Electrical Manufacturers, 1875–1900: A Study in Competition, Entrepreneurship, Technical Change, and Economic Growth*. Technology and Society. New York: Arno Press, 1972.

Perlgut, Mark. *Electricity across the Border: The U.S.-Canadian Experience*. Montreal: Canadian-American Committee, 1978.

Phillips, Sarah T. *This Land, This Nation: Conservation, Rural America, and the New Deal*. Cambridge: Cambridge University Press, 2007.

Platt, Harold L. *The Electric City: Energy and the Growth of the Chicago Area, 1880–1930*. Chicago: University of Chicago Press, 1991.

Pope, Daniel. *Nuclear Implosions: The Rise and Fall of the Washington Public Power Supply System*. New York: Cambridge University Press, 2008.

Pratt, Joseph A. *A Managerial History of Consolidated Edison, 1936–1981*. New York: Consolidated Edison Company of New York, 1988.

Raushenbush, Hilmar Stephen, and Harry Wellington Laidler. *Power Control*. New York: New Republic, 1928.

Regehr, T. D. *The Beauharnois Scandal: A Story of Canadian Entrepreneurship and Politics*. Toronto: University of Toronto Press, 1989.

Reisner, Marc. *Cadillac Desert: The American West and Its Disappearing Water*. Rev. and updated ed. New York: Penguin Books, 1993.

Righter, Robert W. *The Battle over Hetch Hetchy: America's Most Controversial Dam and the Birth of Modern Environmentalism*. New York: Oxford University Press, 2005.

Rimby, Susan. *Mira Lloyd Dock and the Progressive Era Conservation Movement*. University Park: Pennsylvania State University Press, 2012.

Rose, Mark H. *Cities of Light and Heat: Domesticating Gas and Electricity in Urban America*. University Park: Pennsylvania State University Press, 1995.

Rowland, John. *Progress in Power: The Contribution of Charles Merz and His Associates to Sixty Years of Electrical Development, 1899–1959*. London: Privately published for Merz and McLellan, 1961.

Rudolph, Richard, and Scott Ridley. *Power Struggle: The Hundred-Year War over Electricity*. 1st ed. New York: Harper and Row, 1986.

Russell, Andrew L. *Open Standards and the Digital Age: History, Ideology, and Networks*. Cambridge Studies in the Emergence of Global Enterprise. Edited by Louis Galambos and Geoffrey Jones. New York: Cambridge University Press, 2014.

Schewe, Phillip F. *The Grid: A Journey through the Heart of Our Electrified World*. Washington, DC: J. Henry Press, 2007.

Schiffer, Michael B. *Power Struggles: Scientific Authority and the Creation of Practical Electricity before Edison*. Cambridge, MA: MIT Press, 2008.

Schivelbusch, Wolfgang. *Disenchanted Night: The Industrialisation of Light in the Nineteenth Century*. New York: Berg, 1988.

Seifer, Marc J. *Wizard: The Life and Times of Nikola Tesla: Biography of a Genius.* Secaucus, NJ: Citadel, 1996.

Singer, Bayla Schlossberg. "Power to the People, the Pennsylvania–New Jersey–Maryland Interconnection, 1925–1970." PhD diss., University of Pennsylvania, 1983.

Stromberg, Joseph. "Atomic Cowboys: The South Texas Project and the Decline of Nuclear Power." PhD diss., University of Houston, 2012.

Sweeney, James L. *The California Electricity Crisis.* Stanford: Hoover Institution Press, 2002.

Tobey, Ronald C. *Technology as Freedom: The New Deal and the Electrical Modernization of the American Home.* Berkeley: University of California Press, 1996.

Vietor, Richard H. K. *Energy Policy in America since 1945: A Study of Business Government Relations.* Studies in Economic History and Policy. New York: Cambridge University Press, 1984.

Walker, J. Samuel. *Three Mile Island: A Nuclear Crisis in Historical Perspective.* Berkeley: University of California Press, 2004.

Walker, J. Samuel, and the US Nuclear Regulatory Commission. *A Short History of Nuclear Regulation, 1946–1999.* Washington, DC: US Nuclear Regulatory Commission, 2000.

Wellock, Thomas Raymond. *Critical Masses: Opposition to Nuclear Power in California, 1958–1978.* Madison: University of Wisconsin Press, 1998.

Wellock, Thomas Raymond. *Preserving the Nation: The Conservation and Environmental Movements, 1870–2000.* The American History Series. Wheeling, IL: Harlan Davidson, 2007.

Wellstone, Paul David, and Barry M. Casper. *Powerline: The First Battle of America's Energy War.* 1st ed. Minneapolis: University of Minnesota Press, 2003.

White, Richard. *The Organic Machine.* New York: Hill and Wang, 1995.

Worster, Donald. *Rivers of Empire: Water, Aridity, and the Growth of the American West.* 1st ed. New York: Pantheon, 1985.

Yates, Joanne. *Structuring the Information Age: Life Insurance and Technology in the Twentieth Century.* Baltimore: Johns Hopkins University Press, 2005.

Index

Page numbers followed by *f* or *t* refer to figures and tables, respectively.

Printed in the United States
by Baker & Taylor Publisher Services